The Mechanobiology in Atherosclerosis

Guixue Wang

The Mechanobiology in Atherosclerosis

Basic, Frontier and Advance

重庆大学出版社

Springer

Guixue Wang
Key Laboratory for Biorheological Science
and Technology of Ministry of Education &
National Local Joint Engineering
Laboratory for Vascular Implants
College of Bioengineering
Chongqing University
Chongqing, China

JinFeng Laboratory
Chongqing, China

ISBN 978-981-97-6511-9 ISBN 978-981-97-6512-6 (eBook)
https://doi.org/10.1007/978-981-97-6512-6

Jointly published with Chongqing University Press Co., Ltd.
The print edition is not for sale in China (Mainland). Customers from China (Mainland) please order the print book from: Chongqing University Press Co., Ltd.

This Springer imprint is published by the registered company Springer Nature Singapore Pte Ltd.
The registered company address is: 152 Beach Road, #21-01/04 Gateway East, Singapore 189721, Singapore

Foreword

Cardiovascular diseases (CVDs) have been the leading cause of death for both urban and rural residents worldwide since the early 20th century. According to the World Health Organization's data from 2017, coronary artery disease was responsible for nearly one-third of the global deaths. In China, cardiovascular disease is the leading cause of various diseases among urban and rural residents, with rural areas accounting for 46.66% and urban areas accounting for 43.81%. The incidence rate is on the rise, making research on the pathogenesis and prevention of CVD a major and urgent issue in the field of life science and medical research.

Atherosclerosis, which tends to occur in bends, branches, and bifurcations of blood vessels, serves as the pathophysiological basis of CVD. However, the mechanism of the formation and development of atherosclerosis has not been fully elucidated. Autopsy and clinical studies have revealed that the biomechanical response of vascular endothelial cells plays a crucial role in the localization, progression, and stability of atherosclerosis. Additionally, multiple clinical studies have shown that changes in the local biomechanical environment after intravascular stent implantation impact intimal hyperplasia and in-stent restenosis. Therefore, exploring and deciphering the mechanobiological response mechanism of vascular wall cells is essential for developing effective strategies and targets for the prevention and treatment of atherosclerosis. Moreover, this study provides new insights for reducing the risks associated with in-stent restenosis, developing new stents and implants, and translating basic research achievements into clinical practice. This, in turn, promotes the development of new diagnostic and therapeutic methods in the field.

This monograph is based on the research conducted in our laboratory and combines it with the current progress in biomechanics and mechanical biology of atherosclerosis. This review is divided into eight sections covering topics such as lipoprotein concentration polarization, the stress response of vascular endothelial cells, biomechanical mechanisms of cardiovascular and cerebrovascular development, biomechanical mechanisms of atherosclerotic plaque formation, biomechanical mechanisms of in-stent restenosis, and mechanobiological mechanisms of drugs that regulate cell viability to inhibit vascular restenosis. This review systematically discusses the mechanobiological mechanisms of atherosclerosis and intravascular stent restenosis, as well as the strategies for their prevention and treatment. It also introduces the latest frontiers and progress in these fields. We hope that this monograph will not only serve as a reference for scholars in atherosclerosis research regarding biomechanics and mechanical biological mechanisms but also as a valuable resource for clinical cardiovascular doctors and vascular stent developers.

The writing and publication of this monograph have been supported by JinFeng Laboratory Chongqing, China; the Key Laboratory of Biorheological Science and Technology of the Ministry of Education at the School of Bioengineering, Chongqing University, the National Local Joint Engineering Laboratory for Vascular Implant Development, the Public Experimental Center of Chongqing National Bioindustry Base, and Chongqing University Press. I express our highest respect and heartfelt gratitude to all the institutes and individuals who provided care and support during the process of writing and publishing this monograph. I am particularly grateful to professor Xiaoyan Deng at Beihang University, China and my over twenty postdoctoral and doctoral students who have already graduated or are currently studying, such as Dr. D. H. Wei, Dr. Y. Shen, Postdoc. Y. Wang, Postdoc. LL Fan, Postdoc. X. Xie, Dr. Y. P. Zhao, Dr. K. Zhang, Postdoc. T. Zhang, Postdoc. D. X. Lei, Postdoc. R. L. Du, Ms. H. P. Zhang, Mr. W. X. Qin, Mr. E. D. Qin, Ms. Z. C. Xu, Dr. Y. Wang, Dr. L. Wen, Dr. J. X. Xu, Dr. J. X. Wang, Ms. W. Liao, Mr. Z. Q. Hu, Dr. J. Y. Huang, Ms. C. C. Deng, Dr. C. J. Tang, for their outstanding contributions in writing, translating, and proofreading this monograph.

Their dedication is sincerely appreciated. Additionally, I extend our respect and gratitude to the numerous scholars both domestically and internationally who have made significant contributions to this field. I also appreciate the authors and publishers of the cited and referenced literature in this monograph, as well as all the friends who assisted in its publication. Special thanks are extended to the Key/General Projects of the National Natural Science Foundation of China (11332003, 120032007, 11572064, 30970721, 31370949, 31971242), the International Science and Technology Cooperation Key Project of the Ministry of Science and Technology of China (2004DFA06400), the National Scientific Support Program Key Project of China (2012BAI18B02), and the National Key Research and Development Program of China (2016YFC1102305) for their financial support.

Due to the author's limited expertise and the tight completion schedule, there may be errors, omissions, and oversights in this monograph. We sincerely request that readers provide guidance, criticism, and corrections without hesitation.

November 2023

Preface I

Atherosclerosis (AS) and its systemic pathological changes pose a significant threat to human health. Over the years, as life expectancy has increased in China, cardiovascular and cerebrovascular diseases have emerged as major health concerns and leading causes of death and disability. The rapid advancement of modern basic medical research has led to breakthroughs in various emerging disciplines, including genomics, proteomics, metabolomics, bioinformatics, and single-cell genomics. These advancements paved the way for in-depth research on vascular physiology and pathology.

Understanding the rules and characteristics of hemodynamics in the development and progression of atherosclerosis is crucial. Likewise, finding effective ways to improve and repair the damage to the vascular intima and tissue caused by atherosclerosis, as well as designing and efficiently implanting vascular stents to address vascular stenosis or obstruction, are pressing challenges. Meeting these challenges requires a comprehensive monograph that systematically summarizes current international frontiers and research progress in the field.The "Mechanobiological Basis and Frontiers of Atherosclerosis" serves as a scientific foundation for addressing these questions. Biomechanics emerged as an independent discipline in the 1960s, aiming to merge the principles and methodologies of mechanics with those of biological science. Its goal is to understand the quantitative laws governing life processes and leverage them to maintain and enhance human health. Over the past half-century, biomechanics has made remarkable progress, advancing our understanding of various aspects of life movement and expanding into medical fundamentals, clinical practice, and rehabilitation.

Mechanobiology, in particular, stands out as a frontier field that explains the biomechanical mechanisms underlying growth, development, and pathogenesis of diseases. New technologies for disease diagnosis, treatment, and repair/regeneration have been developed.

This monograph provides a systematic introduction and elaboration of the hemodynamic causes of atherosclerosis, covering its etiology, mechanical biology, biomaterials, and clinical medicine. Notably, beyond basic research on atherosclerosis, further research should be performed on the design of vascular stent implantation, vascular repair mechanisms, and patient prognosis.

The monograph is organized into eight sections that comprehensively summarize the latest research achievements. These include the distribution and adsorption of blood lipids, the mechanical response of endothelial cells and vascular injury, vascular development, and the maintenance of homeostasis, as well as vascular restenosis after stent implantation and drug coating intervention. New progress, ideas, and achievements in the field of mechanical biology related to arteriosclerosis have been systematically combined. With clear ideas, innovative content, and thorough and articulate descriptions, this monograph offers high reference value to individuals engaged in basic vascular research and clinical work in related fields.

This monograph encompasses the insights, experiences, and accomplishments of Prof. Guixue Wang's nearly thirty years of research in vascular mechanics and biology. Professor Wang is a meticulous and diligent expert in the field of biomedical engineering. He has been recognized as an "Elsevier Highly Cited Chinese Researchers" in his field for several consecutive years. His outstanding achievements in basic scientific research on the hemodynamic causes of atherosclerosis and vascular stent implantation repair are commendable. To ensure the monograph's quality, the professor at your university, along with proofreading experts, meticulously read, reviewed, and corrected every word and sentence, investing significant effort.

I believe that the publication of this monograph will undoubtedly enrich readers' perspectives on vascular mechanics and biology and make significant contributions to the advancement of scientific research in this field in China.

December 2023

Academician of the Chinese Academy of Sciences
Former Chairperson, the Cardiovascular Branch of the Chinese Medical Association
Professor and President, Biomedical Research Institute, Fudan University
Chairperson of the China Cardiovascular Health Alliance
Shanghai, China

Preface II

I am delighted to share the exciting news that Professor Guixue Wang's monograph titled "Mechanobiological Basis and Frontiers of Atherosclerosis" will soon be published by Springer Press. Congratulations to Prof. Guixue Wang on this achievement.

Atherosclerotic disease is a major health concern that poses a serious threat to human well-being. In China, the proportion of deaths caused by cardiovascular and cerebrovascular diseases has increased from 12.07% in 1957 to 44% in 2020. Therefore, research on the prevention, diagnosis, and treatment of atherosclerotic diseases has become a critical focus for our government. Both clinical studies and autopsies have shown that atherosclerotic lesions are more common in specific areas, such as the branches and bends of the abdominal aorta, carotid artery, coronary artery, and peripheral artery (e.g., femoral artery). These areas, referred to as focal atherosclerotic lesions, are primarily characterized by low shear stress or flow separation. Consequently, understanding the biomechanical and mechanobiological mechanisms of atherosclerosis is essential for the pathogenesis, treatment, and management of cardiovascular and cerebrovascular diseases.

Professor Guixue Wang is a renowned academic leader in the field of atherosclerosis biomechanics in China. In the first edition (2004) and the second edition (2009) of the book "Atherosclerotic Cardiovascular Disease—Basis and Clinic" published by the Science Press, which I edited in chief, he was responsible for writing the chapter "Hemodynamic Changes and Atherosclerosis". His recent monograph "Mechanobiological Basis and Frontiers of Atherosclerosis" encompasses various topics, including the theory of lipoprotein concentration polarization in the formation of atherosclerosis, biomechanical mechanisms underlying the occurrence and progression of atherosclerosis, mechanical biological mechanisms of endothelial cell damage repair, mechanical signal transmission, treatment strategies, and drug development. By encompassing these areas, the monograph provides a comprehensive and detailed overview based on the forefront of cardiovascular biomechanics and mechanical biology science and technology. Its aim is to summarize and showcase the notable research achievements of Prof. Guixue Wang's research team and other experts in the field, providing valuable references on pathophysiology, biomechanics, biomedical engineering, and medicine for postgraduates and young scientific and technological workers interested in atherosclerosis.

This monograph combines fundamental knowledge with cutting-edge developments, encompassing both familiar and novel content. Each chapter reflects the author's extensive knowledge and research experience. With its high reference value, especially for readers engaged in interdisciplinary research, this monograph is expected to be well received by a wide range of readers.

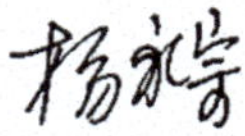

November 2023

Former Chairperson of the Chinese Branch of the International Atherosclerosis Society
Former Chairperson of the Atherosclerosis Committee of the Chinese Society of Pathophysiology
Former President of Hengyang Medical University, China
Hengyang, China

Preface III

Cardiovascular diseases (CVDs) pose the most serious threat to human health today. The mortality rate due to CVD has occupied the first place for many years among both urban and rural residents in China. Atherosclerosis, in particular, is the leading cause of cardiovascular diseases such as coronary heart disease and hypertension. The cardiovascular circulatory system functions as a typical mechanical system, and abnormalities in hemodynamics within this system are closely associated with the occurrence and progression of cardiovascular diseases.

In recent years, research on biomechanics has advanced and expanded to the cellular and molecular levels, giving rise to a new field known as mechanical biology. Consequently, understanding the biomechanical causes and mechanobiological mechanisms underlying cardiovascular diseases is crucial. This understanding will contribute to a deeper understanding of the pathogenesis, clinical diagnosis, and treatment of these diseases.

The development of biomechanics as a new interdisciplinary field in China can be attributed to the direct involvement and strong support of renowned Chinese-American scholar Y.C. Fung, also known as the "Father of Biomechanics." In 1979, Professor Y.C. Fung organized an advanced workshop on biomechanics at Chongqing University, thereby initiating and fostering the development of biomechanics at the university. Over the past 40 years, Chongqing University has become a crucial center for talent cultivation, scientific research, and social service in the field of biomechanics in China and even globally. The university houses a key laboratory in biomechanics science and technology under the Ministry of Education and a research base group covering the complete chain from fundamental research to applied innovation and achievement translation. Notable establishments within this group include the National "111" Plan Innovation Introduction Base for Biomechanics and Tissue Repair Engineering, the National Bioindustry Base Public Experimental Center in Chongqing, and the National Local Joint Engineering Laboratory for Vascular Implant Development.

From a biomechanical standpoint, exploring how mechanical factors regulate and influence the development, function, and reconstruction of the cardiovascular system is highly important. Moreover, understanding the mechanical factors and biomechanical effects associated with cardiovascular diseases and elucidating the biomechanical mechanisms of atherosclerosis at the individual, tissue, cell, and molecular levels can contribute to establishing new mechanisms and theories for vascular repair. Furthermore, the identification of potential therapeutic targets can guide the design and improvement of cardiovascular intervention materials, promote advancements in clinical diagnosis and treatment technologies and related biomaterials, and aid in the development of medical device-related industries.

Professor Guixue Wang's monograph, "Mechanobiological Basis and Frontiers of Atherosclerosis" focuses on cardiovascular mechanical biology. It comprehensively presents and analyses the research achievements and latest cutting-edge progress of his research team, as well as domestic and international peers in the field. The monograph covers topics such as the hemodynamic effects of cardiovascular and cerebrovascular systems, biomechanical mechanisms of atherosclerosis occurrence and development, mechanical and biological effects and mechanisms of endothelial cell damage repair, and new treatment strategies and drug development for atherosclerosis.

As the president of the National Local Joint Engineering Laboratory for Vascular Implant Development and the Key Laboratory for Biorheological Science and Technology of the Chinese Ministry of Education, Prof. Guixue Wang has dedicated his career to the study of biomechanics and developmental biology of cardiovascular and cerebrovascular diseases. He is a renowned scholar with significant achievements and influence in the fields of biomechanics and mechanobiology. He has been featured multiple times on the Elsevier Highly Cited Chinese Researchers and Global Top 100000 Scientists List in the field of biomedical engineering. The publication of his monograph will serve as an important reference for graduate students, young scientific and technological workers, and researchers engaged in cardiovascular physiology and pathophysiology research, particularly in biomedical engineering, basic medicine, and clinical medicine related to atherosclerotic diseases. Additionally, this study will play a vital role in advancing research and development in cardiovascular biomechanics and mechanical biology.

November 2023

Former President, Chinese Society of Biomedical Engineering
Chairperson of the Biomechanics
Professional Committee of the Chinese Society of Mechanics/
Chinese Society of Biomedical Engineering
Fellow of the American Academy of Medicine and Biotechnology
Recipient of the National Outstanding Youth Fund of China
Distinguished Professor "Changjiang Scholars" by the Ministry of Education, China
Director of the Key Laboratory of Biomechanics and Mechanical
Biology of the Ministry of Education, China
Beihang University
Beijing, China

Synopsis

Epidemiological surveys show that the incidence, disability, and mortality rates of cardiovascular and cerebrovascular diseases rank first among all diseases and will continue to increase. Atherosclerotic heart disease is one of the important causes of cardiovascular disease-related mortality. Taking the focal mechanical biological mechanism of atherosclerotic plaque formation and rupture as the research theme, the authors provide direct evidence for early theory of "lipid concentration polarization" through the systematic study of the whole animal at the cellular-molecular level. Using biomechanical theories and methods, the author enriches and develops the "lipid concentration polarization theory" through in vitro and in vivo studies, elucidating the focal nature of atherosclerosis from a new perspective. The scientific hypothesis that "angiogenesis is the occurrence of vulnerable plaques in the proximal end of narrowed blood vessels" was also proposed for the first time. Through cellular and molecular biology methods, the cellular and molecular mechanisms underlying focal rupture of atherosclerotic plaques were further explored from the perspectives of tissue and cellular mechanical biological characteristics and mechanical signal transduction. This monograph is based on the research achievements of the author's research team and domestic and foreign peers in the field of atherosclerosis biomechanics and mechanical biology for more than 20 years. The monograph is suitable for researchers, educators, and students in the fields of biomechanics, hemodynamics, mechanobiology, and atherosclerotic diseases for reference and reading.

Contents

1 Lipoprotein Concentration Polarization and Atherosclerosis 1
1.1 The Overview of Lipoprotein Concentration Polarization 2
1.2 In Vitro Measurement of Lipid Concentration Polarization on the Surface of Tunica Intima 3
1.2.1 Numerical Simulation and Measurement Methods 4
1.2.2 Numerical Simulation and Measured Results 6
1.3 In Vivo Observation of Lipid Concentration Polarization on the Surface of Tunica Intima—Zebrafish Model 12
1.3.1 Zebrafish Model 12
1.3.2 Fluorescence Intensity Analysis 13
1.3.3 Distribution of Low-Density Lipoprotein Concentration in Straight Sections 13
1.3.4 Distribution of Low-Density Lipoprotein Concentration at Vessel Bifurcation 14
1.4 Lipoprotein Concentration Polarization Promotes Atherosclerosis—Rabbit Carotid Artery Cuff Model 16
1.4.1 Rabbit Carotid Artery Cuff Model 16
1.4.2 Numerical Simulation of Local Narrow Flow Field 17
1.4.3 Numerical Simulation of Low-Density Lipoprotein Concentration Distribution in Local Stenosis Vessels 18
1.4.4 Experimental Measurement of Low-Density Lipoprotein Concentration Distribution in Local Stenotic Vessels 18
1.4.5 The Effect of Polarization of Low-Density Lipoprotein Concentration on Local Stenotic Vascular AS Lesions 20
1.5 Research Progress and Prospects 22
1.5.1 The Development of Low-Density Lipoprotein Targeting Therapy 22

1.5.2 Numerical Simulation of Low Density Lipoprotein Concentration Polarization Study ... 23
References ... 25

2 Low Shear Stress Regulates Biological Behavior of Vascular Endothelial Cells to Promote Atherosclerosis ... 27
2.1 The Injury of Vascular Endothelial Cell is a Key Factor in the Process of Atherosclerosis ... 28
2.2 In Vitro Model of Shear Stress Regulating the Biological Characteristics of Vascular Endothelial Cells ... 29
2.3 Animal Model of Shear Stress Regulating the Biological Characteristics of Vascular Endothelial Cells ... 31
2.3.1 Hemodynamic Model for Local Stenosis of the Common Carotid Artery ... 31
2.3.2 Low Oscillatory Shear Stress Animal Model (Local Ligation Model of Common Carotid Artery) ... 33
2.3.3 Controllable Shear Stress Animal Model ... 34
2.4 The Cumulative Effect of Cell Membrane Tensile Stress on the Secretion of ET-1 in Large Vascular Endothelial Cells ... 40
2.5 Low Shear Stress Promotes the Production and Secretion of Chemokines in Large Vascular Endothelial Cells ... 43
2.6 Research Progress and Prospects ... 44
References ... 46

3 Hypothesis of Hemodynamics for Vascular Endothelial Cell Injury Repair ... 49
3.1 Effects of Atherosclerotic Plaque Formation on Hemodynamics ... 50
3.2 Effects of Stent Implantation on Vascular Injury and Hemodynamics ... 51
3.3 Cell Sources for Vascular Injury Repair ... 52
3.4 Surface Modification of Stents for Promoting Vascular Endothelial Repair ... 57
3.4.1 Research on Surface Modification of Vascular Stents ... 58
3.4.2 The Modified Intravascular Stents of Transgenic Endothelial Cell ... 59
3.4.3 CD133 Antibody Capture EPCs Scaffold ... 61
3.4.4 Molecular Mechanism Study on Surface Materials-Mediated Re-Endothelial Repair of Scaffold ... 62
3.5 Research Progress and Prospects ... 65
References ... 68

4 Mechanobiological Mechanism of Cardiovascular and Cerebrovascular Development ... 73
4.1 The Effect of Hemodynamics on Cardiovascular and Cerebrovascular Development and Its Mechanical-Molecular Mechanism ... 74

4.1.1 The Influence of Hemodynamics on Vascular Development ... 75
4.1.2 Molecular Mechanisms of Hemodynamic Regulation of Vascular Development ... 78
4.1.3 Hemodynamic Regulation of the Mechanical Conduction Mechanism of Vascular Development ... 87
4.1.4 The Effect of Hemodynamics on Cerebrovascular Development and Its Mechanism ... 89
4.2 Effect of Microgravity on Vascular Development and Its Mechanical and Molecular Mechanisms ... 92
4.2.1 Effects of Microgravity on Blood Vessel Development ... 93
4.2.2 Molecular Mechanisms of Microgravity Affecting Vascular Development ... 94
4.3 Research Progress and Prospects ... 99
4.3.1 Morphogenesis of the Vascular System ... 99
4.3.2 Stress on Vascular ECs ... 101
4.3.3 The Formation of the BBB ... 101
4.3.4 Hemodynamics and the BBB ... 101
4.3.5 Prospects of Mechanical Developmental Biology Research ... 103
References ... 104

5 Mechanobiological Mechanisms of Atherosclerotic Plaque Formation ... 109
5.1 Shear Stress and Atherosclerosis ... 109
5.1.1 Shear Stress and Atherosclerosis-Associated Endothelial Dysfunction ... 109
5.1.2 Shear Stress and Atherosclerosis-Related Inflammation ... 110
5.1.3 Relationship Between Shear Stress and Angiogenesis in Atherosclerosis ... 112
5.1.4 The Effect of Blood Flow Shear Stress on the Formation of Atherosclerosis ... 113
5.1.5 Impact of Shear Stress-Regulated Gene Expression on Atherosclerosis ... 119
5.1.6 Stress Regulation of Transcription Factors in Atherosclerosis and Anti-Atherosclerosis ... 120
5.1.7 The Clinical Implications of Shear Stress for the Treatment of Clinical Atherosclerosis ... 123
5.2 Hemodynamic Changes and the Stability of Atherosclerotic Plaques ... 124
5.2.1 Characteristics of Vulnerable Plaques ... 124
5.2.2 Hemodynamic Characteristics at the Plaque Site and Their Impact on Plaque Development ... 125

5.2.3 Mechanical Mechanisms of Atherosclerotic Plaque Rupture 132
5.3 Research Progress and Prospects 132
5.3.1 Atherosclerosis, Shear Stress, and Endothelial Function 133
5.3.2 Force Sensors and Sensitive Transcription Factors in Atherosclerosis 133
5.3.3 Synergistic Effects of Hemodynamic Factors and Systemic Risk in Atherosclerosis Formation 135
References 140

6 Mechanobiological Mechanism by Which Id1 Regulates Lipid Absorption in Atherosclerotic Plaque Formation 143
6.1 Proteomic Analysis of Low Oscillatory Shear Stress Model Blood Vessels 144
6.2 Low Oscillatory Shear Stress Causes Lipid Accumulation to Promote Atherosclerotic Plaque Formation 150
6.3 Id1 Participates in the Lipid Absorption Process Caused by Low Oscillatory Shear Stress 155
6.4 Id1 Participates in the Process of Endothelial Lipid Absorption Induced by Low Oscillatory Shear Stress by Regulating LDLR Expression 157
6.5 Research Progress and Prospects 169
6.5.1 Id1 Affects Endothelial Cell Uptake of LDL 169
6.5.2 Id1 Participates in the Formation of Atherosclerosis 170
References 171

7 Mechanical Mechanisms of In-Stent Restenosis and Atherosclerosis Following the Implantation of Intravascular Stents 175
7.1 Construction of the Rabbit Carotid Artery Stenosis Model and Stent Implantation Experiment 176
7.1.1 Experimental Animal Model 176
7.1.2 Animal Model Processing Results 179
7.2 Analysis of Mechanical Behavior During the Stent Expansion Process 182
7.2.1 Finite Element Model of Stent Expansion 183
7.2.2 Numerical Calculation Results Analysis 189
7.3 Hemodynamic Analysis Based on OCT Imaging 193
7.3.1 3D Model Reconstruction 193
7.3.2 Hemodynamic Analysis 194
7.3.3 Analysis of Vascular Pathological Morphology in the Stent Segment 198
7.4 Research Progress and Prospects 199
7.4.1 Animal Models for Stent Implantation 200
7.4.2 Medical Imaging Evaluation of Stent Implantation 201

7.4.3 Finite Element Modeling of Stent Implantation 202
7.4.4 Mechanical Microenvironment of Arterial Atherosclerosis After Stent Implantation 203
References 205

8 Mechanobiology Mechanism for the Inhibitory Effect of Vascular Restenosis by Drug Regulation of Cell Viability 211
8.1 ATO is an Endothelium-Friendly Anti-Stent Restenosis Drug 212
8.2 ATO Inhibited the Proliferation of VSMCs and Showed an Effect on Their Phenotypic Transformation 213
8.2.1 ATO in Vivo Promotes the Transformation of VSMCs from Synthetic to Contractile Types 215
8.2.2 Cell Experiment of ATO Promoting the Transformation of VSMCs from Synthetic to Constrictive Type 215
8.3 F-Actin Enhances the Contraction Morphology of VSMCs and Increases the Elastic Modulus of Cells 218
8.4 The HiPO-YAP Pathway Mediates ATO to Regulate the Phenotypic Transformation of VSMCs 218
8.4.1 Hippo/YAP is Predicted to Participate in ATO Regulation of VSMC Phenotype by RNA Transcriptome Sequencing 221
8.4.2 YAP Mediates ATO to Regulate the Phenotypic Transformation of VSMCs 223
8.5 Research Progress and Prospects 223
References 227

Glossary of Terms 231

About the Author

Dr. Guixue Wang is a Distinguished Professor of Biomedical Engineering and former Dean of the College of Bioengineering, Chongqing University, China. He served as the director of the Key Laboratory of the Chinese Ministry of Education, the founding director of the Public Experiment Center of National Biological Industrialization Base at Chongqing, and the founding director of the National Local Joint Engineering Laboratory for Vascular Implants of China. His areas of research interests include cardio-cerebrovascular biomechanics and biomedical materials. Other duties include: Director of Atherosclerosis Branch under Chinese Society of Pathophysiology, Director of Materiobiomechanics Branch of the Chinese Society for Biomaterials, the Council Member of World Association for Chinese Biomedical Engineers. He was selected as the Elsevier "Most Cited Chinese Researchers in Biomedical Engineering" in 2019, 2020, 2021, 2022 and 2023; World's Top 2% Scientists in 2023; and a Fellow of International Union of Societies for Biomaterials Science and Engineering (FBSE) in 2023.

Chapter 1
Lipoprotein Concentration Polarization and Atherosclerosis

Atherosclerosis (AS) mainly occurs in large and medium-sized arteries, which is characterized by the formation of lipid rich vascular lesions. The occurrence and development of AS are related to multiple risk factors such as high concentration of low-density lipoprotein (LDL), low concentration of high-density lipoprotein (HDL), hypertension, diabetes, smoking, gender, genetic susceptibility, stress, lack of exercise and aging. Among them, LDL is the main transporter of cholesterol in the blood, which can pass through the endometrium and accumulate under the endometrium, oxidized to oxidized low density lipoprotein (Ox-LDL) (Fernández-Friera et al. 2017). Ox-LDL damages vascular endothelial cells, promotes endothelial cells to secrete chemokines, recruits blood monocytes to migrate into the endothelium, differentiates into macrophages, and then devours Ox-LDL to form macrophage derived foam cells. A large number of lipid rich foam cells accumulated to form early AS lesions. As the lesion progresses, AS lesions protrude and bleed into the lumen, leading to severe stenosis, reduced blood flow velocity, and ischemic symptoms (such as angina). In addition, when the unstable AS plaque ruptures, the blood directly contacts with the plaque or thrombogenic substances at the vascular wall, forming thrombus, leading to blockage of the plaque rupture and downstream vessels, leading to serious clinical cardiovascular events such as heart disease (coronary occlusion) and stroke (cerebral artery occlusion).

Clinical and autopsy studies have shown that AS lesions do not occur randomly in the arterial system, but rather tend to occur in specific locations, such as the branches and bends of the abdominal aorta, carotid artery, coronary artery, and peripheral artery (such as femoral artery), indicating the focal nature of AS lesions. Hemodynamic analysis shows that there is a sharp change in blood flow in the prone area of AS, which is prone to flow separation and eddy currents. As early as 1872, Rindfleisch pointed out that AS lesions are formed at the site of "changes in blood pressure and blood flow", and shear stress is closely related to AS lesions. Subsequent research suggests that the focal nature of AS lesions is closely related to two factors:

G. Wang, *The Mechanobiology in Atherosclerosis*,
https://doi.org/10.1007/978-981-97-6512-6_1

① Blood flow shear stress.
② The disruption of blood flow will slow local blood flow and result the accumulation of AS lipids such as LDL on the surface of vascular endothelial cells and enhanced diffusion across the intima.

In previous studies, researchers have attached great importance on shear stress, mainly focusing on the effects of blood flow shear stress on vascular remodeling, endothelial cell morphology, cell metabolism, physiological and pathological processes of the vascular system, signal transduction and gene expression of vascular endothelial cells, as well as the secretion of bioactive substances and molecules, and their regulatory mechanisms (Baeyens et al. 2016; Davies 2009; Souilhol et al. 2020). However, a large amount of evidence suggests that complex and multi-scale arterial wall material transport plays an important role in the occurrence and development of AS lesions (Mahmoudi et al. 2021). Research on early AS lesions has shown that there is a large accumulation of lipids such as cholesterol under the endometrium, which mainly come from plasma lipoprotein, especially LDL. Wiklund et al. found that LDL infiltrated the artery wall mainly through the leakage of the endothelial cell junction, and entered the artery intima through a receptor independent way (Wiklund et al. 1985). The surface lipids of the vascular lumen directly contact with vascular endothelial cells, and the transmembrane permeation of lipids depends on the concentration of lipids at the blood/arterial wall interface.

The arterial wall is permeable and has a certain filtration rate (Wang et al. 2003). Therefore, we believe that there is a concentration polarization phenomenon (mass transfer phenomenon in engineering) that causes AS lipids in the arterial system. Theoretical predictions indicate that the lipid concentration at the blood/wall interface is higher in the blood flow region with flow recirculation, flow stagnation, and low wall shear rate than in the laminar flow region. The difference in surface lipid concentration within the vascular lumen may be a key factor in the focal nature of AS, and may be more important than the effect of blood flow shear stress.

1.1 The Overview of Lipoprotein Concentration Polarization

We assume that fluids with dissolved solute flows through membrane. If the membrane has a higher resistance to solute than the fluid, the solute will be rejected by the membrane and accumulate on the surface of the membrane. When the concentration of solute on the membrane surface is higher than that in the fluid, the solute on the membrane surface will diffuse into the fluid. When the solute in the fluid accumulates towards the membrane surface area and the solute on the membrane surface diffuses towards the fluid, as well as the solute transport across the membrane reaches equilibrium, the solute steady-state distribution state is reached. Under this steady state, the solute concentration on the membrane surface remains higher than that in the fluid bulk solution, which is called concentration polarization. The accumulated solute layer is called concentration polarization layer (Fig. 1.1).

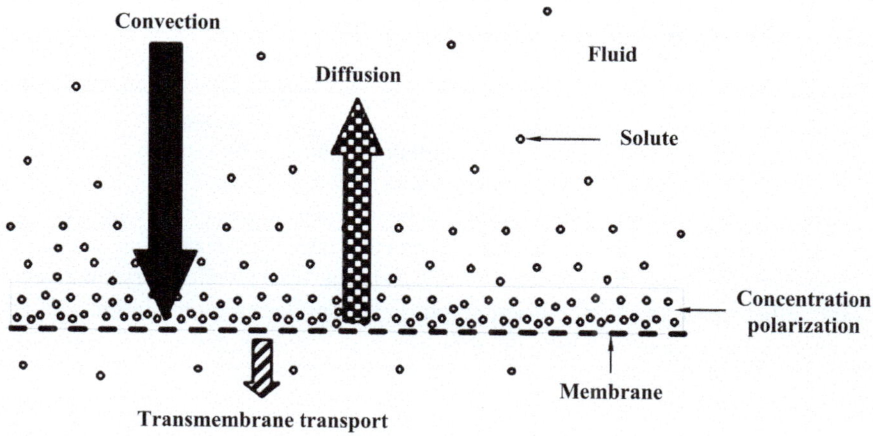

Fig. 1.1 Schematic diagram of the occurrence of concentration polarization. When the convection towards the membrane (solid arrow), the diffusion from the membrane (grid arrow) and the transmembrane transport are balance, the homeostasis is reached (stripe arrow). Adapted from Vincent and Weinberg (2014)

The phenomenon of concentration polarization is commonly seen in the fields of science and engineering, especially when it comes to filtration processes. Since the convection rate of LDL in blood towards the surface of the intima cavity is far greater than the rate of transport from the inner surface of the vessel to the intima, a concentration polarization layer rich in LDL will be formed near the surface of the tunica intima of the vessel. The LDL concentration close to the tunica intima depends on the thickness of the concentration polarization layer and the rate of LDL transport across the intima. The thickness of the concentration polarization layer depends on the local blood flow pattern, especially the local wall shear stress. The rate of endometrial transport depends on the hydraulic permeability of the endometrium, which is closely related to the magnitude and direction of shear stress on the surface of the blood vessel wall. Therefore, the blood flow pattern and blood flow shear stress are closely related to the polarization of LDL concentration and the LDL concentration on the surface of tunica intima (Naiki et al. 1999).

1.2 In Vitro Measurement of Lipid Concentration Polarization on the Surface of Tunica Intima

To confirm the hypothesis of lipid concentration polarization, the transport of LDL in the straight segment of the artery was analyzed through numerical simulation under steady-state and pulsatile flow conditions, and experimentally validated (Fatouraee et al. 1998; Deng et al. 2003).

1.2.1 Numerical Simulation and Measurement Methods

To simplify the analysis, the following assumptions are made:

① The fluid (blood) is homogeneous, incompressible, and Newtonian fluid with a viscosity (v) of 0.035 g/cm s, mass density (ρ) It is 1.05 g/cm^3.
② A blood vessel is a straight cylindrical tube with a uniform inner diameter.
③ The blood vessel wall can penetrate plasma at a rate of 10–6 cm/s.
④ The convection and diffusion flux of LDL into the vessel wall is very small, and the influence on the LDL concentration on the surface of tunica intima can be ignored. For a given variable, the flow and mass transfer caused by convection and diffusion can be described using the following formula:

$$\frac{\partial^2 \psi}{\partial x^2} - \frac{1}{y}\frac{\partial \psi}{\partial y} + \frac{\partial^2 \psi}{\partial y^2} = y\eta \tag{1.1}$$

$$\frac{\partial \eta}{\partial t} + \frac{\partial u\eta}{\partial x} + \frac{\partial v\eta}{\partial y} = \frac{2}{\text{Re}}\left(\frac{\partial^2 \eta}{\partial x^2} + \frac{1}{y}\frac{\partial \eta}{\partial y} + \frac{\partial^2 \eta}{\partial y^2} - \frac{\eta}{y^2}\right) \tag{1.2}$$

$$\frac{\partial c}{\partial t} + \frac{\partial uc}{\partial x} + \frac{1}{y}\frac{\partial yvc}{\partial y} = \frac{2}{\text{Re}\ Sc}\left[\frac{\partial^2 c}{\partial x^2} + \frac{1}{y}\frac{\partial}{\partial y}\left(y\frac{\partial c}{\partial y}\right)\right] \tag{1.3}$$

Navier Stokes Eqs. (1.1) and (1.2) in the form of flow and vorticity functions are used in flow simulation. Among them ψ is a dimensionless flow function, x is the dimensionless axial coordinate, and y is the dimensionless radial coordinate, η is the dimensionless vortex function, t is the dimensionless time, v is the dimensionless radial velocity component, c is the dimensionless LDL concentration, u is the dimensionless axial velocity component, *Re* is the Reynolds number, *Sc* is the Schmidt number (v/D), and D is the diffusion coefficient. Vortex function η *(x, y, t)* (Eq. (1.2)) and Stream Function ψ *(x, y, t)* (Eq. (1.2)) is defined as:

$$\eta = \frac{\partial v}{\partial x} - \frac{\partial u}{\partial y},\quad \upsilon = -\frac{1}{y}\frac{\partial \psi}{\partial x},\ u = -\frac{1}{y}\frac{\partial \psi}{\partial y} \tag{1.4}$$

Since n, r, and D are constants, Eqs. (1.1) and (1.2) are uncoupled from the mass transfer Eq. (1.3); Therefore, they can be solved separately.

The boundary conditions are as follows:

$$u(x,1,t) = 0,\ \ v(x,1,t) = V_\omega/\overline{u}_0,\ \ u(0,y,t) = u_{\omega m},\ v(0,y,t) = 0,\ c(0,y,t) = 1 \tag{1.5}$$

In numerical simulation, the artery is a semi permeable membrane with Vw filtration rate. Based on assumption (1.4), the boundary conditions of the mass transfer Eq. (1.3) can be written as:

$$V_{\omega}C_{\omega} = \frac{2\overline{u}_0 C_0}{\mathrm{Re}\, Sc}\left(\frac{\partial c}{\partial y}\right)_{\omega} \tag{1.6}$$

In this chapter's study, to obtain the LDL concentration of Cw at the blood/wall interface. The finite difference method is used to solve Eqs. (1.1) and (1.3). The non-uniform grid is used in the solution process, and the number of nodes is minimized while the grid maintains sufficient accuracy. In the area closest to the tunica intima, the grid spacing decreases. The blood flow waveform used in the calculation is shown in Fig. 1.2.

In the actual measurement, bovine serum albumin was used, and the measured system is shown in Fig. 1.3. The 10–15 cm long common carotid artery comes from adult hybrid dogs, and the perfusion solution is Krebs solution containing 1.0 mg/mL bovine serum albumin. Krebs solution was prepared as follows (final concentration): NaCl: 118 mmol/L; KCl: 4.7 mmol/L; $NaHCO_3$: 25 mmol/L; KH_2PO_4: 1.2 mmol/L; $MgSO_4$: 1.2 mmol/L; $CaCl_2$: 2.5 mmol/L; Glucose: 11 mmol/L, adjust the pH value of the solution to 7.4. Adjust the height of the storage tank and the resistance of the outlet pipe to obtain the flow rate through the blood vessels. Set the pressure inside the container to either 100 mmHg (normal pressure) or 200 mmHg (hypertension) in two pressure environments.

After the flow rate stabilizes (approximately 20 min after starting), use the sampling needle of a micro syringe to extract a liquid sample at the boundary of the solution/arterial wall. The diameter of the micro injector sampling needle is 0.28 mm, the length is 50 mm, and the total volume sampled is 10 μL with an accuracy of ±0.2 μL. Insert the sampling needle into the inner cavity of the test artery through a specially designed distal cannula, and carefully place the needle horizontally on the inner surface of the arterial cavity. During the experiment, the infusion solution in the overflow top water tank was continuously stirred to prevent albumin precipitation. Samples were taken 10 times at each flow rate.

The improved McKnight colorimetric method was used to determine the concentration of albumin in the sampling solution, which is very sensitive and can be used to detect proteins as low as 0.1 mg. The albumin concentration measured in the sample

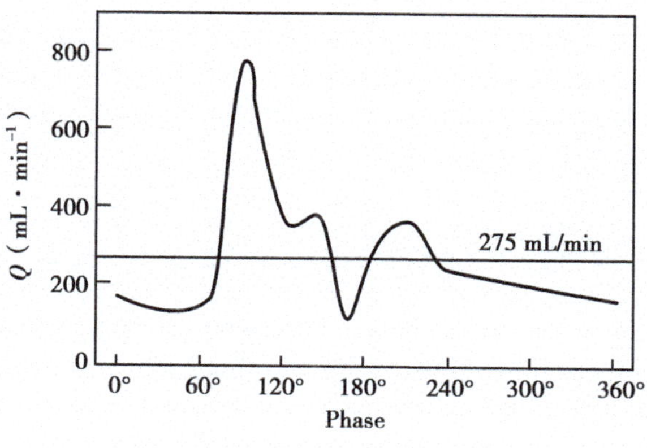

Fig. 1.2 Pulse flow rate in numerical simulation. The waveform is from the human common carotid artery. Adapted from Deng and Wang (2003)

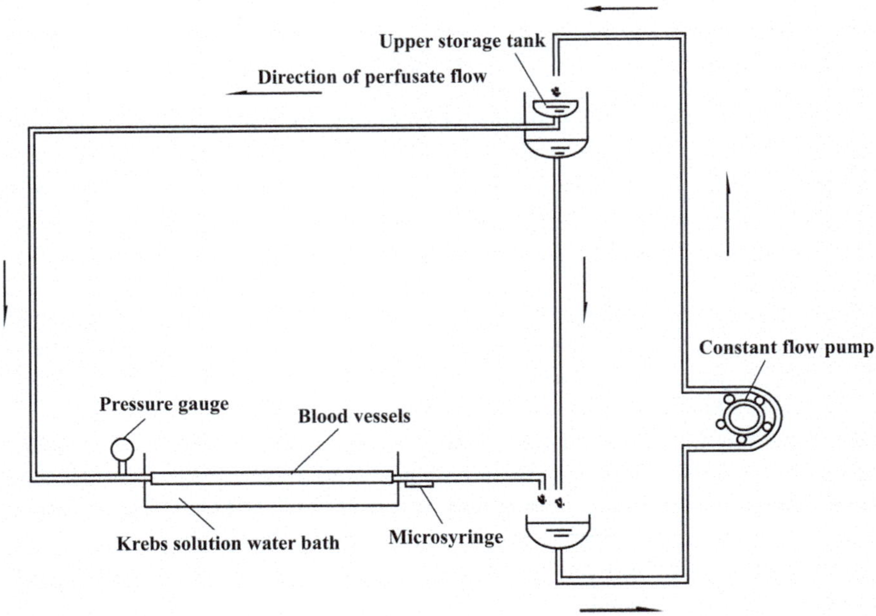

Fig. 1.3 Schematic diagram of the experimental perfusion system. Adapted from Deng and Wang (2003)

is used as the surface albumin concentration C_w on the inner membrane of the test artery, and the total concentration C_0 of the perfusion solution is standardized and expressed as a function of wall shear stress C_w.

1.2.2 Numerical Simulation and Measured Results

(1) Numerical Simulation Results

In the numerical simulation, the parameters of the human common carotid artery were used, with an inner diameter of 7.0 mm, an average flow rate of 275 mL/min, a heart rate of 70/min, Re of 250, and a Womersley number of 5.2. LDL filtration rate is 10^{-6} to 10^{-5} cm/s and Sc is 1.67×10^5–6.67×10^5, corresponding to LDL diffusion coefficient of $5.0 \times 10^{-8} - 2.0 \times 10^{-7}$ cm^2/s.

The results showed that under the conditions of steady and pulsating flow, due to the filtration of the arterial wall, a lipid accumulating boundary layer was formed on the surface of the tunica intima of the vessel, and its LDL concentration was significantly higher than that of the lumen. Under pulsatile flow conditions, the LDL concentration on the surface of tunica intima changes periodically with time. As shown in Fig. 1.4, the average value within a cardiac cycle is slightly higher than the average value under steady-state blood flow conditions. The Sc value (diffusion coefficient D of lipids) has a significant impact on C_w/C_0. When Sc is 1.67 × At 105, C_w/C_0 (C_w: time average value of LDL concentration on the cavity surface) is only about 1.054. However, when the Sc value is 6.6×10^5, C_w/C_0 increased to 1.38.

Under physiological flow conditions (average Re = 250), different Sc and filtration rate affect the LDL concentration on the surface of tunica intima. With the increase of filtration rate V_w, the time average LDL concentration C_w/C_0 on the surface of tunica intima increased linearly. In normal arteries (with a physiological filtration rate of 4.0) $\times 10^{-6}$ cm/s), the LDL concentration on the surface of tunica intima is 5 ~ 14% higher than that in the lumen (Fig. 1.5).

At low wall shear rate, C_w/C_0 drops sharply with the increase of wall shear rate, indicating that at low wall shear rate, LDL concentration on the surface of tunica intima is more sensitive to changes in flow conditions (Fig. 1.5).

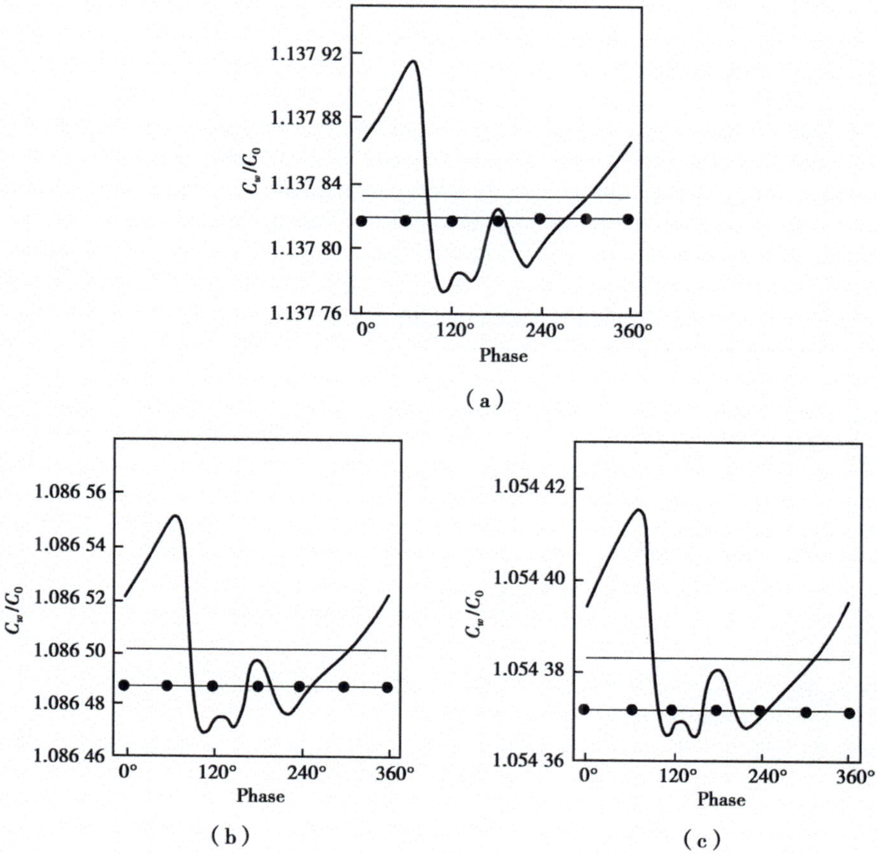

Fig. 1.4 LDL concentration on the surface of the vascular intima during a cardiac cycle. The solid horizontal line represents the time average value of the surface LDL concentration in a cardiac cycle, and the dotted line is the value under the condition of steady flow. $V_w = 4.0 \times 10^{-6}$ cm/s. **a** $Sc = 6.6 \times 10^5$; **b** $Sc = 3.3 \times 10^5$; **c** $Sc = 1.6 \times 10^5$. Adapted from Deng and Wang (2003)

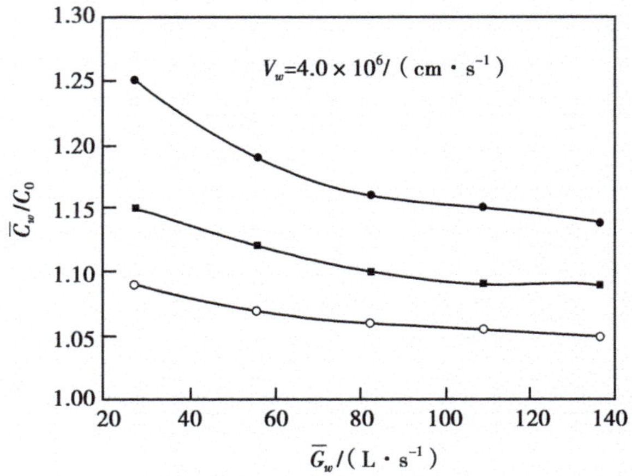

Fig. 1.5 The relationship between the LDL concentration on the surface of vascular intima and the wall shear rate. Adapted from Deng and Wang (2003)

(2) Actual Experimental Results

The albumin concentration on the surface of tunica intima was measured under steady flow condition. The experimental results showed that the albumin concentration G_w on the surface of tunica intima was always higher than the concentration C_O in the lumen, indicating that concentration polarization occurred in the canine carotid artery. Similar to numerical simulation, at low wall shear rates, the relative concentration G_w/C_O with the vascular lumen sharply decreases. By comparing the data under two different pressure levels, it can be seen that the filtration rate V_w has a very obvious effect on the surface albumin concentration of tunica intima. When $G_w = 0$ and 185 s^{-1}, when $V_w = 8.9 \pm 1.7 \times 10^{-6}$ cm/s, C_w is 65% and 15% higher than C_0, respectively. When $V_w = 4.8 \pm 0.6 \times 10^{-6}$ cm/s, C_w is only 42 and 5% higher than C_0 (Fig. 1.6).

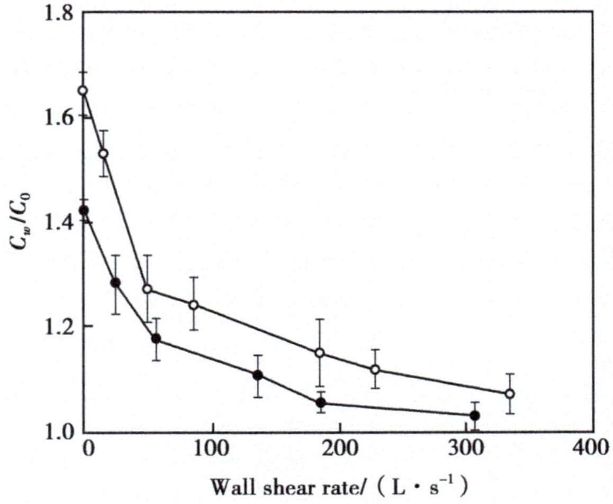

Fig. 1.6 Intima surface albumin concentration versus wall shear rate in canine carotid artery.·Pressure = 100 mmHg ($V_w = 4.8 \pm 0.6 \times 10^{-6}$ cm/s); °pressure = 200 mmHg ($V_w = 8.9 \pm 1.7 \times 10^{-6}$ cm/s). Adapted from Deng and Wang (2003)

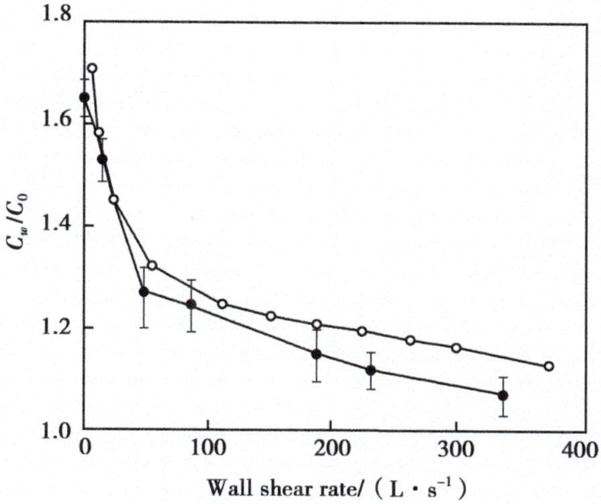

Fig. 1.7 Comparison of experimental (·) and theoretical (°) results of 200 mmHg (Vw = 8.9 ± 1.7 × 10^{-6} cm/s). In the numerical simulation, the diffusion coefficient D of BSA is 8.7 × 10^{-8} cm^2/s. Adapted from Deng and Wang (2003)

To verify the numerical simulation, under the pressure of 200 mmHg ($V_w = 8.9 \pm 1.7 \times 10^{-6}$ cm/s), theoretical calculations were conducted using flow parameters corresponding to the experiment, and the experimental and computational results were compared. The results showed that the theoretical results were very consistent with the experiment, indicating that concentration polarization did indeed occur in the arterial system, and numerical simulation can accurately analyze this mass transfer phenomenon (Fig. 1.7).

In another experiment, the size of the sampling probe was changed from 31 to 27 (outer diameter = 0.43 mm) for sampling, and the results showed that the size of the sampling probe may affect the albumin concentration of the retrieved liquid sample. As shown in Fig. 1.8, the smaller the sampling needle (i.e., the closer the sampling needle is to the intima of the artery), the higher the albumin concentration of the liquid sample. The transmural pressure of 100 mmHg (Fig. 1.8a) is more pronounced than that of 200 mmHg (Fig. 1.8b), indicating that concentration polarization occurs in a thin layer very close to the arterial intima. When the needle is thicker, some of the liquid samples extracted may come from the outside of the concentrated protein layer.

In recent years, researchers have increasingly focused on the transport of substances in the circulatory system and the interaction between blood cells and vascular walls. They have proposed the concepts of "residence time" and "deposition" of AS causing substances in vascular walls, and believe that the "residence time" and "deposition" of AS causing substances in vascular walls are related to AS lesions and thrombosis. Obviously, the concentration polarization of macromolecules in the arterial system will affect the "residence time" and "deposition" of AS substances in the vascular wall.

Concentration polarization is a well-known engineering phenomenon. Due to the semi permeable membrane of the arterial wall, water can penetrate, so we believe that concentration polarization of AS lipids can occur in the arterial system (Deng

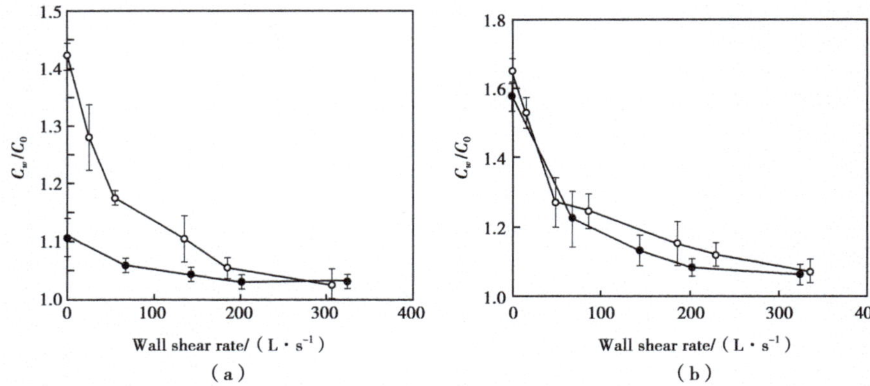

Fig. 1.8 The effect of the size of the injection needle on the albumin concentration of the recovered liquid sample. Data from gauge 31; °: results from gauge 27. **a** Pressure = 100 mmHg; **b** pressure = 200 mmHg. The thickness of the protein concentrate layer on vascular intima depends on the filtration rate. The higher the filtration rate contributes the thicker the filter layer. When the layer is thick enough [in the case of this figure], the difference of the albumin concentration from needles 27 and 31 was decreased. Adapted from Deng and Wang (2003)

et al. 2003). Using 19 nm fluorescent microspheres as tracer particles, Naiki and Karino observed concentration polarization on the monolayer surface of cultured endothelial cells (Naiki & Karino 2000). Numerical simulation also demonstrated that concentration polarization occurs in the curved segment of the aorta.

The distribution of LDL on the surface of tunica intima was analyzed by numerical simulation. The results showed that the LDL concentration on the surface of tunica intima was 5 ~ 14% higher than that in the arterial lumen under pulsatile flow conditions. The LDL concentration on the surface of tunica intima is closely related to the wall shear rate and filtration rate, but the pulsation of blood flow has little effect on the LDL concentration on the surface of tunica intima.

Although the LDL concentration on the surface of tunica intima changes with time in the cardiac cycle, its time average value is almost the same as that under the steady flow condition.

To validate the numerical analysis results, bovine serum albumin was used as a tracer macromolecule to directly measure the concentration of surface proteins on the canine carotid artery intima in vitro. This experiment clearly shows that the concentration of albumin on the surface of tunica intima is higher than that in the lumen, indicating that the concentration of macromolecular lipids is indeed polarized in the arterial system. At low shear rate, the relative concentration of albumin C_w/C_0 on the surface of tunica intima decreased sharply with the increase of shear rate, which was consistent with the results of numerical simulation. The experiment shows that the filtration rate V_w has a significant effect on the albumin concentration on the surface of tunica intima. The larger the V_w is, the higher the albumin concentration on the surface of tunica intima is. This indicates that the increase of albumin concentration on the surface of tunica intima is not caused by the absorption of albumin by endothelial cells, but by the concentration polarization phenomenon. The most interesting thing is that the size of the sampling needle significantly affects the concentration of albumin in the extracted liquid sample. This result indicates that concentration polarization occurs in a thin layer very close to the surface of the arterial intima. The thickness of the polarization layer of macromolecules depends on the filtration rate, and the higher the filtration rate, the thicker the polarization layer of AS material.

Current theoretical and experimental studies have shown that the lipid concentration at the blood/vascular wall interface is influenced by local flow fields. The recirculation and stagnation of flow in the low wall shear rate region are higher than those in the high wall shear rate laminar flow region. Within the normal physiological filtration rate range ($V_w = 4.8 \pm 0.6 \times 10^{-6}$ cm/s), At $G_w = 185\ \text{s}^{-1}$, C_w is only 5% higher than C_0 at 10^{-6} cm/s. However, when $G_w = 0\ \text{s}^{-1}$, C_w is 42% higher than C_0. Considering that AS is a chronic and progressive disease, the local change of AS lipid concentration caused by the surface of tunica intima is very important for the occurrence and development of AS. If this material transport phenomenon does exist in the human cycle, it may play an important role in the localization of AS, and its impact on AS is at least as important as the effect of shear stress, or even greater than the effect of shear stress.

Clinical observations indicate that AS only occurs in the arterial system and not in the venous system. It is difficult to explain this phenomenon solely by considering the effect of wall shear stress. The shear stress of the venous system is much lower than that of the arterial system, and there are also many branches and bifurcations in the venous system, where blood flow is significantly disrupted and slow recirculation is formed. However, it is easy to explain from the perspective of macromolecular material transport. Firstly, compared to veins, arteries are high-pressure systems, under which lipids can more easily penetrate the arterial walls. Secondly, the arterial wall is much thicker than the venous wall, so the infiltrated lipids can hardly pass through the arterial wall, but are excreted through the lymphatic system within the outer membrane (the lymphatic system in the arterial wall is only located within the outer membrane layer of the arterial wall, serving as a drainage channel for blood vessels). Therefore, lipids may accumulate within the arterial wall. The vein is a low-pressure system. More importantly, its walls are very thin, so the infiltrated lipids easily penetrate the walls and are excreted through the lymphatic system. Therefore, lipids hardly accumulate in the venous system. Interestingly, if a venous segment is implanted into the arterial system (arterial bypass surgery), AS can also occur in the implanted venous segment, which is even faster and can be well explained from the perspective of macromolecule transport.

It must be mentioned that the shear stress caused by flow can affect the permeability of the arterial wall, which is one of the important factors in the formation of AS. Although under normal arterial permeability, blood flow itself may cause a local increase in lipid concentration at the blood/vessel wall interface, further research is needed to address whether elevated lipid concentration can lead to AS. Although most scientists are committed to studying the effects of shear stress, we believe it is necessary to clarify the correlation between material transport in the arterial system and AS. As is well known, after bypass surgery, intimal hyperplasia occurs in the transplanted vein, which accelerates the occurrence of AS and leads to the failure of arterial transplantation. Clinical studies have shown that the vein segment implanted in the arterial system undergoes arterialization, meaning that the blood vessel wall of the vein segment thickens, thereby reconstructing an adaptive response to arterial pressure. As a result, the infiltrated AS causing lipids will accumulate abnormally in the venous wall, thereby accelerating the progression of AS lesions. Anastomotic pseudoaneurysm is another issue in arterial reconstruction and bypass surgery, which is related to the accumulation of lipids in the walls of artificial vascular prostheses. The accumulation of lipids in the walls of vascular prostheses can lead to calcification and accelerate the oxidative degradation of the material, affecting its overall mechanical properties. Therefore, research on lipid transport in the arterial system not only helps to better understand the role of hemodynamic factors in AS, but also has important guidance for the effective prevention and treatment of clinical vascular diseases.

1.3 In Vivo Observation of Lipid Concentration Polarization on the Surface of Tunica Intima—Zebrafish Model

Zebrafish is often used as an animal model due to its relatively low maintenance costs, rapid proliferation, and ease of genetic manipulation. In addition, the optical transparency of zebrafish embryos combined with advanced imaging technology makes it possible to directly observe complex life phenomena at the entire biological level. Zebrafish has become a powerful tool for in vivo research in the cardiovascular field and lipoprotein biology (Fang et al. 2014; Schlegel 2016; Stoletov et al. 2009; Vedder et al. 2020). Here, laser scanning confocal microscope (LSCM) was used to observe the polarization of LDL concentration in zebrafish.

1.3.1 Zebrafish Model

Flk1: green fluorescent protein zebrafish (Flk1: GFP), endothelial cells express GFP (provided by developmental biology Laboratory of Tsinghua University). 10 octacosyl-3, 3, 30, 30 tetramethyl indole carbonyl anthocyanin LDL (DiI LDL, 200 mg ml^{-1}, Biomedical Technologies Inc.) was injected into the circulating blood of GFP embryos 48 h after fertilization (h.p.f.) by microinjection. The control group was injected with phosphate buffer saline (PBS) at the injection site shown in Fig. 1.9. Anesthetized zebrafish embryos [exposed to 0.02% Tricaine for a short period of time], awakened with fresh water, and then placed in a fish tank. Euthanasia of zebrafish was achieved by prolonged exposure to 0.02% Tricaine.

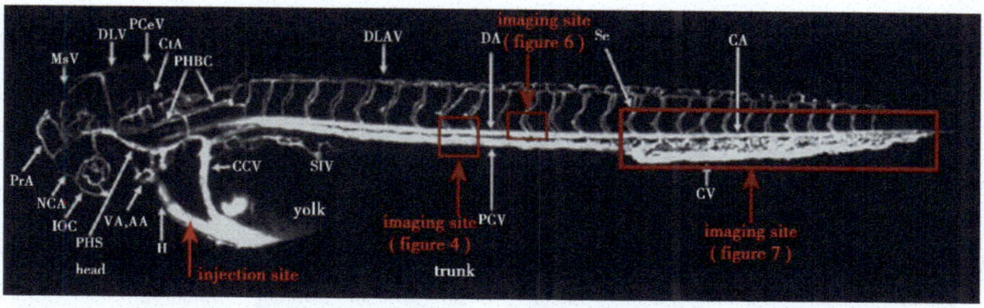

Fig. 1.9 The map of injection and imaging site. Adapted from Xie et al. (2013)

The embryos of zebrafish are fixed with 1.5% low melting point agarose to maintain normal blood circulation. Use Leica LSCM (SP5) to screen for LDL migration rates in straight and bifurcated vascular segments. The excitation wavelength of green fluorescence is 488 nm, and the excitation wavelength of red fluorescence is 561 nm. Record using a 10x objective lens, with 512 scans per scan × A focal plane of 512 square pixels.

1.3.2 Fluorescence Intensity Analysis

We collect samples from five different positions (distinguishing different colors) at the surface of the vascular lumen to the center of the vessel (Fig. 1.10). Analyze the red fluorescence intensity at different positions of the vascular lumen using IMAGE-PRO PLUS v.6.0 software.

1.3.3 Distribution of Low-Density Lipoprotein Concentration in Straight Sections

By injecting DiI-LDL into Flk1: GFP embryonic blood, fluorescence distribution was observed at 48 h.p.f using LSCM at 52 h.p.f. The quantitative measurement of red fluorescence intensity shows that the LDL concentration in arteries (Fig. 1.11a, c) and veins (Fig. 1.11a, c) shows a gradient distribution from the vascular wall to the axis. These results indicate that there is a concentration polarization of LDL in the vascular system.

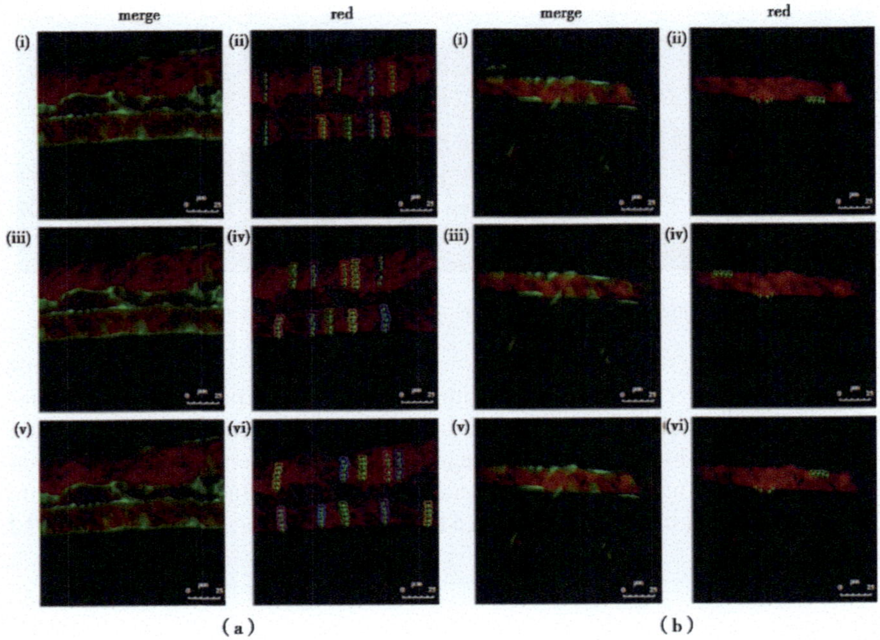

Fig. 1.10 Red fluorescence intensity analysis. Flk1: GFP embryos at 52 h.p.f. and DiI LDL embryos at 48 h.p.f. The endothelial cells were green and DiI LDL was red. **a** Red fluorescence intensity analysis of straight vessel segment. Fluorescence intensity analysis diagram from three time points [(i) $t = 0$ s, (iii) $t = 2.22$ s and (v) $t = 8.88$ s]. (ii, iv, vi); the samples collected at three time points were marked with a small grid. **b** The red fluorescence intensity of the bifurcation region was analyzed. [(i) $t = 1.674$ s, (iii) $t = 2.511$ s and (v) $t = 13.671$ s]. The samples collected at three time points were marked with a small grid and fluorescence intensity was analyzed (ii, iv, vi). Adapted from Xie et al. (2013)

1.3.4 Distribution of Low-Density Lipoprotein Concentration at Vessel Bifurcation

The change of flow field caused by the change of vessel geometry will significantly affect the LDL concentration distribution. Previous studies have shown that flow field disturbances occur in the bifurcation vascular segment (Fan et al. 2019). To support this hypothesis, DiI-LDL was injected into Flk1: GFP embryos, with blood circulating at 48 h.p.f and LDL fluorescence distribution observed by LSCM at 52 h.p.f. The quantitative measurement of red fluorescence intensity indicates that the LDL concentration on the forked cavity surface is significantly higher than that on the flat surface (Fig. 1.12).

At 72 h.p.f, significant lipid deposition was observed in the tail vein (Fig. 1.13). At the same site, LDL infiltrates under the endothelium (Fig. 1.13, arrow).

Our experiment shows that lipid deposition occurs in the tail vein of zebrafish embryos, and LDL infiltrates into the subendothelial layer of the tail vein. In the embryonic stage of zebrafish, the great arteries and veins are directly connected,

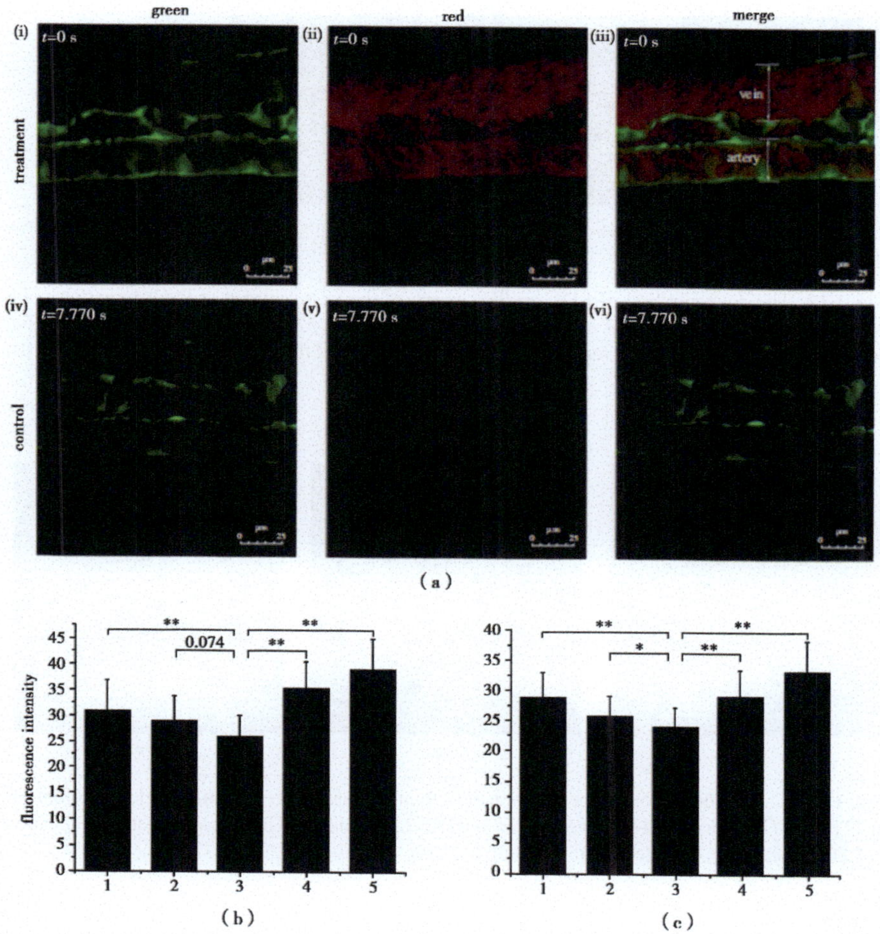

Fig. 1.11 LDL concentration polarization in the vascular system. **a** Flk1: 52 h.p.f. GFP embryos; (i–iii) 48 h.p.f. DiI—LDL embryos. The endothelial cells were green and DiI—LDL was red. Red fluorescence intensity of vein (**b**) and artery (**c**). Numbers 1–5 indicated the position of the sample in the vessel (In this figure a ii, iv, v): 1 and 5 presented the lumen surface of the vessel; 2–4 presented from the lumen surface to the vessel center. 3 presented the vascular center (*$p < 0.05$; **$p < 0.01$). Adapted from Xie et al. (2013)

rather than indirectly connected through a capillary network, and lipid deposition in the veins is closely related to the sharp changes in vascular geometry. In adult zebrafish, lesions are only found in the dorsal aorta and not in the caudal vein. Why AS lesions only occur in arteries? We believe that arteries are high-pressure systems, so lipids can more easily penetrate the blood vessel wall. Secondly, due to the thick arterial wall, lipids are easily blocked in the arterial wall, making it difficult for them to penetrate and be excreted by the lymphatic system. The venous wall is a very thin

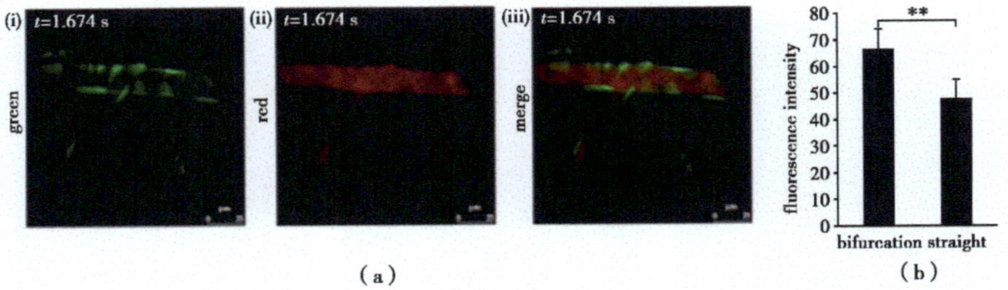

Fig. 1.12 The concentration of LDL on the surface of the vessel lumen at the bifurcation point is significantly higher than that in the straight section. **a** Flk1: 52 h.p.f. GFP embryo; endothelial cells were green, DiI LDL was red. (b) Statistical analysis of red fluorescence intensity. $^{**}p < 0.01$. Adapted from Xie et al. (2013)

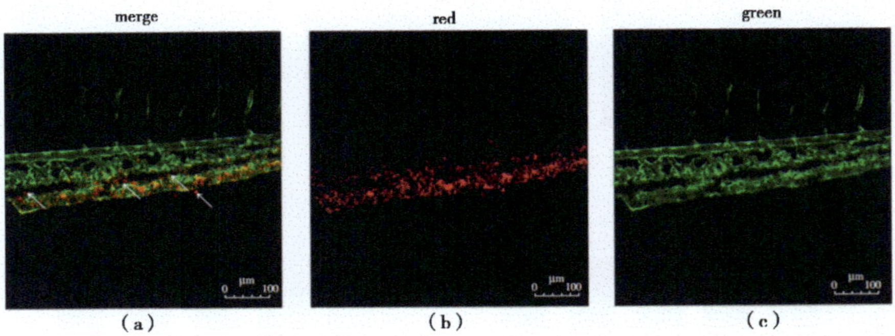

Fig. 1.13 The accumulation and penetration of LDL in the vascular system. **a–c** Flk1: GFP embryos at 72 h.p.f.; The endothelial cells were green and DIL LDL was red. The arrow indicates the location of LDL penetration. Adapted from Xie et al. (2013)

low-pressure system, and lipids are easy to seep out of the venous wall and are easily excreted by the lymphatic system.

1.4 Lipoprotein Concentration Polarization Promotes Atherosclerosis—Rabbit Carotid Artery Cuff Model

1.4.1 Rabbit Carotid Artery Cuff Model

A local vascular stenosis model was obtained using a silicone tube loop surgery on the carotid artery of New Zealand white rabbits (Getz & Reardon 2011; Winkel et al. 2015). After anesthesia with midazolam (0.2–0.4 mL/kg), a fixed incision was made along the midline of the neck between the thyroid and sternum, with a length of 4 cm. The subcutaneous tissue and neck muscles were dissected layer by layer, exposing the trachea, and separating the middle segment of the left common carotid artery (LCCA) from the left cervical sheath by about 2 cm. Establish local stenosis of LCCA using an 8 mm long medical silicone tube (stenosis degree of 40%), ligate with surgical silk thread, and use the right common carotid artery as a sham surgical control. Analgesics (with a concentration of 0.3 mg/kg buprenorphine) were administered every 8 h within 24–48 h after surgery, accompanied by broad-spectrum antibiotics. The postoperative high-fat group was given a high-fat diet containing 1% cholesterol.

1.4.2 Numerical Simulation of Local Narrow Flow Field

Establish the geometric modeling of the local stenosis artery (Fig. 1.14). It is assumed that the vessel wall is rigid, and the blood is incompressible Newtonian fluid, its viscosity is $\mu = 3.5$ mPa/s, and the blood density is: $\rho = 1.05 \times 10^3$ kg/m^3, the Navier Stokes equation for steady flow is:

$$\nabla u = 0$$

And

$$\rho(\nabla u)u = -\nabla p + \mu \nabla^2 u$$

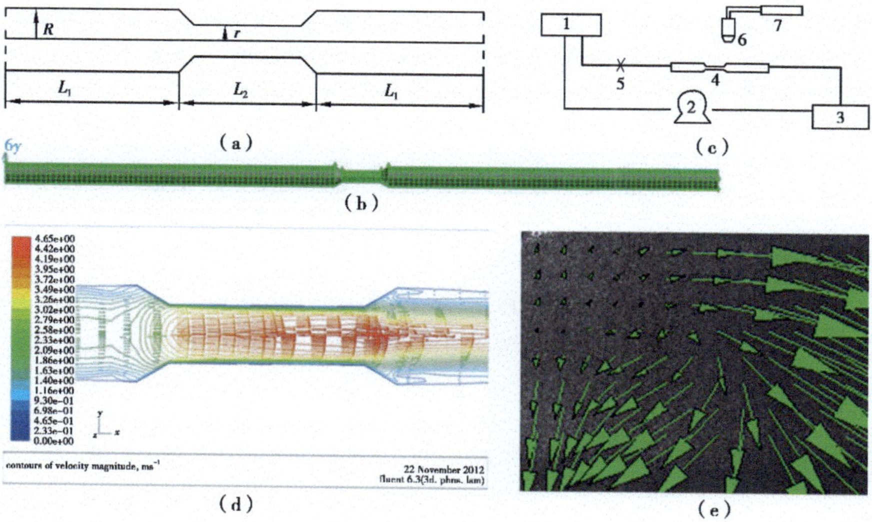

Fig. 1.14 Flow field analysis of local stenotic artery. **a** The geometric model of the local stenotic artery. The stenotic vessels were axisymmetric. The radius of the stenotic part r was 0.075 cm, and the diameter of the straight part R was 0.125 cm. (R-r)/R is the degree of stenosis (40%), L1 = 5 cm and L2 = 0.8 cm. **b** Three dimensional model and mesh generation of stenotic vessels. **c** Particle image velocimetry (PIV) system schematic diagram. The distribution of flow field in the stenosis area was determined by simulation (**d**) and PIV measurement (**e**). Adapted from Xie et al. (2013)

P is the blood pressure gradient, Δu is the velocity divergence, $\Delta^2 u$ is the Laplacian Equation Vector for u.

$$u = 2u_0\left[1 - \left(\frac{r}{R}\right)^2\right], \quad x = 0, \; v = 0, \; w = 0$$

u, v, w are the velocity vectors in the x, y, and z directions, respectively, u_0 is the entry velocity, and r is the radius of the narrow vessel cross-section, R is the radius of a straight blood vessel cross-section.

Then, a three-dimensional model of the stenosis vessel was established using GAMBIT software (with a stenosis degree of 40%). The diameter of the vascular inlet is 0.25 cm, which is the same as the diameter of the rabbit carotid artery (Fig. 1.14a). GAMBIT software divides the three-dimensional model mesh (Fig. 1.14b).

1.4.3 Numerical Simulation of Low-Density Lipoprotein Concentration Distribution in Local Stenosis Vessels

The numerical simulation results indicate that due to the presence of seepage through the arterial wall, LDL will accumulate on the surface area of the vessel wall at the distal end of the stenosis, forming a region with a sharp increase in concentration at the distal end of the stenosis. There is also a peak of LDL, which decreases with the increase of inlet flow velocity and becomes more pronounced with the increase of stenosis degree. However, when the inlet flow rate is constant, this peak value changes with the degree of stenosis. At Re of 250, the peak LDL concentration at the distal end of 40% stenosis vessels is 153% of the body concentration, which is higher than 30% stenosis (129%) and 50% stenosis (141%) (Fig. 1.15).

1.4.4 Experimental Measurement of Low-Density Lipoprotein Concentration Distribution in Local Stenotic Vessels

On the basis of numerical simulation proving the polarization of lipoprotein concentration, LSCM measured the concentration distribution of LDL in carotid artery of New Zealand white rabbits in z-axis in vitro. The experimental test was conducted under steady flow conditions, with LSCM scanning along the longitudinal axis from the inner wall surface of the blood vessel to the axial plane of the vessel. By comparing the fluorescence intensity ratio of the inner wall surface of the distal end of the narrow vessel to the central plane of the vessel, it can be seen that the concentration on the inner wall surface of the distal end of the narrow vessel is higher than the concentration on the body, and the wall concentration decreases with an increase in inlet flow

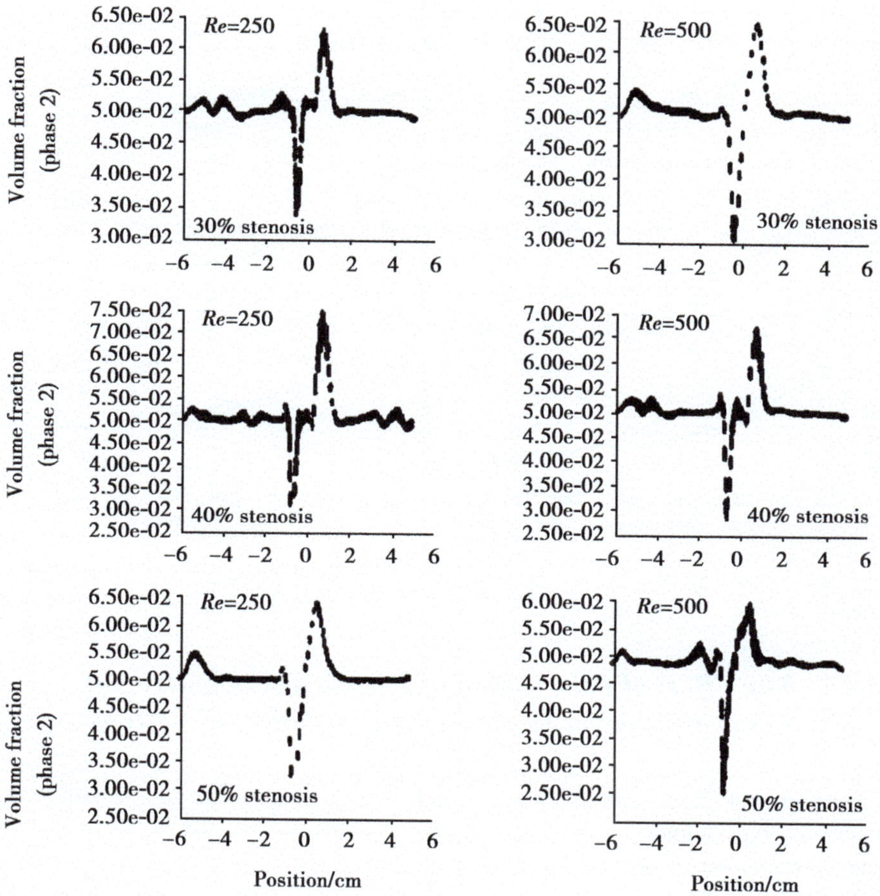

Fig. 1.15 Numerical simulation of LDL concentration distribution in local stenotic vessels. Adapted from Wei et al. (2007)

velocity; When the degree of stenosis is 40%, the wall concentration at the distal end of the stenosis vessel is (147 ± 17)% of the lumen concentration, while when Re = 500, the wall concentration at the distal end of the stenosis vessel is (132 ± 10)% of the lumen concentration, and the wall concentration is related to the degree of stenosis. Under the same inlet flow rate conditions, although the wall concentration of different stenosis degrees also changes, the cavity surface concentration of LDL is always higher than the LDL concentration of the fluid body, which is basically consistent with the numerical simulation results (Fig. 1.16).

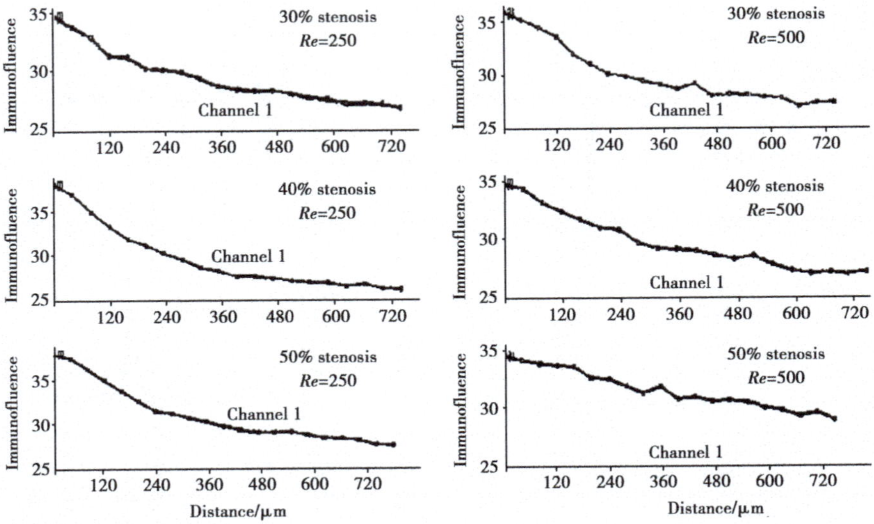

Fig. 1.16 The distribution of LDL in the z-axis direction of the rabbit carotid artery. Adapted from Wei et al. (2007)

1.4.5 The Effect of Polarization of Low-Density Lipoprotein Concentration on Local Stenotic Vascular AS Lesions

The results of LDL concentration detection in New Zealand white rabbits with local stenosis of the common carotid artery on a high-fat diet showed that their serum LDL concentration was significantly higher than that of the normal diet control group (Fig. 1.17a). A large amount of lipid accumulation was found in both the normal diet and the high fat diet stenosis model group (Fig. 1.17c (iii) and (iv)), and the high fat diet stenosis model group was more pronounced (Fig. 1.17c (iv)), while no significant lipid accumulation was observed in the control group (Fig. 1.17c (i) and (ii)). And the total cholesterol content in the stenosis model group was significantly higher than that in the control group, and the total cholesterol content in the high-fat diet group was also higher than that in the normal diet group (Fig. 1.17b).

A silicone tube was placed on the straight segment of the rabbit carotid artery to create a local stenosis model of the common carotid artery. Through computer simulation and PIV measurement, the results indicate the presence of LDL concentration polarization in the recirculation zone of the distal end of narrow blood vessels, indicating that flow field disturbances (eddies or flow stagnation) caused by changes in vascular geometry lead to the deposition of LDL carried by the blood flow. And although the serum LDL and cholesterol levels in the high-fat diet group were significantly higher than those in the normal diet group, only the stenosis model group formed AS plaques at 8 weeks. In addition, significant intimal thickening and lipid deposition were observed in the stenosis model group. This may be due to the direct contact of vascular endothelial cells with high concentrations of LDL polarization

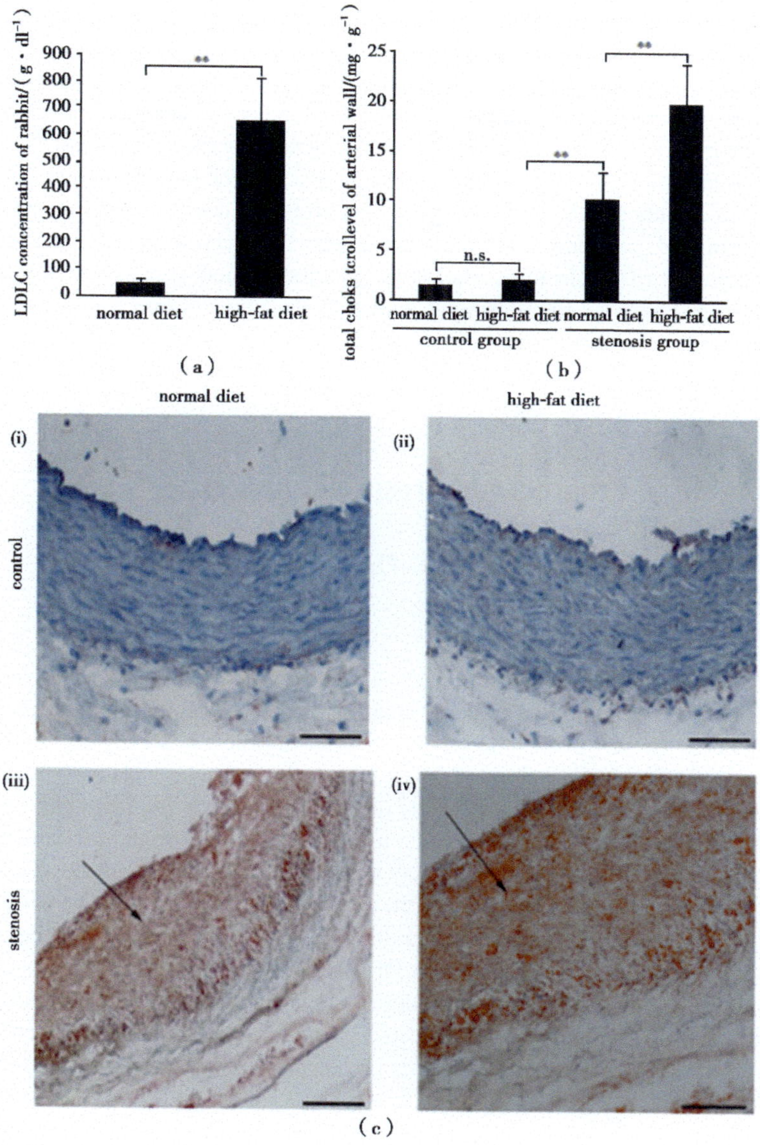

Fig. 1.17 LDL concentration polarization promoted atherosclerosis of rabbit common carotid artery. The local stenosis mode 1 was established by operation. Rabbits were fed with normal diet or high-fat diet for 8 weeks. **a** Serum LDL concentration, $*p < 0.01$. **b** Total cholesterol level $**p < 0.01$. **c** Oil red O staining. Rabbits without local stenosis were fed with normal diet (i) or high-fat diet (ii). Rabbits with local stenosis were fed with normal diet (iii) or high fat diet (iv). Adapted from Xie et al. (2013)

layer, which damages vascular endothelial cells and promotes the occurrence and development of AS lesions.

Although a large number of studies have shown that LDL concentration polarization exists in the arterial system and is related to the occurrence and development of AS, there is still a lack of strong direct evidence for the relationship between LDL concentration polarization and the occurrence, development, plaque stability, and clinical events of AS. Moreover, there is currently a lack of effective means to intervene in LDL concentration polarization, thereby intervening in the occurrence and progression of AS lesions. Therefore, developing new research methods and methods to explore the relationship between LDL concentration polarization and the occurrence and development of AS, as well as its regulatory mechanisms, may provide new ideas and strategies for the effective prevention and treatment of AS.

1.5 Research Progress and Prospects

1.5.1 The Development of Low-Density Lipoprotein Targeting Therapy

Genome wide association study (GWAS) has shown that AS lesions are associated with single nucleotide polymorphisms in genes encoded by lipid related proteins or nearby genes. These results indicate that plasma lipid levels are closely related to the occurrence and development of AS. Research on mouse models has shown that knocking out the LDL receptor (LDLR) or apolipoprotein E (ApoE) gene leads to an increase in plasma cholesterol levels, ultimately leading to AS susceptibility in C57BL/6 mice, indicating that lipids play a crucial role in the pathogenesis of AS.

Research suggests that plasma lipoprotein damage to vascular endothelial cells and accumulation under dysfunctional endothelium is the initiating link in the occurrence and development of AS, with LDL accumulation under the endometrium closely related to the occurrence and development of AS. The LDL accumulated under the intima, on the one hand, ingests LDL in the form of cholesterol complex or crystal through the micro endocytosis or phagocytosis of macrophages; On the other hand, LDL is oxidized by reactive oxygen species (ROS) and Fe^{2+} to form Ox-LDL, which is uptake by macrophages through Toll like receptor pathways or scavenger pathways. The continuous influx of cholesterol ultimately exceeds the metabolic capacity of macrophages, forming intracellular lipid droplets and cholesterol microcrystals, activating inflammasomes and triggering inflammatory reactions. Reducing the accumulation of cholesterol in macrophages and increasing the efflux of cholesterol can help suppress the inflammatory response of AS plaques, and even promote plaque regression. Therefore, reducing LDL levels and inhibiting inflammatory responses represent the basic treatment strategies for current anti AS. Among them, HMG-CoA reductase (statins) inhibits endogenous cholesterol synthesis, lowers LDL cholesterol, and has multiple anti-inflammatory effects, which can prevent, delay, and even reverse AS lesions. However, it should be noted that even after receiving high-dose statin therapy, more than 20% of patients with acute coronary syndrome will relapse within 30 months.

Recent studies have found that LDL, as a related autoantigen, can drive the autoimmune response in AS plaques (Wolf et al. 2019). CD4 + T cells in human AS plaque recognize Ox LDL by binding to major histocompatibility complex (MHC) presenting peptide epitopes of ApoB-100. Anti Ox LDL antibodies produced by B cell-derived plasma cell were detected in the serum of people or animals with AS. Therefore, the immune mechanism of LDL promoting AS deserves in-depth exploration and research. Unlike cytokine antibodies, immunotherapy and vaccination have antigen specificity and can prevent host defense. The antiatherosclerosis effect of LDL vaccines is also a field and strategy worth exploring.

1.5.2 Numerical Simulation of Low Density Lipoprotein Concentration Polarization Study

LDL is one of the most important biochemical factors in AS and has always been a key molecular focus in research related to AS. The arterial wall can be seen as a porous medium composed of cells, and the number and distribution of gaps between vascular endothelial cells determine the material transport conditions of the porous model in numerical simulation (Avgerinos & Neofytou 2019). At present, there are three types of numerical simulation models for LDL mass transfer in arterial walls, namely, no wall model, homogeneous wall model, and multi-layer wall model. Different models have different boundary conditions for fluid and solute. In the wall free model, the blood flow in the vascular cavity is independent of the material transfer mechanism within the wall, so only necessary boundary conditions need to be considered. This model only defines the blood flow velocity at the inlet and the constant filtration rate of the pipe wall, which is relatively simple. It is often used to discuss the LDL concentration and its distribution in the whole lumen, but cannot analyze the solute concentration distribution in the wall. The homogeneous wall model assumes that the vascular wall is a single-layer homogeneous porous medium, with relatively constant solute convection and diffusion. The concentration at the inlet and outlet of the lumen, as well as between the middle and outer membranes, is also relatively constant. The multi-layer wall model is a very comprehensive model that considers the different characteristics of each arterial layer as a membrane or porous medium, thus considering the heterogeneity of the vascular wall, which can better analyze the distribution of LDL within the vascular wall (Roustaei et al. 2018). The four-layer wall model defines the endothelium, inner membrane, inner elastic plate, and middle membrane as uniform porous media, ignoring the boundary effects of different porous layers. The filtration rate is continuous at the interface between the lumen, inner membrane, inner elastic plate, and middle membrane.

Darcy law, Darcy Forchheimer model, Brinkman model, Vafai and Tien, and the Brinkman Forchheimer Darcy equation have been used in material transport models to describe biological phenomena. Darcy's law states that there is a linear relationship between flux and pressure gradient through porous media. The Darcy Forchheimer model is an improved model to explore the relationship between LDL and blood vessel wall based on liquid Reynolds number permeability. The Brinkman model considers porous media and has a non-smooth boundary condition on the surface. All the above mathematical models simplify the biochemical processes related to the formation and progression of AS lesions. But in reality, LDL is oxidized to Ox LDL by oxygen free radicals under the endothelium, causing oxidative stress. On the one hand, it induces endothelial cells to express adhesion molecules such as vascular cell adhesion molecule-1 (VCAM-1), intercellular cell adhesion molecule-1 (ICAM-1) and P-selectin; On the other hand, Ox LDL stimulates EC and smooth muscle cells to secrete monocyte chemoattractant protein-1 (MCP-1) and monocyte colony-stimulating factor (M-CSF), and then causes adhesion and transmembrane migration of leukocytes (mainly monocytes and T cells) in circulating blood. Therefore, in the future, more complex models need to be constructed to simulate the near wall transport of LDL, monocytes, and MCP-1, and to explore the interactions between other lipid components, cell types, and biochemical signals. Monocytes and LDL play a crucial role in the formation of AS plaques. Previous studies have used discrete Lagrangian models and aggregated continuum models to explore the transmembrane migration of monocytes and their mechanisms.

Research suggests that blood flow, vascular wall deformation, and filtration rate interact and affect the LDL concentration on the vascular wall. Due to the close correlation between LDL concentration and arterial geometry, accurate characterization of arterial geometry is necessary for the analysis of mass transfer. Introducing more practical geometric shapes will facilitate the assessment of the interrelationships between biological processes such as hemodynamics and inflammation, as well as their impact on the occurrence and development of AS lesions.

The arterial wall is a viscoelastic deformable tissue, and the deformation of the arterial wall and the shear stress of the blood flow interact closely with each other. Hemodynamic factors affect the filtration rate of LDL in the blood and also affect the dynamic characteristics of the vascular wall. These interactions between blood flow and vascular wall deformation play a role in LDL concentration polarization and AS formation. The porous high elasticity model is suitable for characterizing the viscoelasticity of arterial walls, as well as the pressure gradient and filtration flow within the walls. Therefore, the porous high elasticity arterial wall model will be an important model for analyzing material transport in vascular walls.

Although significant efforts have been made so far, numerical simulations cannot fully predict the polarization of LDL concentration and its impact on the occurrence and development of AS, as well as its regulatory mechanisms. The development of more complex prediction tools depends on more accurate parameters involved in material transport, such as permeability, porosity, and diffusivity. Close cooperation in fields such as medicine, engineering, physics, and chemistry will lead to more accurate parameters involved in the transport of substances within arterial walls, thereby improving modeling accuracy in arterial characteristics, hemodynamics, biochemical processes, and boundary conditions (Avgerinos & Neofytou 2019; Libby et al. 2011). This will be the direction of future efforts.

References

Avgerinos NA, Neofytou P. Mathematical modelling and simulation of atherosclerosis formation and progress: a review[J]. Ann Biomed Eng, 2019, 47(8): 1764–1785.

Baeyens N, Bandyopadhyay C, Coon BG, Yun S, Schwartz MA. Endothelial fluid shear stress sensing in vascular health and disease [J]. J Clin Invest, 2016, 126(3): 821–828.

Davies PF. Hemodynamic shear stress and the endothelium in cardiovascular pathophysiology[J]. Nat Clin Pract Cardiovasc Med, 2009, 6(1): 16–26.

Deng XY, Wang GX. Concentration polarization of atherogenic lipids in the arterial system[J]. Science in China (Series C), 2003, 46(2): 153–164.

Deng XY, Wang GX, Yang Y. Experimental simulation of model platelet adhesion to a semi-permeable wall exposed to flow disturbance[J]. Chinese Science Bulletin. 2003,48(22): 2421–2426.

Fan Z, Liu X, Zhang P, Gu J, et al. Transfer of low-density lipoproteins in coronary artery bifurcation lesions with stenosed side branch: numerical study[J]. Comput Math Methods Med, 2019, 2019: 5297284.

Fang L, Liu C, Miller YI. Zebrafish models of dyslipidemia: relevance to atherosclerosis and angiogenesis[J]. Transl Res, 2014, 163(2): 99–108.

Fatouraee N, Deng X, Champlain A, et al. Concentration polarization of low density lipoproteins (LDL) in the arterial system[J]. Ann N Y Acad Sci, 1998, 858: 137–146.

Ference BA, Ginsberg HN, Graham I, et al. Low-density lipoproteins cause atherosclerotic cardiovascular disease. Evidence from genetic, epidemiologic, and clinical studies. A consensus statement from the European Atherosclerosis Society Consensus Panel. Eur Heart J, 2017; 38(32): 2459–2472.

Fernández-Friera L, Fuster V, López-Melgar B, et al. Normal LDL-cholesterol levels are associated with subclinical atherosclerosis in the absence of risk factors[J]. J Am Coll Cardiol, 2017, 70(24): 2979–2991.

Getz GS, Reardon CA. Animal models of atherosclerosis[J]. Arterioscler Thromb Vasc Biol, 2012, 32(5): 1104.

Libby P, Ridker PM, Hansson GK. Progress and challenges in translating the biology of atherosclerosis[J]. Nature, 2011, 473(7347): 317.

Mahmoudi M, Farghadan A, McConnell DR, et al. The story of wall shear stress in coronary artery atherosclerosis: biochemical transport and mechanotransduction[J]. J Biomech Eng, 2021, 143(4): 041002.

Naiki T, Karino T. Visualization of flow-dependent concentration polarization of macromolecules at the surface of a cultured endothelial cell monolayer by means of fluorescence microscopy[J]. Biorheology, 2000, 37(5–6): 371–384.

Naiki T, Sugiyama H, Tashiro R, et al. Flow-dependent concentration polarization of plasma proteins at the luminal surface of a cultured endothelial cell monolayer[J]. Biorheology, 1999, 36(3): 225–241.

Roustaei M, Nikmaneshi MR, Firoozabadi B. Simulation of Low Density Lipoprotein (LDL) permeation into multilayer coronary arterial wall: interactive effects of wall shear stress and fluid-structure interaction in hypertension[J]. J Biomech, 2018, 67: 114–122.

Schlegel A. Zebrafish models for dyslipidemia and atherosclerosis research[J]. Front Endocrinol (Lausanne), 2016(7): 159.

Souilhol C, Serbanovic-Canic J, Fragiadaki M, et al. Endothelial responses to shear stress in atherosclerosis: a novel role for developmental genes[J]. Nat Rev Cardiol, 2020, 17(1): 52–63.
Stoletov K, Fang L, Choi SH, et al. Vascular lipid accumulation, lipoprotein oxidation, and macrophage lipid uptake in hypercholesterolemic zebrafish[J]. Circ Res, 2009, 104(8): 952–960.
Vedder VL, Aherrahrou Z, Erdmann J. Dare to compare. Development of atherosclerotic lesions in human, mouse, and zebrafish[J]. Front Cardiovasc Med, 2020, 7: 109.
Vincent PE, Weinberg PD. Flow-dependent concentration polarization and the endothelial glycocalyx layer: multi- scale aspects of arterial mass transport and their implications for atherosclerosis[J]. Biomech Model Mechan, 2014, 13(2): 313–326.
Wang GX, Deng XY, Robert Guidoin. Concentration polarization of macromolecules in canine carotid arteries and its implication for the localization of atherogenesis[J]. Journal of Biomechanics, 2003,36(1): 45–51.
Wei DH. The effect of low-density lipoprotein concentration polarization on the formation of atherosclerosis and its mechanism [D] Chongqing University, 2006.
Wei DH, Wang GX, Tang CJ. Concentration polarization of low density lipoprotein at the distal end of carotid stenosis promotes atherogenesis[J]. Acta Physiologica Sinica, 2007, 59(6): 831–839.
Wiklund O, Carew TE, Steinberg D. Role of the low density lipoprotein receptor in penetration of low density lipoprotein into rabbit aortic wall. Arteriosclerosis. 1985, 5(2): 135–41.
Winkel LC, Hoogendoorn A, Xing R, et al. Animal models of surgically manipulated flow velocities to study shear stress-induced atherosclerosis[J]. Atherosclerosis, 2015, 241(1): 100–110.
Wolf D, Ley K. Immunity and inflammation in atherosclerosis[J]. Circ Res, 2019, 124(2): 315–327.
Xie X, Tan J, Wei DH, et al. In vitro and in vivo investigations on the effects of low-density lipoprotein concentration polarization and haemodynamics on atherosclerotic localization in rabbit and zebrafish[J]. J R Soc Interface, 2013, 10(82): 20121053.

Chapter 2
Low Shear Stress Regulates Biological Behavior of Vascular Endothelial Cells to Promote Atherosclerosis

Cardiovascular disease is a serious threat for human health. At present, the death of cardiovascular disease is the first death factor of all diseases in China. AS is the main pathophysiology basis of cardiovascular disease. The lesions of AS first show that the vascular endothelial cells are damaged, the permeability of the vascular intima is increased, the lipid in the blood enters the intima for oxidative modification and accumulation, and induces the phenotypic changes of vascular smooth muscle cells and the foam of macrophages, leading to the formation and development of AS plaque (Gomez et al. 2012; Petsophonsakul et al. 2019; Zhang et al. 2018).

A large number of studies have shown that although the formation and development of AS lesions are related to systemic risk factors such as Hypercholesterolemia, aging, smoking, etc., the AS lesions are localized at the lesion site. The lesion site of AS is more likely to occur in areas where the geometric shape of blood vessels in the arterial system changes rapidly, such as the bending and branching of blood vessels (Li et al. 2014). The blood flow field in these areas with rapidly changing geometric shapes undergoes significant changes, from laminar flow to turbulent flow, and the blood flow shear stress changes from physiological laminar shear stress to low shear stress or oscillatory shear stress (Ajami et al. 2017).

Vascular endothelial cells are single-layer epithelial cells that come into direct contact with blood on the luminal surface of blood vessels. These cells perceive changes in the speed and direction of blood flow, integrate biomechanical signals, and convert them into biochemical signals, thereby triggering a series of physiological and pathological reactions.

G. Wang, *The Mechanobiology in Atherosclerosis*,
https://doi.org/10.1007/978-981-97-6512-6_2

2.1 The Injury of Vascular Endothelial Cell is a Key Factor in the Process of Atherosclerosis

Vascular endothelial cells are located on the inner surface of blood vessels, directly in contact with blood and sensing changes in the blood flow field (Zeng et al. 2018). The intima of blood vessels is a selective permeable interface composed of endothelial cells, which regulates the transportation and exchange of fluid and macromolecules between blood and vascular walls through a system of intercellular vesicles and intercellular junction complexes. When this selective barrier function is lost, it will promote the adhesion and migration of Monocyte in circulating blood. Normal functioning endothelial cells have anticoagulant and fibrinolytic properties, which can effectively inhibit coagulation cascade reactions, platelet adhesion, and antithrombotic formation.

Shear stress occurred by blood flow plays an important role in vascular homeostasis. Vascular endothelial cells are sensitive to the mechanical environment and participate in the biomechanical response of vascular endothelial cells induced by shear stress and/or stretching through various mechanical sensors (Baeyens et al. 2016). From the perspective of biological engineering, the single endothelial cell that forms the lining of each part of the cardiovascular system is a local biomechanical transducer, which can sense various forces generated by blood flow and convert them into biological reactions. For example, integrin, platelet endothelial cell adhesion molecule-1 (PECAM-1), VE-cadherin and Vascular endothelial growth factor receptor are mechanical sensors that can be activated by shear stress and tension, and mediate a variety of biological behaviors (Brown et al. 2016). Shear stress induces phosphorylation of PECAM-1, which activates Src, Rho and PI3K to mediate cytoskeleton remodeling and integrin activation, and finally leads to changes in endothelial cell arrangement (long axis aligned along the blood flow direction) and nuclear adaptation (Budatha et al. 2018); PECAM-1 regulated by Transcriptional factors (such as NF-κB)) The function of regulating the phenotype of vascular endothelial cells (pro-phenotype or anti-AS phenotype) (Chen et al. 2001). Tensile strain can cause polymerization of Actin in cytoskeleton and extracellular matrix and NF-κB activation. Compared to pathological low stretching (2%) or no stretching, normal levels (8%) of multidirectional stretching can significantly reduce the activation of NF-κB in vascular endothelial cells. In addition, research has shown that physiological shear stress only momentarily activates NF-κB. Pathological low shear stress can lead to continuous activation of NF- κB in vascular endothelial cells (Chen et al. 2015).

In the straight section of blood vessels, the blood flow is laminar, which helps to maintain the anti-inflammatory and antithrombotic properties of the endothelium. Changes in the geometric shape of blood vessels (such as bifurcation, curvature) or disease conditions (such as AS) can greatly affect the biological function of vascular endothelial cells. Physiological blood flow and shear stress make cytoskeleton protein consistent with the direction of blood flow, thus reducing blood flow resistance (Cox et al. 2016). Pathological blood flow and shear stress can activate vascular endothelial cells, transition from an antithrombotic phenotype to a prothrombotic phenotype, and promote platelet adhesion. In addition, low shear stress can cause a large release of inflammatory factors in vascular endothelial cells, reduce nitric oxide production, lead to LDL oxidation and apoptosis of vascular endothelial cells (Chen et al. 2018).

Endothelial cell dysfunction is a key part of AS, which is also the earliest detectable pathological change in the progression of AS lesions (Gimbrone et al. 2016; Kumar et al. 2019). The damaged vascular endothelial cells recruit Monocyte from the circulating blood into the subintima, and these Monocyte differentiate into macrophages and ingest a large amount of oxidized lipoproteins to become macrophage derived Foam cell (a marker of early AS lesions); Activated endothelial cells and macrophages secrete a variety of chemokines and growth factors and act on adjacent smooth muscle cells, promoting smooth muscle cells in the middle membrane to change from contractile to synthetic, and proliferate and synthesize extracellular matrix under the intima, producing fibrous plaques (Chistiakov et al. 2015). In the process of pathological changes, the AS plaques undergo structural remodeling to form fibrous caps, covering the necrotic core composed of oxidized lipoprotein, cholesterol crystals and cell debris, forming atherosclerotic plaques. Inflammatory cells (activated macrophages and T cells, natural killer T cells, dendritic cells) in atherosclerotic plaques further promote the inflammatory phenotype of vascular endothelial cells and develop into unstable or vulnerable plaques. The rupture of unstable or vulnerable plaques releases highly thrombogenic components in the necrotic core, leading to AS embolism and serious clinical events (Timmins et al. 2017). Even though stable plaques (characterized by thick fibrous caps, low levels of lipids, and inflammatory cell content) typically do not cause AS events, they gradually erode the vascular lumen, causing it to gradually narrow, leading to ischemic symptoms (Gomez et al. 2012).

2.2 In Vitro Model of Shear Stress Regulating the Biological Characteristics of Vascular Endothelial Cells

In the arterial system, vascular endothelial cells continuously response to shear stress caused by blood flow, which strongly affects the morphology and function of vascular endothelial cells (Chiu et al. 2009). The in-depth study on how vascular endothelial cells respond to shear stress will help us better understand how changes in mechanical environment affect the morphology, structure and function of vascular endothelial cells, and the relationship between the mechanical conduction barrier of vascular endothelial cells and the occurrence and development of cardiovascular diseases (Chiu et al. 2011).

The parallel plate flow chamber is a commonly used extracorporeal circulation device that explores the impact of shear stress on the biological characteristics of vascular endothelial cells. This system controls the shear stress on the surface of vascular endothelial cells by adjusting the flow rate of the circulating fluid, The loading microscope can dynamically observe the vascular endothelial cells in the flow field in real time and record the morphological response of endothelial cells to shear stress and the changes of adhesion behavior of monocytes and vascular

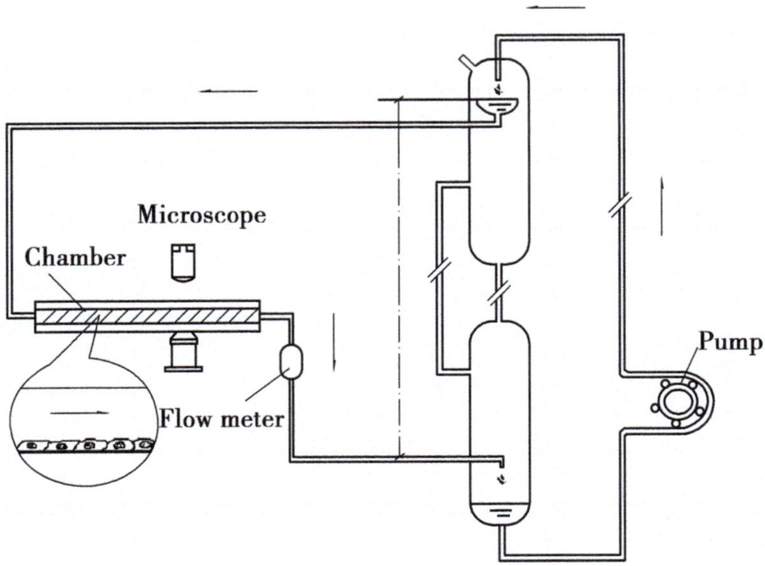

Fig. 2.1 Improved parallel plate flow chamber system. Adapted from Wang et al. (2000)

endothelial cells under the flow field (Civelek et al. 2011; Iring et al. 2019). Due to its simplicity and ease of operation, the system has high practicality. Based on the commonly used parallel plate flow chamber, the system can be modified and improved according to actual needs. This device (Fig. 2.1) is an improvement on the shear stress device proposed by Frangos et al. and consists of three main modules:

(1) Perfusion module: the perfusion module is composed of a liquid storage tank, a micro flowmeter and a Peristaltic pump, which is connected by a polytetrafluoroethylene pipe in the middle. The Peristaltic pump drives the perfusion fluid from the lower tank of the storage fluid to the small cup in the upper tank of the storage fluid, and the perfusion fluid in the small cup in the upper tank enters the lower tank through the connecting system, the flow chamber and the flowmeter. The flow field on the surface of the vascular endothelial cells in the flow chamber of the system is a steady laminar flow.
(2) Flow chamber module: composed of two pieces of organic glass, gaskets, stainless steel frame, and cell slides. Open holes at both ends of the organic glass, with the inlet aperture (0.15–0.20 cm) slightly larger than the outlet aperture (0.10–0.15 cm), and the thickness of the gasket is 0.02 cm. Before use, rinse the organic glass and gasket with tap water, then rinse repeatedly with distilled water, soak in 75% alcohol for sterilization, clean with sterile physiological saline, and air dry. Cell slides are sterilized by ultraviolet irradiation. Other components (such as storage tanks and polytetrafluoroethylene pipes) are washed with distilled water, dried, and sterilized under high pressure. In the experiment, the slides covered with cells were placed in the flow chamber to fix the flow chamber system and prevent leakage. The circulating liquid was driven by the Peristaltic

pump to flow on the surface of the slides covered with cells. During perfusion, place the flow chamber and polytetrafluoroethylene tube in a 37 °C incubator, and connect them to the gas cylinder through the storage tank joint to maintain a humidity of 95% and a CO_2 content of 5% for the circulating liquid.

(3) Recording module: composed of a phase contrast microscope, camera, monitor, and video recorder.

The adhesion and morphological changes of endothelial cells on cell slides under flow conditions were observed by using a phase contrast microscope (Liu et al. 2015). The images were captured and saved by a monitor and video recorder.

① Shear stress on endothelial cells on the surface of flow chamber cell slides τ (N/cm^2) is calculated by Eq. (2.1):

$$\tau = \frac{6\mu Q}{wh^2} \tag{2.1}$$

Among them μ Is the liquid viscosity (N s/cm^2), Q is the flow rate (cm^3/s), h is the flow chamber height (cm), and w is the chamber width (cm).

② The Reynolds number (Re) of the flow in the flow chamber is calculated using the following equation:

$$\mathrm{Re} = \frac{hv\rho}{\mu} \tag{2.2}$$

Among them v Is the average speed (cm/s), ρ Is the density of the circulating liquid (g/cm^3), μ Same equation as h (2.1). The average speed v is:

$$v = \frac{Q}{wh} \tag{2.3}$$

2.3 Animal Model of Shear Stress Regulating the Biological Characteristics of Vascular Endothelial Cells

2.3.1 Hemodynamic Model for Local Stenosis of the Common Carotid Artery

Caro et al. found during autopsy that AS is more common in areas where the geometric shape of blood vessels in the human arterial system undergoes sharp changes, such as bifurcation, curvature, and stenosis, especially in the outer wall and inner wall of arterial branches, carotid sinus, and abdominal aorta (rather than the ascending main artery). These areas are all low shear stress areas (Ni et al. 2010).

In order to simulate the impact of low shear stress on the formation and development of AS lesions and its mechanism of action, a low shear stress model of local

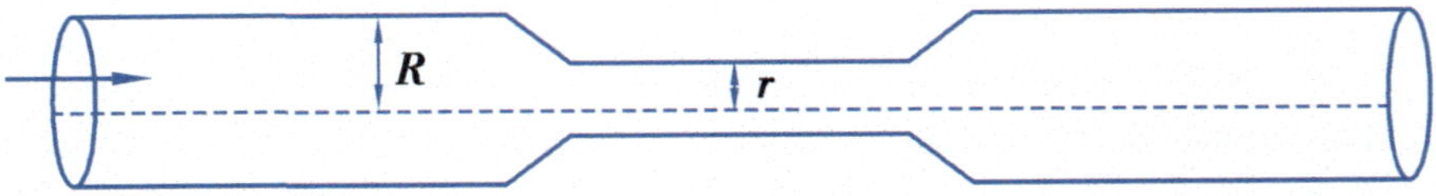

Fig. 2.2 Low shear stress model of local stenosis of the common carotid artery. R: The diameter of straight blood vessel, r: the diameter of vessel in stenosis segment

stenosis of the common carotid artery was established using a silicone tube local cuff method, and its hemodynamic characteristics were simulated and measured using numerical simulation and PIV methods (Wei et al. 2010).

The model used is an axisymmetric circular tube with an inner diameter of 0.15 cm, which is equivalent to the diameter of the common carotid artery in rabbits (Liu et al. 2009). Control the degree of stenosis with (R − r)/R, where r is the inner diameter of the stenosis segment and R is the inner diameter of the straight segment (Fig. 2.2). Assuming that the blood vessel wall is rigid and the blood is an incompressible Newtonian fluid with viscosity μ At 3.5 mPa/s, blood density $\rho = 1.05 \times 10^3$ kg/m^3.

The numerical simulation results of the flow field in the posterior segment of the stenosis show that the flow field in the posterior segment of the stenosis becomes turbulent, and there are vortices formed at the posterior end of the stenosis. The intensity of the vortices increases with the increase of the inlet flow velocity, and the flow field is also more complex. At the same flow rate, the downstream vortex intensity is positively correlated with the degree of stenosis (Pedrigi et al. 2015).

The numerical simulation results of the wall shear stress in the posterior segment of stenosis indicate that the shear stress in the posterior segment of stenosis changes sharply (Reneman et al. 2006). Performance is as follows: in the narrow posterior segment, the wall shear stress first rapidly decreases until it reaches zero (forming a local low shear stress area). Then, as it slowly rises again, the wall shear stress returns to its normal value under undisturbed conditions after a certain length of narrow distance.

Experimental research, theoretical analysis, and numerical simulation are three basic methods for studying fluid motion characteristics. In order to further confirm the flow field characteristics of the posterior segment of narrow blood vessels, the PIV method was used to measure the flow field of the posterior segment of narrow blood vessels. PIV technology is a non-contact transient full flow field testing technology with fast dynamic response and high spatial resolution. By post-processing flow velocity data, some important parameters in hemodynamics, such as shear force, vorticity, etc., can be obtained, which can be well applied to flow field analysis of steady and unsteady flows. At the same time, the principle of PIV technology is simple, which is to add tracer particles to the flow field to generate scattering effect on light. Optical methods are used to record the position of particles in the flow field at different times, obtain the displacement of particles, and then calculate the motion speed of particles. Based on the particle's following ability to the flow field, the particle speed represents the motion speed of the fluid at the corresponding position in the flow field where it is located. In foreign countries, PIV flow field testing

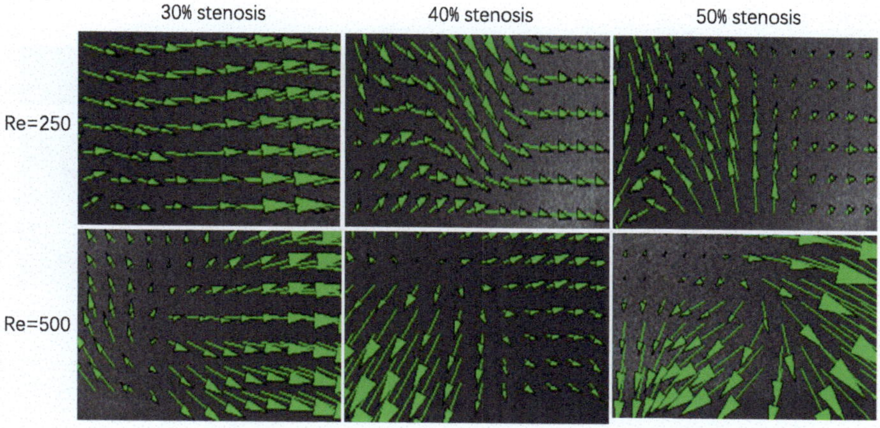

Fig. 2.3 Flow field distribution at the distal end of the stenosis with different degrees of stenosis and different Re. Adapted from Wei et al. (2010)

technology has been successfully applied to the study of flow fields in different cardiovascular models, including vascular bifurcation models.

The results of PIV testing indicate that the flow field in the straight section of the narrow tube has a clear hierarchy of streamline, which is laminar flow; The flow field in the rear section of the narrow pipe undergoes a sharp change, with significant changes in the velocity vector field and streamline, resulting in turbulent flow; When setting the inlet flow rate, the higher the degree of stenosis, the more complex the flow field in the rear section, and the easier it is to generate stagnation and reflux. At Re = 250, the flow field in the posterior segment with 30% stenosis does not show significant changes, while the flow field in the posterior segment with 50% stenosis is significantly disordered, with the formation of vortices and secondary flow. When setting the degree of stenosis, the inlet flow velocity affects the distribution and characteristics of the flow field in the posterior segment of the stenosis (Wei et al. 2010). For a 30% narrow circular tube, Re = 500 has a more complex rear end flow field than Re = 250, forming obvious eddies and secondary flows (Fig. 2.3).

2.3.2 *Low Oscillatory Shear Stress Animal Model (Local Ligation Model of Common Carotid Artery)*

Establish a low oscillatory shear stress animal model using the method of Nam et al. In this model, ApoE $^{-/-}$ mice (8 weeks old) were anesthetized by intraperitoneal injection of 8% (wt/vol) chloral hydrate, fixed supine, and disinfected with 75% alcohol or iodine on the surface of the neck skin; After hair removal, carefully separate the left carotid artery, and then ligate the external carotid artery (ECA), internal carotid artery (ICA), and occipital artery (OA) separately, leaving only the

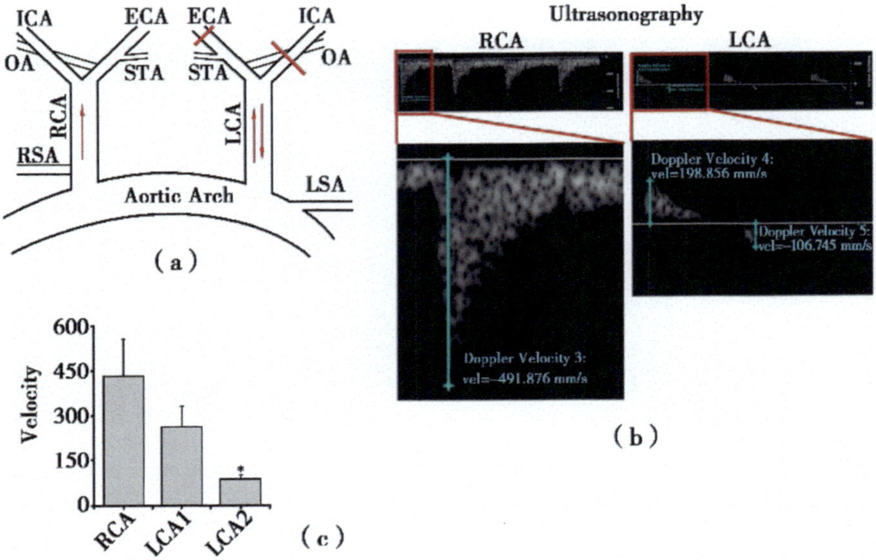

Fig. 2.4 Local ligation of the common carotid artery produces low oscillatory shear stress. **a** Schematic illustration of local ligation of carotid artery. **b** After 24 h of local ligation of the carotid artery, the velocity of blood flow was detected by ultrasound. **c** The quantitative analysis of velocity, LCA1 and LCA2 are the maximum positive and negative velocities, respectively (n = 3). Compared with RCA, $*p < 0.05$. Adapted from Zhang et al. (2018)

superior thyroid artery (STA) unobstructed. The right carotid artery was used as a sham surgical control, where the right carotid artery was separated but not ligated (Fig. 2.4). The Vevo 2100 high-resolution small animal ultrasound instrument detects carotid blood flow velocity using an ultrasound frequency of 24–30 MHz. The results indicate that the anterior segment of local ligation is a low shear stress area.

2.3.3 *Controllable Shear Stress Animal Model*

A silicone tube constriction and ligation surgery was performed on the left common carotid artery of the rabbit (Fig. 2.5) to create a local stenosis vascular model. The blood flow field will undergo significant changes, resulting in the formation of a region with reduced local shear stress (Liu et al. 2009).

Rabbits were anesthetized by intravenous injection of 3% pentobarbital sodium solution (30 mg/kg) until the eyelash reflex disappeared and fixed in a supine position. Cut an incision approximately 4 cm long from the thyroid cartilage to the sternum along the midline of the neck, passively separate subcutaneous tissue and neck muscle groups layer by layer, and expose the trachea. Find the left carotid artery sheath, separate the middle segment of the left common carotid artery with a length of about 2 cm, place the sterilized scale perpendicular to the separated blood vessel on the same

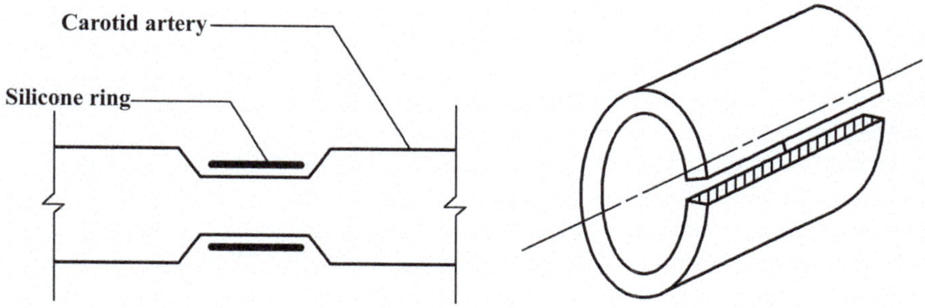

Fig. 2.5 Schematic diagram of rabbit common carotid artery stenosis model. Adapted from Liu et al. (2009)

horizontal plane, take photos with a digital camera, and quickly input the photos into the computer (Liu et al. 2009; Qiu et al. 2017). In Windows drawing software, based on the distance ratio between the scale on the scale next to the blood vessel in the photo and the diameter of the blood vessel, the diameter of the blood vessel can be quickly and accurately measured, thereby calculating the circumference of the silicone tube required for preset stenosis. Select a medical silicone tube with a longitudinal incision of 2 mm in inner diameter, cut it to the required circumference as needed, with a length of about 8 mm. Disinfect it with iodophor 3 times, rinse it with sterile physiological saline, and place it on the left common carotid artery. Triple ligation and fixation with surgical silk thread No. 4. Separate the right common carotid artery using the same method, without using a silicone tube as a sham surgical control, and suture the incision layer by layer (Fig. 2.6). According to the intraoperative measurement of the diameter of the left common carotid artery, silicone tubes with 30%, 40%, and 50% stenosis degrees were used for ligation surgery, creating different flow fields in the local blood vessels and subjecting endothelial cells to different shear stresses.

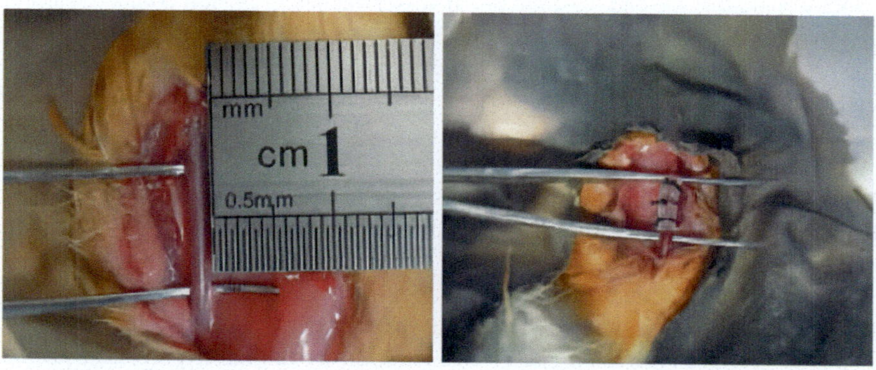

(a) Before the left carotid artery loop **(b) After the left carotid artery loop**

Fig. 2.6 Rabbit left common carotid artery coarctation loop. **a** Before the left carotid artery loop. **b** After the left carotid artery loop. Adapted from Liu et al. (2009)

After surgery, penicillin 400000 U/d intramuscular injection was administered for 3–5 days to prevent infection. Feed a daily ration of 150 g of feed for 8 weeks. According to the experimental groups, four types of feed were fed: 0% cholesterol feed (regular feed), 0.5% cholesterol feed (0.5% cholesterol + 5% lard + basic feed), 1% cholesterol feed (1% cholesterol + 5% lard + basic feed), and 1.5% cholesterol feed (1.5% cholesterol + 5% lard + basic feed). By feeding feed containing different concentrations of cholesterol, different polarization levels of LDL concentration in the bloodstream system are formed. Using GE's LOGIQ400 color Doppler flow imaging (CDFI), various hemodynamic parameters were measured in rabbits fed for 8 weeks. Firstly, the two-dimensional color blood flow distribution characteristics of the carotid artery were detected to determine the direction of blood flow. Then, the end diastolic blood flow diameter (Dr), end systolic blood flow velocity (VSP), end diastolic blood flow velocity (VED), average blood flow velocity (Vm), resistance index (RI), and pulsatile index (PI) were measured. The left common carotid artery is the narrow side, and the right side is used as the control. Measure the relevant parameters of each side of the blood vessel 10 times, and finally take the average value. Blood samples were collected from the ear vein of rabbits before and 4 and 8 weeks after surgery, and serum total cholesterol (TC), triglycerides (TG), high-density lipoprotein (HDL), and low density lipoprotein (LDL) were measured using Sapphire 600 automatic biochemical analyzer. The FASCO-2050A fully automatic hemorheological rapid tester was used to measure the whole blood viscosity of each group of animals at 8 weeks after surgery.

According to the formula

$$\tau_m = \frac{4 \cdot \eta \cdot V_m}{D_r} \tag{2.4}$$

Calculate the wall shear stress (WSS) of blood vessels, τ_m is the average shear stress of the blood flow, η Is the blood viscosity value, V_m is the average blood flow velocity, and D_r is the end diastolic blood vessel diameter. The left common carotid artery (LCCA) with the same degree of stenosis increases blood viscosity and shear stress with the increase of dietary cholesterol concentration; For the group fed with feed containing the same cholesterol concentration, the average blood flow rate of LCCA was lower in the 40% stenosis group, and the shear stress was also lowest at that stenosis level; LCCA underwent silicone tube constriction surgery and had AS lesions such as intimal thickening, resulting in a significant decrease in vessel diameter and higher shear stress compared to the control side of the right common carotid artery (RCCA) ($p < 0.01$) (Table 2.1).

Scanning electron microscopy examination of bilateral common carotid artery specimens from each group of experimental animals revealed varying degrees of AS lesions in the LCCA of each group, characterized by intimal hyperplasia, plaque formation, disordered arrangement and detachment of local endothelial cells, and proliferation of fibrous tissue on the surface of plaques. The endothelial cells of the

Table 2.1 The shear stress value of the left and right common carotid arteries of each group of animals/(dyn cm^{-2})

Cholesterol Concentration in Feed	30% Stenosis		40% Stenosis		50% Stenosis	
	LCC	RCC	LCC	RCC	LCC	RCC
0% Cholesterol	13.33 ± 1.32^a	7.76 ± 1.89	12.16 ± 1.2^a	7.49 ± 0.78	14.14 ± 1.43^a	7.29 ± 1.08
0.5% Cholesterol	14.82 ± 2.54^a	8.28 ± 1.06	12.88 ± 1.47^a	7.63 ± 1.17	15.60 ± 2.44^a	8.25 ± 1.07
1% Cholesterol	17.47 ± 2.03^a	10.06 ± 0.96	15.71 ± 1.26^a	8.87 ± 1.88	19.93 ± 2.12^a	11.06 ± 1.47
1.5% Cholesterol	20.64 ± 2.18^a	$12.02 + 2.23$	19.08 ± 2.03^a	11.17 ± 1.82	23.47 ± 2.68^a	13.89 ± 2.43

Note [a] Compared with RCCA with the same degree of stenosis, $p < 0.01$

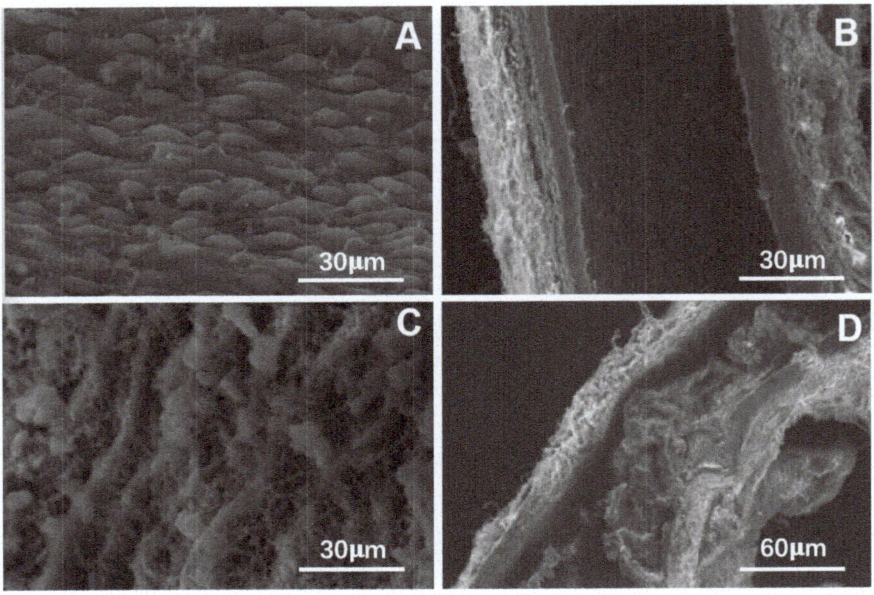

Fig. 2.7 Scanning electron microscope observation of bilateral common carotid artery lesions. **a** endothelial cells of RCCA arrange neatly, and no plaque formed in the group of 30% stenosis + 0% cholesterol. **b** Smooth lumen of RCCA and without plaque hyperplasia in the group of 40% stenosis +1% cholesterol. **c** Fibrous proliferation on the surface of LCCA plaque and no obvious endothelial cells in the group of 40% stenosis +0.5% cholesterol. **d** Significant proliferative plaques and lumen stenosis in the group of 40% stenosis +1% cholesterol. Adapted from Liu et al. (2009)

RCCA on the control side were arranged neatly, and no significant plaque formation was observed. Among each group of animals, the 40% stenosis group showed particularly significant lesions (Fig. 2.7).

The pathological section was observed under the optical microscope. It was found that the LCCA of animals in each group had different degrees of AS lesions, and a large number of foam cells, mucus degeneration, lipid aggregation, elastic fibers and vascular wall smooth muscle cells were in disorder, vascular intima degeneration and necrosis, blood thrombus formation, plaque formation and other lesions appeared on the local vascular wall; On the control side, the intima of RCCA was smooth and intact, with no plaque formation, and only focal foam cells appeared in some local vessels. Among all groups of animals, the 40% stenosis group showed particularly significant lesions (Fig. 2.8).

Establishing a stable and reliable animal model is the foundation of animal experiments and an important link in conducting related research. Currently, the methods for making carotid artery stenosis models include surgical silk thread ligation, balloon injury, dry air injury, chemical burning of the intima, etc. However, the above methods can cause damage to the vascular intima or wall. For the determination of stenosis degree, most of them use detection methods such as postoperative digital subtraction angiography (DSA) or intravascular ultrasound (IVUS), which can better determine

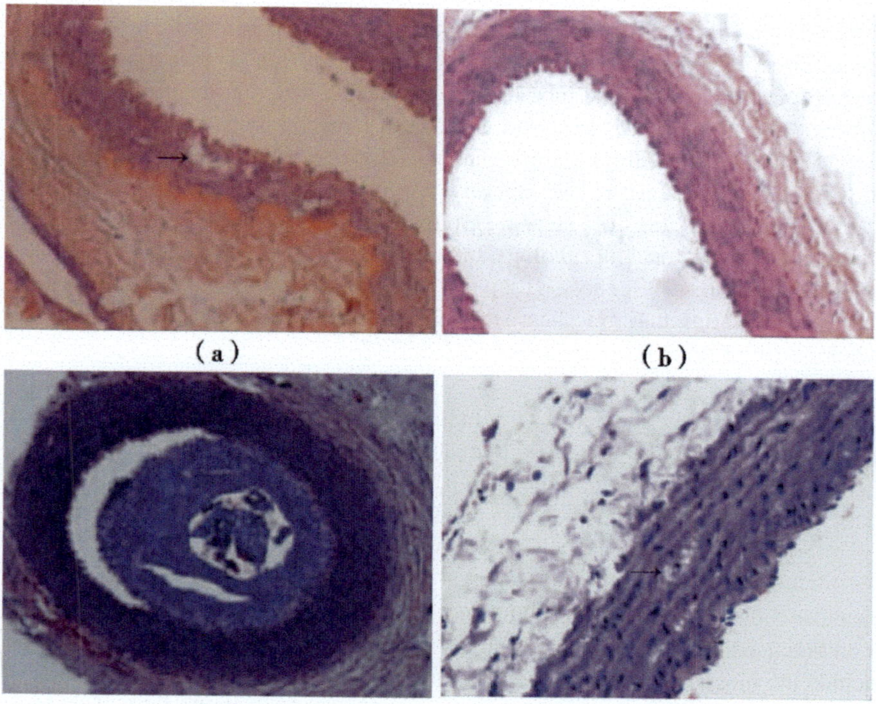

(a) (b) (c) (d)

Fig. 2.8 Pathological observation of carotid artery. **a** Plaque formation in LCCA lumen of 40% stenosis +0.5% cholesterol (HE × 100), ↑: Lipid aggregation and cholesterol crystals in the intima. **b** The RCCA in the group of 40% stenosis +0.5% cholesterol (HE × 100). (c) The thrombosis was formed and recanalized after mechanization in the group of 50% stenosis +1.5% cholesterol. **d** The RCCA in the group of 30% stenosis +1.5% cholesterol (HE × 100), ↑: Foam cells scattered among the smooth muscle cells. Adapted from Liu et al. (2009)

the stenosis degree of narrow blood vessels. However, there are drawbacks such as multiple procedures, complex operations, and the inability to set the stenosis degree arbitrarily. In this chapter's experiment, we used a silicone tube to ligate the left common carotid artery, which has the advantages of good biocompatibility and no damage to the vascular intima. In addition, the model adopts a method of combining digital photography during surgery with computer software measurement, which can accurately measure the diameter of blood vessels. Based on the size of the common carotid artery diameter, a silicone tube that meets the requirements is made on site, making the determination of stenosis simple, fast, and more accurate, thus enabling the preset stenosis to be accurately controlled. This method is stable, reliable, simple, and easy to implement. The length and degree of stenosis of the narrowed blood vessels can be freely set and regulated, and can quickly create an AS animal model for in vivo research on the relationship between shear stress changes and AS lesions.

In this chapter's experiment, we used silicone tube ligation of the left common carotid artery in rabbits and a high-fat diet after surgery to quickly create an AS animal model. By applying silicone tube ligation with three different degrees of stenosis, different flow fields were formed in the local narrowed blood vessels, causing the blood vessel wall to withstand different shear stresses, making local vascular shear stress a controllable factor. The experimental results showed that regardless of the type of animal feed, the 40% stenosis group had the lowest shear stress in the same feed group. Pathological morphology and scanning electron microscopy results also confirmed that the 40% stenosis group had the most significant AS lesions (Figs. 2.7 and 2.8), indicating that low shear stress induced and promoted the formation and development of AS.

2.4 The Cumulative Effect of Cell Membrane Tensile Stress on the Secretion of ET-1 in Large Vascular Endothelial Cells

Vascular endothelial metabolism is closely related to the functional status of vascular endothelial cells. The various metabolic functional changes of the endothelium include the biosynthesis and degradation of vasoactive substances, the synthesis of scavenging reactive oxygen species, the transportation and metabolism of lipoproteins, the secretion and remodeling of extracellular matrix components, the biosynthesis of various growth factors, cytokines and hormonal substances, as well as prostaglandins and other potent carotenoids (Rosenson et al. 2016; Wentzel et al. 2012).

The shear stress caused by flow acts on the endothelial cell membrane, and shear stress is closely related to the function of vascular endothelial cells. Research suggests that the internal stress or membrane tension of endothelial cells is directly related to changes in endothelial cell metabolism and ultrastructure. Theoretical analysis shows that cell membrane tension or tensile stress may accumulate due to shear stress caused by flow. When the osmotic pressure or static pressure difference between cells and blood is zero or greater (in vivo environment), cell membrane tension will accumulate in the direction opposite to the blood flow, that is, membrane tension will accumulate upstream. Under certain conditions, the tensile stress in the cell membrane can grow in the opposite direction to the blood flow and accumulate to a very high level, even three or more orders of magnitude higher than the flow shear stress on the membrane surface. This phenomenon is called cell membrane tension accumulation (CMTA).

Some scholars have used interference microscopy to measure the three-dimensional geometric shape of human vascular endothelial cells under flow shear stress conditions (Steward et al. 2015). The results indicate that there is indeed a phenomenon of tensile stress accumulation in the endothelial cell membrane, which depends on the blood endothelial interface and may have a profound impact on integrins or ion channels, which is of great significance for metabolism and biological functions (Finney et al. 2017). Therefore, the phenomenon of membrane tension accumulation should be considered when studying the mechanical behavior of endothelial cells (Wang et al. 2003).

Due to the difficulty of directly measuring cell membrane tension, we believe that if tension or tensile stress does accumulate and increase against blood flow, the average endothelial cell membrane tensile stress of blood vessels under the same shear stress environment is related to vessel length, and the metabolism and secretion rate of bioactive substances in endothelial cells may also be related to vessel length. In this study, a paired experiment was designed to validate this hypothesis by measuring the secretion of endothelin-1 (ET-1) in the monolayer of endothelial cells exposed to stable laminar shear stress (Wang et al. 2000).

Obtain the umbilical vein of the baby within 4 h after delivery and rinse the blood with sterile physiological saline. Under sterile conditions, divide a umbilical vein into three venous segments (A, B, and C). The length of the venous segment is 10 cm (A), 15 cm (B), and 15 cm (C), respectively. The infusion adopts the aforementioned parallel plate flow chamber system. After connecting the venous segment with the perfusion system, the flow rate is measured by a flow monitor, and the entire system is maintained at 37 °C and pH 7.4. The shear stress of sections A and B is 0.48 N/m^2, while the shear stress of section C is 0.73 N/m^2. During the 24 h perfusion process, 1 mL of liquid sample is taken from the bottom reservoir every 2 h, and an equal amount of new culture medium is added immediately after the sample is extracted, keeping the total amount of perfusion fluid in the entire system unchanged. Radioimmunoassay was used to determine the content of ET-1 (Wang et al. 2002). The cumulative secretion of ET-1 measured was fitted to the nonlinear regression equation using Marquart nonlinear regression analysis and statistical software package DATA PROCESS SYSTEM (DPS), and paired t-tests were used for statistical analysis. Under a shear stress of 0.48 N/m^2 for 24 h, the average secretion rate of ET-1 in a 15 cm long vein (B) was significantly higher than that in a 10 cm long vein segment ($p < 0.01$) (Table 2.2).

At the same level of shear stress (0.48 N/m^2), the ET-1 secretion rate of both venous segments decreased linearly with time (Fig. 2.9).

When subjected to different levels of shear stress, the rate of ET-1 secretion varies significantly among venous segments (B and C) with the same length. Within 24 h, the average secretion rate of C-segment ET-1 under shear stress of 0.73 N/m^2 was 46.289 29 ± 0.1180 pg/cm^2 h, and the average secretion rate of B-segment ET-1

Table 2.2 The average secretion rate of ET-1 in the vein

Segment	Length/cm	Shear stress/(N m^{-2})	ET-1 secretion rate [Average ± SEM. (pg cm^{-2} h^{-1})]
A	10	0.48	30.627 4 ± 0.491 2
B	15	0.48	34.915 4 ± 0.983 0
C	15	0.73	46.289 2 ± 0.118 0

Duration of flow shear = 24 h

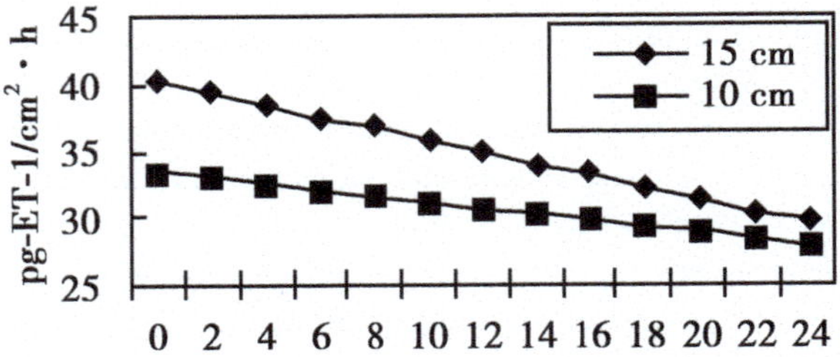

Fig. 2.9 The relationship between the secretion rate of ET-1 and the length of the vein. Adapted from Wang et al. (2003)

under shear stress of 0.48 N/m^2 was 34.91549 ± 0.9830 pg/cm^2 h, with the former being nearly 30% higher than the latter ($p < 0.01$).

Under the condition of 0.73 N/m^2, the ET-1 secretion rate of segment C remained stable throughout 24 h, which was significantly different from that of segment B under shear stress of 0.48 N/m^2 (Fig. 2.10).

Due to the difficulty of directly measuring tensile stress, the hypothesis of membrane tension accumulation can be inferred by utilizing the correlation between the release of bioactive substances and the length of endothelial cell monolayers. Measure the secretion of ET-1 in human umbilical vein segments of different lengths (10 and 15 cm, respectively). Place two venous segments in the same flow field and apply the same shear stress (0.48 N/m^2) for 24 h. The results showed that at the same shear stress level (0.48 N/m^2), there was a significant difference in the secretion rate of ET-1 between a 10 cm long venous segment (A) and a 15 cm long venous segment (B).

Exposed to a shear stress of 0.48 N/m^2 for 24 h, the rate of ET-1 secretion in segment B was 34.91549 ± 0.9830 pg/cm^2 h, which was nearly 14% higher than the average rate of 30.62749 ± 0.4912 pg/cm^2 h in segment A. When at different shear stress levels, the rate of ET-1 secretion varies significantly among venous segments with the same length (B and C). Due to the proportional relationship between tensile stress and shear stress, high shear stress causes high tensile stress, which generates more ET-1 (Wang et al. 2016). These data confirm that longer blood vessel segments, or those subjected to higher shear stress, accumulate higher tensile stress on the cell membrane, and the accumulation of higher tensile stress in the cell membrane leads to more secretion of ET-1.

Therefore, the CMTA phenomenon does occur in vascular endothelial cells and affects their metabolism (Wang et al. 2003; White et al. 2011).

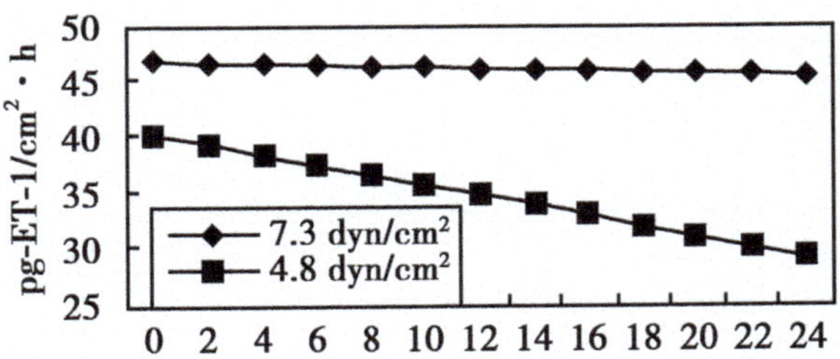

Fig. 2.10 The relationship between the secretion rate of ET-1 and the flow shear time. Adapted from Wang et al. (2003)

2.5 Low Shear Stress Promotes the Production and Secretion of Chemokines in Large Vascular Endothelial Cells

Chemokines are a large class of small molecule cytokines that were first discovered due to their ability to promote cell chemotactic migration. Subsequently, it was found that chemokines play a crucial role in cell recruitment, development, and homeostasis regulation, especially in the migration and homing of stem cells. Recent studies have shown that the expression imbalance of chemokines is involved in various pathological processes. Among numerous chemokines, stromal cell derived factor-1 (SDF-1) is an important chemokine, classified as the "CXC" subfamily due to the separation of two cysteine near the N-terminus by one other amino acid. Numerous studies have shown that SDF-1 binds to specific receptors (CXC motif chemokine receptor type 4, CXCR4), activates G protein, initiates a series of signaling molecules, and participates in various physiological and pathological processes (Wu et al. 2006).

Research has shown that SDF-1 is highly expressed in the liver, heart, pancreas, and spleen. Constitutionally expressed in liver cells contributes to the maintenance of liver homeostasis and damage repair response. In ischemic damaged myocardial tissue, SDF-1 repairs the tissue through chemotactic stem cells, but it also causes myocardial damage through interactions with inflammatory factors, indicating that SDF-1 plays a dual role in the repair of myocardial injury. During the occurrence and development of cancer, SDF-1 is closely related to tumor angiogenesis, cell proliferation, and chemotherapy tolerance.

Recent studies have shown that the capture of white blood cells, angiogenesis, migration and proliferation of smooth muscle cells during the formation of AS plaques are closely related to the role of chemokines, and SDF-1 is closely related to the occurrence and development of AS. Research has found that SDF-1 is only highly expressed in AS plaques, while it is not found in normal arterial walls. High expression of SDF-1 was also detected during the process of restoring coverage after arterial intimal injury in $ApoE^{-/-}$ mice. The expression of SDF-1 was up-regulated in the atherosclerotic lesions caused by transplantation rejection in the mouse model, and was expressed in monocytes, damaged endothelial cells and activated platelets closely related to the occurrence and development of atherosclerosis.

Research has shown that the expression of chemokines is significantly upregulated in areas with turbulent blood flow fields such as shear stress, flow separation, and secondary flow, and plays an extremely important role in mediating the adhesion of monocytes, platelets, and damaged vascular endothelial cells, arterial intimal hyperplasia, and the occurrence and development of atherosclerosis. In a low shear stress animal model established using the carotid artery cuff method (experimental model as described earlier, Fig. 2.2), it was found that the expression of SDF-1 and its receptor CXCR4 was significantly upregulated in AS lesions induced by low shear stress. Further research has found that Ox-LDL, an independent risk factor for AS, upregulates the expression of SDF-1 mRNA and protein in vascular endothelial cells, and upregulates the expression of CXCR4 in THP-1 derived monocytes, thereby promoting monocyte chemotaxis, adhesion to vascular endothelial cells, and cross endothelial migration (Qiu et al. 2011). These results indicate that low shear stress may upregulate the expression of SDF-l, promote adhesion and transmembrane migration between monocytes and damaged vascular endothelial cells in the circulating blood, and mediate low shear stress-induced AS (Souilhol et al. 2020).

2.6 Research Progress and Prospects

As a carrier of blood flow, blood vessels not only play a role in transporting nutrients, but also can sense various signals and respond, maintaining the entire blood circulation system in a stable state. Blood flow shear stress plays a key role in the formation, progression and rupture of atherosclerosis (Samady et al. 2011). Atherosclerosis is formed in the low shear stress area, while atherosclerotic plaque rupture mainly occurs in the high shear stress area. High shear stress can induce upregulation of NO and endothelial growth factor in endothelial cells in the proximal region, leading to the formation of microvessels in plaques, which are newly formed thin-walled microvessels within the plaques (Koskinas et al. 2013; Qiu et al. 2017). Under normal physiological conditions, the general vascular adventitia layer and the outer layer of the medial membrane only exist to nourish the vascular branch system. In the process of atherosclerosis, the situation becomes more complex, and the vascular network becomes more abundant, and even can extend from the outer membrane to the vascular intima layer, providing repair nutrients and oxygen to the core damage area. Various stimulators are activated in the process of atherosclerotic plaque lesions to promote angiogenesis. However, new blood vessels are fragile and immature, only enclosing a layer of endothelial cells without a complete protective basement membrane. There are also no pericytes and smooth muscle cells covering the blood vessels. Due to the immaturity of new blood vessels, plaque rupture mediates inevitable bleeding within the plaque. In addition, inflammatory cell infiltration in pathological vessels, lipoprotein deposition and hemorrhage in plaques are also the main causes of atherosclerosis. Reducing the leakage of the vascular system caused by high shear stress in the vascular environment can establish a microenvironment more conducive to the survival of smooth muscle cells, thereby hindering the formation of vulnerable plaques. The coordination of risk factors such as hypercholesterolemia, hyperlipidemia, hypertension, and hyperglycemia will cause abnormal vascular shear stress environment, which can induce the pathological characteristics of early atherosclerosis (Stone et al. 2011). The changes in blood lipids, blood sugar, and blood pressure in the human body with three high diseases can also cause changes in the forces exerted on blood vessel wall cells. Clinical cases show that the shear stress of the diseased part of the blood vessel is much higher than that of the surrounding normal vascular area, and the dilatational remodeling of the blood vessel often occurs in the diseased area, and the dilatational remodeling of the blood vessel is a risk factor for vulnerable plaque rupture (Stone et al. 2012). At the same time, the high tension of the diseased part of the blood vessel may also regulate the occurrence and development of atherosclerosis.

Under physiological conditions, blood flow shear stress can cause changes in the morphology of cells and regulate their force transmission pathways. Among them, laminar shear stress induces the production of nitric oxide (NO) and reduces the production of reactive oxygen species (ROS) and the increase of apoptosis by regulating the expression and phosphorylation of eNOS (eNOS phosphorylation active sites including Ser1177, Ser635, and inhibitory sites Ser116 and Thr497), while the opposite is true under turbulent flow conditions (Stone et al. 2012). At the same time, shear force also regulates gene transcription involving various processes, including inflammation, cell cycle, apoptosis, cell migration, protein oxidation, and unfolded protein reactions. For example, perturbed flow can regulate cell proliferation, morphology, and the occurrence of inflammation (Back et al. 2019). These shear stress receptor molecules include bone morphogenetic protein 4 (BMP4) and transforming growth factor- β (TGF β)、 WNT signaling pathway and its other components, genes related to endothelial mesenchymal transition (EndMT), hypoxia inducible factor 1 α (HIF1 α), Vascular endothelial growth factor (VEGF), glycolytic related genes, and angiopoietin related receptors. At the same time, shear stress can also affect the methylation level of vascular endothelial cells, and blood flow can control the expression of multiple genes through epigenetic regulation of whole genome DNA methylation levels, such as the methylation of HOX family genes, Hippo YAP related signaling pathways, and transcriptional co activators with PDZ binding motifs, WNT, and Notch (Dunn et al. 2014; Jiang et al. 2015).

Endothelial cells are very sensitive to shear stress, and prolonged exposure to laminar shear stress can also activate the vascular endothelial growth factor (VEGF) receptor to stimulate serine/threonine kinase Akt (also known as protein kinase B or Rac kinase) phosphorylation on Ser473 (Fan et al. 2014). The activation of Akt in turn phosphorylates eNOS at Ser1177, leading to eNOS activation and NO production. At the same time, laminar shear stress can also regulate the mechanical signal complex of PECAM-1/VE cadherin/VEGFR2 to induce Akt phosphorylation and activate eNOS to release anti AS signals (Hong et al. 2019). This process can mediate the arrangement and elongation of endothelial cells along the flow direction, leading to the reconstruction of the intracellular skeleton structure. In addition, after sensing shear stress, the surface of endothelial cells undergoes deformation, leading to ion displacement, thereby activating downstream intracellular signaling pathways and affecting gene and protein expression, regulating the structure and function of endothelial cells. On the contrary, disturbance flow will promote NF- κ The B related inflammatory signaling pathway leads to the expression of leukocyte adhesion molecules, including VCAM-1 and ICAM-1, as well as chemokines, including CCL2, leading to the activation of inflammatory signals. In recent studies, it has been reported that Piezo1, as a responsive receptor for blood flow shear stress, can induce endothelial cells to release ATP, thereby activating the Gq/G11 coupled purinergic P2Y2 receptor and exacerbating AS (Chen et al. 2017). Therefore, laminar shear stress upregulates the expression of AS protective genes and antiapoptotic genes. However, when blood vessels are under low shear stress or oscillatory flow conditions, endothelial cell-mediated vasodilation will be damaged and cellular inflammation will occur and aggravate atherosclerosis (Feaver et al. 2010).

Overall, endothelial cells can sense changes in the mechanical environment, and uniform shear stress can induce the expression of endothelial protective factors; When the blood flow status is abnormal, such as in uneven laminar flow, turbulence, and turbulent flow, the secretion of endothelial protective cytokines by cells decreases, and endothelial lesions promote the formation of AS lesions (Yang et al. 2018; Zhou et al. 2014). At the same time, under the stimulation of mechanics and various factors, blood vessels are prone to rupture, and the infiltrated red blood cells enter the interior of the plaque. As a large number of red blood cells enter the plaque, they will recruit inflammatory factors and cells to exacerbate the increase in plaque area (Suvorava et al. 2018; Yun et al. 2016).

References

Ajami NE, Gupta S, Maurya MR, et al. Systems biology analysis of longitudinal functional response of endothelial cells to shear stress[J]. Proc Natl Acad Sci USA, 2017, 114(41): 10990–10995.

Aulak KS, Barnes JW, Tian L, et al. Specific O-GlcNAc modification at ser-615 modulates eNOS function[J]. Redox Biol, 2020, 36: 101625.

Back M, Yurdagul AJr, Tabas I, et al. Inflammation and its resolution in atherosclerosis: mediators and therapeutic opportunities[J]. Nat Rev Cardiol, 2019, 16(7): 389–406.

Baeyens N, Bandyopadhyay C, Coon BG, et al. Endothelial fluid shear stress sensing in vascular health and disease[J]. J Clin Invest, 2016, 126(3): 821–828.

Bondareva O, Tsaryk R, Bojovic V, et al. Identification of atheroprone shear stress responsive regulatory elements in endothelial cells[J]. Cardiovasc Res, 2019, 115(10): 1487–1499.

Brown AJ, Teng Z, Evans PC, et al. Role of biomechanical forces in the natural history of coronary atherosclerosis[J]. Nat Rev Cardiol, 2016, 13(4): 210–220.

Budatha M, Zhang J, Zhuang ZW, et al. Inhibiting integrin α5 cytoplasmic domain signaling reduces atherosclerosis and promotes arteriogenesis[J]. J Am Heart Assoc, 2018, 7(3): e007501.

Chen BP, Li YS, Zhao Y, et al. DNA microarray analysis of gene expression in endothelial cells in response to 24-h shear stress[J]. Physiol Genomics, 2001, 7(1): 55–63.

Chen J, Green J, Yurdagul AJ, et al. αvβ3 integrins mediate flow-induced NF-KB activation, proinflammatory gene expression, and early atherogenic inflammation[J]. Am J Pathol, 2015, 185(9): 2575–2589.

Chen X, Qian S, Hoggatt A, et al. Endothelial cell-specific deletion of P2Y2 receptor promotes plaque stability in atherosclerosis-susceptible apoe-null mice[J]. Arterioscler Thromb Vasc Biol. 2017, 37(1): 75–83.

Chen Z, Oliveira S, Zimnicka AM, et al. Reciprocal regulation of eNOS and caveolin-1 functions in endothelial cells[J]. Mol Biol Cell, 2018, 29(10): 1190–1202.

Chistiakov DA, Orekhov AN, Bobryshev YV. Contribution of neovascularization and intraplaque haemorrhage to atherosclerotic plaque progression and instability[J]. Acta Physiol (Oxf), 2015, 213(3): 539–553.

Chiu JJ, Usami S, Chien S. Vascular endothelial responses to altered shear stress: pathologic implications for atherosclerosis[J]. Ann Med, 2009, 41(1): 19–28.

Chiu JJ, Chien S. Effects of disturbed flow on vascular endothelium: pathophysiological basis and clinical perspectives[J]. Physiol Rev, 2011, 91(1): 327–387.

Civelek M, Manduchi E, Riley RJ, et al. Coronary artery endothelial transcriptome in vivo: identification of endoplasmic reticulum stress and enhanced reactive oxygen species by gene connectivity network analysis[J]. Circ Cardiovasc Genet, 2011, 4(3): 243–252.

Cox CD, Bae C, Ziegler L, et al. Removal of the mechanoprotective influence of the cytoskeleton reveals PIEZO1 is gated by bilayer tension[J]. Nat Commun, 2016, 7: 10366.

Dunn J, Qiu H, Kim S, et al. Flow-dependent epigenetic DNA methylation regulates endothelial gene expression and atherosclerosis[J]. J Clin Invest, 2014, 124(7): 3187–3199.

Fang Y, Wu D, Birukov KG. Mechanosensing and mechanoregulation of endothelial cell functions[J]. Compr Physiol, 2019, 15, 9(2): 873–904.

Fan R, Tang D, Yang C, et al. Human coronary plaque wall thickness correlated positively with flow shear stress and negatively with plaque wall stress: an IVUS-based fluid-structure interaction multi-patient study[J]. Biomed Eng Online, 2014, 13(1): 32.

Feaver RE, Gelfand BD, Wang C, et al. Atheroprone hemodynamics regulate fibronectin deposition to create positive feedback that sustains endothelial inflammation[J]. Circ Res, 2010, 106(11): 1703–1811.

Finney AC, Stokes KY, Pattillo CB, et al. Integrin signaling in atherosclerosis[J]. Cell Mol Life Sci, 2017, 74(12): 2263–2282.

Gimbrone MA, Garcia-Cardena G. Endothelial cell dysfunction and the pathobiology of atherosclerosis[J]. Circ Res, 2016, 118 (4): 620–636.

Gomez D, Owens GK. Smooth muscle cell phenotypic switching in atherosclerosis[J]. Cardiovasc Res, 2012, 95(2): 156–164.

Hong FF, Liang XY, Liu W, et al. Roles of eNOS in atherosclerosis treatment[J]. Inflamm Res, 2019, 68(6): 429–441.

Iring A, Jin YJ, Albarrán-Juárez J, et al. Shear stress-induced endothelial adrenomedullin signaling regulates vascular tone and blood pressure[J]. J Clin Invest, 2019, 129(7): 2775–2791.

Jiang YZ, Manduchi E, Stoeckert CJ, et al. Arterial endothelial methylome: differential DNA methylation in athero-susceptible disturbed flow regions in vivo[J]. BMC Genomics, 2015, 16: 506.

Koskinas KC, Sukhova GK, et al. Thin-capped atheromata with reduced collagen content in pigs develop in coronary arterial regions exposed to persistently low endothelial shear stress[J]. Arterioscler Thromb Vasc Biol, 2013, 33(7): 1494–1504.

Kumar S, Williams D, Sur S, et al. Role of flow-sensitive microRNAs and long noncoding RNAs in vascular dysfunction and atherosclerosis[J]. Vascul Pharmacol, 2019, 114: 76–92.

Li X, Yang Q, Wang Z, et al. Shear stress in atherosclerotic plaque determination[J]. DNA Cell Biol, 2014, 33(12): 830–838.

Liu H, Wang GX, Ye LQ. Establishment of controllable wall shear stress and LDL concentration polarization model in rabbits and their effect on atherosclerosis[J]. Journal of Army Medical University, 2009, 31(12): 1981–1985.

Liu J, Wang Y, Goh WI, et al. Talin determines the nanoscale architecture of focal adhesions[J]. Proc Natl Acad Sci USA, 2015, 112(35): e4864–e4873.

Ni CW, Qiu H, Rezvan A, et al. Discovery of novel mechanosensitive genes in vivo using mouse carotid artery endothelium exposed to disturbed flow[J]. Blood, 2010, 116(15): e66–e73.

Pedrigi RM, Poulsen CB, Mehta VV, et al. Inducing persistent flow disturbances accelerates atherogenesis and promotes thin cap fibroatheroma development in D374Y-PCSK9 hypercholesterolemic minipigs[J]. Circulation, 2015, 132(11): 1003–1012.

Petsophonsakul P, Furmanik M, Forsythe R, et al. Role of vascular smooth muscle cell phenotypic switching and calcification in aortic aneurysm formation[J]. Arterioscler Thromb Vasc Biol, 2019, 39(7): 1351–1368.

Qiu J, Lei D, Hu J, et al. Effect of intraplaque angiogenesis to atherosclerotic rupture-prone plaque induced by high shear stress in rabbit model[J]. Regen Biomater, 2017, 4(4): 215–222.

Qiu J, Wang G, Hu J, et al. Id1-induced inhibition of p53 facilitates endothelial cell migration and tube formation by regulating the expression of beta1-integrin[J]. Mol Cell Biochem, 2011, 357(1–2): 125–133.

Reneman RS, Arts T, Hoeks AP. Wall shear stress—an important determinant of endothelial cell function and structure—in the arterial system in vivo. Discrepancies with theory[J]. J Vasc Res, 2006, 43(3): 251–269.

Rosenson RS, Brewer HB, Ansell BJ, et al. Dysfunctional HDL and atherosclerotic cardiovascular disease[J]. Nat Rev Cardiol, 2016, 13(1): 48–60.

Samady H, Eshtehardi P, McDaniel MC, et al. Coronary artery wall shear stress is associated with progression and transformation of atherosclerotic plaque and arterial remodeling in patients with coronary artery disease[J]. Circulation, 2011, 124(7): 779–788.

Souilhol C, Serbanovic-Canic J, Fragiadaki M, et al. Endothelial responses to shear stress in atherosclerosis: a novel role for developmental genes[J]. Nat Rev Cardiol, 2020, 17(1): 52–63.

Steward R Jr, Tambe D, Hardin CC, et al. Fluid shear, intercellular stress, and endothelial cell alignment[J]. Am J Physiol Cell Physiol, 2015, 308(8): C657–C664.

Stone G, Maehara A, Lansky A, et al. Prospective natural-history study of coronary atherosclerosis[J]. New Engl J Med, 2011, 364 (3): 226–235.

Stone PH, Saito S, Takahashi S, et al. Prediction of progression of coronary artery disease and clinical outcomes using vascular profiling of endothelial shear stress and arterial plaque characteristics: the PREDICTION Study[J]. Circulation, 2012, 126(2): 172–181.

Suvorava T, Cortese-Krott MM. Exercise-induced cardioprotection via eNOs: a putative role of red blood cell signaling[J]. Curr Med Chem, 2018, 25(34): 4457–4474.

Timmins LH, Molony DS, Eshtehardi P, et al. Oscillatory wall shear stress is a dominant flow characteristic affecting lesion progression patterns and plaque vulnerability in patients with coronary artery disease[J]. J R Soc Interface, 2017, 14(127): 20160972.

Wang GX, Luo XD, Ouyang KQ. Effect of shear stress on metabolism by capillary vascular endothelial cells[J]. Acta Biophysica Sinica, 1993, 3(1): 178–185.

Wang GX, Cai SX, Rao XC, et al. The phenomenon of cell membrane tensile stress accumulation and its effect on endothelin-1 secretion by vascular endothelial cells[J]. Colloid Surface B, 2003, (28): 273–278.

Wang G, Cai S, Deng X, et al. Secretory response of endothelin-1 in cultured human glomerular microvascular endothelial cells to shear stress[J]. Biorheology, 2000, 37(4): 291–299.

Wang GX, Cai SX, Wang PQ, et al. Shear-induced changes in endothelin-1 secretion of microvascular endothelial cells[J]. Microvasc Res, 2002, 63(2): 209–217.

Wang Y, Qiu J, Luo S, et al. High shear stress induces atherosclerotic vulnerable plaque formation through angiogenesis[J]. Regen Biomater, 2016, 3(4): 257–267.

Wei DH, Wang GX, Tang CJ. Numerical simulation and PIV assay of flow field at distal stenosis vascular[J]. Medical Science Journal of Central South China, 2010, 38(1): 24–26.

Wentzel JJ, Chatzizisis YS, Gijsen FJ, et al. Endothelial shear stress in the evolution of coronary atherosclerotic plaque and vascular remodelling: current understanding and remaining questions[J]. Cardiovasc Res, 2012, 96(2): 234–243.

White SJ, Hayes EM, Lehoux S, et al. Characterization of the differential response of endothelial cells exposed to normal and elevated laminar shear stress[J]. J Cell Physiol, 2011, 226(11): 2841–2848.

Wu MJ, Wei DH, Wang GX. Stromal cell-derived factor-1 expression in atherosclerotic lesions induced by low shear stress[J]. Chin J Arterioscler, 2006, 12(14): 1028–1030.

Yahagi K, Kolodgie FD, Otsuka F, et al. Pathophysiology of native coronary, vein graft, and in-stent atherosclerosi[J]. Nat Rev Cardiol, 2016, 13(2): 79–98.

Yang Q, Xu J, Ma Q, et al. PRKAA1/AMPKα1-driven glycolysis in endothelial cells exposed to disturbed flow protects against atherosclerosis[J]. Nat Commun, 2018, 9(1): 4667.

Yun S, Budatha M, Dahlman JE, et al. Interaction between integrin α5 and PDE4D regulates endothelial inflammatory signalling[J]. Nat Cell Biol, 2016, 18(10): 1043–1053.

Zeng Y, Zhang XF, et al. The role of endothelial surface glycocalyx in mechanosensing and transduction[J]. Adv Exp Med Biol, 2018, 1097: 1–27.

Zhang K, Chen Y, Zhang T, et al. A novel role of Id1 in regulating oscillatory shear stress-mediated lipid uptake in endothelial cells[J]. Ann Biomed Eng, 2018, 46(6): 849–863.

Zhou J, Li YS, Chien S. Shear stress-initiated signaling and its regulation of endothelial function[J]. Arterioscler Thromb Vasc Biol, 2014, 34(10): 2191–2198.

Chapter 3
Hypothesis of Hemodynamics for Vascular Endothelial Cell Injury Repair

Percutaneous coronary intervention (PCI) is the use of catheters and balloons to clear the narrow blocked coronary lumen, improve myocardial blood perfusion, combined with coronary stent interventional therapy, can quickly and effectively relieve the disease and save the lives of patients. It has become a routine treatment for coronary heart disease. In 1969, Dotter first used metal rings to make vascular stents and successfully implanted them into animals to keep blood flow in the blood vessels unblocked (Dotter 1969). In 1987, Sigwart et al. successfully performed the first coronary stent operation (Sigwart et al. 1987). Subsequently, coronary artery stents were widely recognized as the second milestone in the history of interventional therapy for coronary heart disease. But the implantation of stents raises a new issue: in-stent restenosis (ISR). At the beginning of the 21st century, sirolimus and paclitaxel drug-eluting stents (DES), represented by Johnson & Johnson's Cypher and Boston's TAXUS, were hailed as the third milestone. DES has been shown to significantly reduce ISR, and its effectiveness has been widely recognized (Khan et al. 2012). However, anti-proliferative drugs also delay the re-endothelialization process, increasing the risk of late stent thrombosis, and their safety is still widely questioned (Moss et al. 2010). As is well known, vascular endothelial cells regulate the balance of vascular contraction and relaxation by secreting active substances such as nitric oxide (NO), PGI2, and ET-1, and release coagulation factors to regulate the dynamic balance of coagulation and anti-coagulation, thereby maintaining the normal physiological function of blood vessels (Kolluru et al. 2010). Due to the inevitable mechanical damage caused by stent implantation or expansion to blood vessels, re-endothelial repair of stent segment blood vessels is a key factor in inhibiting ISR and reducing late stent thrombosis (Vyalov et al. 1996).

G. Wang, *The Mechanobiology in Atherosclerosis*,
https://doi.org/10.1007/978-981-97-6512-6_3

3.1 Effects of Atherosclerotic Plaque Formation on Hemodynamics

Under normal physiological conditions, laminar shear stress is an important factor in maintaining healthy blood vessels (Chiu & Chien 2011). It remains constant within a certain range and maintains its direction and size, which can inhibit AS, thrombosis, adhesion of granulocytes, proliferation of vascular smooth muscle cells (VSMCs), vascular reconstruction, injury of vascular endothelial cells (VECs) etc., that play a very important role in the biological processes such as the regulation of vascular tube size, proliferation of vascular wall cells, thrombosis, and inhibition of inflammation (Gimbrone & García-Cardeña 2016). Laminar shear stress stimulates VECs to release various bioactive substances, such as superoxide dismutase (SOD), glutathione, NO, and ET-1 (Li et al. 2014b; Shyy et al. 2011). NO causes relaxation of vascular smooth muscle, while ET-1 causes contraction (Qu et al. 2007). The balance between the two maintains vascular tension within a certain range. NO also inhibits platelet adhesion and excessive synthesis of VSMCs, and release of growth factors by VECs, and prevent the excessive proliferation of VSMCs, promotes growth factor release and prevents excessive proliferation of VSMCs (Chen et al. 2003). Therefore, stable laminar shear stress is important in the formation of anti-AS (Zhou et al. 2013). In normal coronary artery, blood is in a laminar flow state, with a stable flow rate and relatively constant shear stress on endothelial cells, which is conducive to maintaining the normal physiological function of blood vessels (Cong & Kong 2020). The presence of atherosclerotic plaque leads to lumen stenosis. The blood flow velocity at the stenosis of the plaque measured by MRI increases sharply, forming a region with high shear stress at the proximal end of the plaque and relatively low shear stress at the distal end.

Due to the influence of blood flow velocity characteristics, the shear stress range of VECs in arterial and venous vessels is significantly different. The shear stress of the arterial wall is 10–70 dyn/cm^2, while the normal value of shear stress on VECs in the corresponding veins is 1–6 dyn/cm^2, which is significantly lower than that of the arteries. In the vascular system, areas with high shear stress exist at the curvature, bifurcation, and vascular anastomosis of the arch-like arteries, which are caused by turbulence or increased flow velocity due to the anatomical structure or geometric shape of the blood vessels. Moreover, due to the presence of countercurrent, the direction of shear stress may also change. The low shear stress zone is often accompanied by the presence of unstable fluids, such as turbulence, blood reflux areas, and blood "stasis" areas (congestion), which is often be seen at the bifurcation of the major arteries or arteries.

3.2 Effects of Stent Implantation on Vascular Injury and Hemodynamics

The implantation of vascular stents greatly changes the mechanical microenvironment of local blood vessels (Foin et al. 2014). From a macro perspective, due to the dilation effect of the stent on the lumen, the entire stent segment of the blood vessels returns to normal vascular patency, while the high shear stress state of the blood flow at the site of plaque stenosis will also return to relatively low shear stress (Van der Heiden et al. 2013). However, in the local vascular segment where the stent is implanted, the presence of the stent results in a fluid separation zone in the unidirectional flow of blood flow (Rikhtegar et al. 2014). At the interface between the stent and the blood vessel, the fluid undergoes a sharp reversal, resulting in turbulent flow. However, there is high shear stress at the edge and local surface of the stent near the heart (Fig. 3.1).

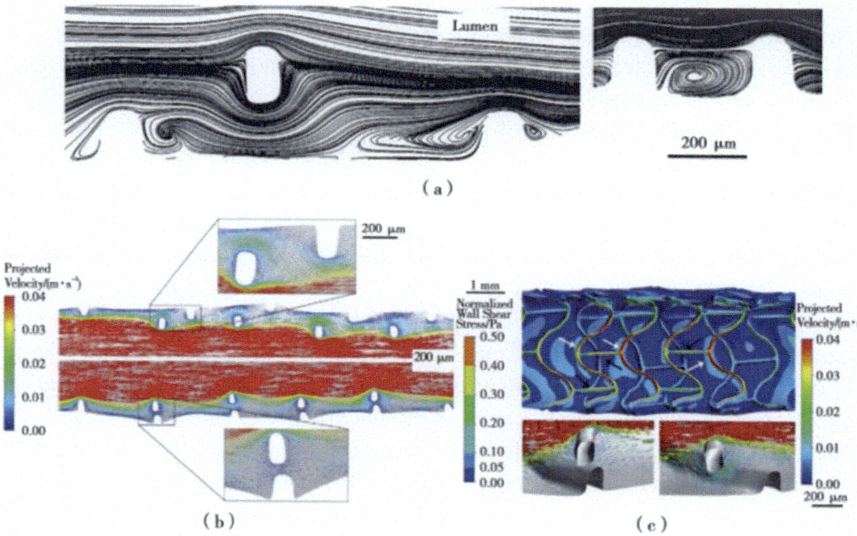

Fig. 3.1 Effect of stent implantation on hemodynamics. ANSYS ICEM CFD simulated the shear force distribution on the vascular wall after stent implantation, showing that the surface of the scaffold wire was a region of high shear force, while the shear force on the vascular wall was low. The scaffold locally causes turbulence, and the proximal end has high shear force, forming eddy currents. Adapted from Farhad et al. (2014)

3.3 Cell Sources for Vascular Injury Repair

Research has shown that endothelial repair after vascular injury not only relies on the proliferation and migration of mature VECs adjacent to the damaged intima, but also closely related to the homing, adhesion, and differentiation of endothelial progenitor cells (EPCs) derived from bone marrow (Fig. 3.2). In 1997, Asahara et al. first published an article in Science magazine discovering EPCs (Asahara et al. 1997). EPCs are precursor cells of VECs, also known as angioblasts, involved in angiogenesis, indicating their important role in cardiovascular disease, ischemic disease, vascular restenosis, wound healing, and other processes (Tesfamariam 2016). It mainly comes from human umbilical vein blood, adult peripheral blood and bone marrow. $CD133^+$ EPCs cultured in vitro can successfully induce differentiation into VECs (Reyes et al. 2002).

The currently reported screening methods for EPCs mainly include density gradient centrifugation and immunomagnetic bead sorting. Its main biomarkers include $CD34^+$, $VEGFR2^+$, etc. $CD34^+$ is not only expressed on hematopoietic stem cells, but also highly expressed on mature endothelial cells. Research has shown that $CD34^+$ cells also have the function of EPCs, and $CD34^+$ monocytes derived from bone marrow and peripheral blood can quickly differentiate into EPCs like cells in the body and contribute to the formation of new blood vessels. In 2003, Rehman et al. believed that EPCs were transdifferentiated $CD14^+$ monocytes, exhibiting the phenotypic characteristics of VECs in the vascular environment (Rehman et al. 2003). CD133/AC133, as a surface antigen of a glycosylated peptide with a molecular weight of 120 kD, contains 5 transmembrane domains (Shmelkov et al. 2005). Gordon et al. pointed out that CD133 is an antigen with unclear biological characteristics, expressed in hematopoietic stem cells (Gordon et al. 2003). As hematopoietic stem cells differentiate and mature, CD133 expression rapidly decreases, but is not expressed in mature endothelial cells and monocytes. Therefore, AC133 is currently a very useful marker for the separation of hematopoietic and endothelial progenitor cells (Yin et al. 1997). In addition, $VEGFR2^+$ expression is an important criterion for identifying cells with the ability to produce endothelial cells, which can originate from hematopoietic stem cells or monocytes. One hypothesis is that the expression of $VEGFR2^+$ is a fundamental requirement for regulating signal transmission caused by its ligand VEGF. However, $CD14^+/VEGFR2^+$, like $CD34^+/VEGFR2^+$, also exhibits other functional features. $VEGFR2^+$ cells exhibit stronger ability to proliferate, migrate, and release growth factors. In addition, EPCs overexpress CD31 on the cell membrane, while organelles of Weibel Palade bodies in the cytoplasm express vWF.

Percoll gradient centrifugation and adhesion methods were used to isolate EPCs from rat bone marrow. The isolated EPCs were identified by immunofluorescence, immunocytochemistry, and flow cytometry, and the specific binding matrix of EPCs was screened; Using single microtubule suction technology to study the adhesion of EPCs to different substrates and their dependence on substrate concentration; Using parallel plate flow chamber technology to study the effect of shear stress on the matrix binding ability of EPCs (Xiao et al. 2011). By screening anti restenosis drugs—total alkaloids of Sophora flavescens—to study their impact on the biological function of EPCs. The research results indicate that EPCs with higher purity

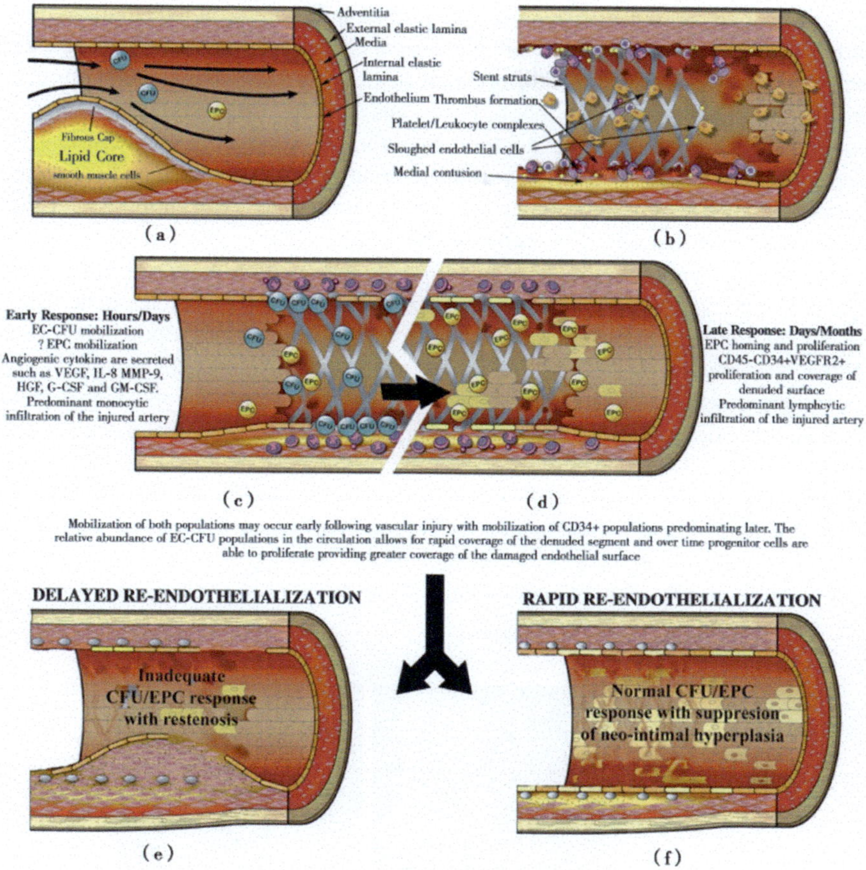

Fig. 3.2 Schematic diagram of the involvement of EPCs and VECs in the vascular repair process after stent implantation. **a** Atherosclerotic plaque formation. **b** The implantation of stents compresses plaques and maintains unblocked blood vessels, but at the same time damages blood vessels and activates inflammatory response and coagulation system. **c** At the early stage of stent implantation, VECs and EPCs were mobilized and activated. **d** At the later stage of implantation, the proliferation and migration of VECs and the homing of EPCs were attached to the surface of the scaffold for endothelial regeneration and repair. The repair process may produce two results; **e** Incomplete repair leads to delayed endothelialization and restenosis; **f** Complete repair and rapid endothelialization of the scaffold surface. Adapted from Padfield et al. (2010)

of VEGFR2$^+$/CD34$^+$/CD133$^+$ can be separated using Percoll gradient centrifugation method (Fig. 3.3a); The adhesion of EPCs on the matrix coated with antibodies VEGFR2, CD34, and CD133 is concentration dependent. EPCs have the strongest adhesion to antibody CD133, and compared to VECs, EPCs exhibit significant specific adhesion to antibody CD133. There is no CD133 antigen on the surface of VECs, and CD133 antibodies have more specificity towards EPCs (Fig. 3.3b). EPCs are subjected to shear stress on the flow chamber lined with CD133, and their retention and proliferation rates are significantly higher than those of the other two types of antibody lining; EPCs were subjected to shear stress on the chamber lined with CD133, resulting in a significant increase in NO expression. A certain degree of shear stress may induce EPCs to differentiate towards ECs. Animal experiments have shown that total alkaloids of Sophora flavescens can effectively inhibit endometrial hyperplasia, and appropriate concentrations of total alkaloids of Sophora flavescens have a proliferative effect on EPCs, without reducing their ability to adhere to specific matrices; In vitro results showed that the capture ability of CD133 antibody on EPCs was stronger than that of CD34 antibody (Wu et al. 2015).

Vascular injury causes the detachment of the endothelial layer, resulting in the exposure of the vascular intimal layer. This is accompanied by changes in hemodynamics and the release of inflammatory factors, activating and mobilizing various cellular components in the body to recruit to the site of injury for repair. In this process, previous views generally believed that endothelial cells involved in vascular repair mainly originate from the proliferation and migration of VECs adjacent to the damaged intima. This viewpoint relies on the observation results of epithelial cell migration and repair of epidermal wounds, which were confirmed in early Fishman's research: air blowing into a segment of the vascular lumen causes the inner wall of the vessel to dry, leading to the loss of endothelial cells in that part of the vessel. Subsequently, rapid proliferation of endothelial cells at the edge of exposed blood vessels

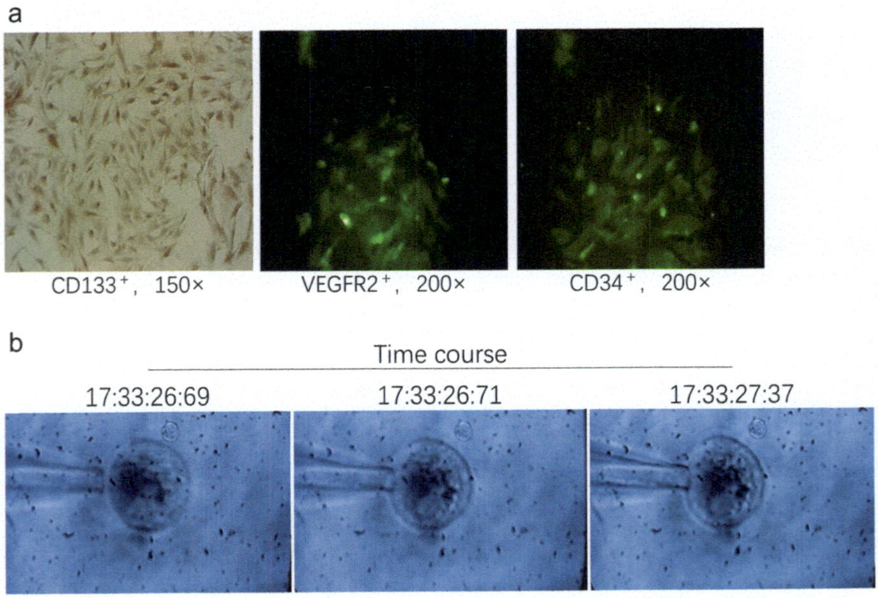

Fig. 3.3 Identification of EPCs and their adhesion to the substrate. **a** Immunohistochemistry combined with flow cytometry was used to determine the EPCs in the bone marrow of rats separated by Percoll gradient centrifugation and adherent method, and the expression of VEGFR2 and CD133 was positive. **b** The microtubule sucking system was used to observe the process of EPCs being pulled away from the adhesive substrate by negative pressure, and then the adhesion force value was calculated. Adapted from Xiao (2007)

was observed, migrating and growing towards the central celless area (Fishman et al. 1975). However, with the discovery of EPCs, new perspectives have been established on the cellular sources and repair mechanisms of vascular repair. Research has found that the newly formed endothelial layer after vascular transplantation mainly comes from the recipient's own bone marrow cells, rather than the donor's blood vessels (Reyes et al. 2002). Subsequently, it was confirmed in animal models such as acute ischemia after myocardial infarction and coronary artery bypass grafting that vascular injury will stimulate the proliferation of EPCs in the blood circulation, leading to their directional differentiation into VECs (Hu et al. 2003). Therefore, it is believed that the mobilization, homing, and adhesion of EPCs are the main processes of vascular endothelial layer repair, which once dominated the research on vascular injury repair (Padfifield et al. 2010). The researchers subsequently developed methods such as CD34, CD133, VE-Cadherin antibody capture EPCs scaffolds (Klomp et al. 2009), and EPCs cell therapy (Pelliccia et al. 2010), which have a significant promoting effect on the re-endothelialization of blood vessels in the stent implantation segment. However, Tsuzaki et al. proposed that no bone marrow-derived cells were found in the regenerated endothelial layer after vascular injury (Tsuzuki 2009). Recently, Hagensen et al. transplanted the blood vessels of Tie2-GFP and Tie2-LacZ transgenic mice into a wild-type mouse with damaged carotid artery model, and observed that the newly formed endothelial layer expressed green fluorescence and x-gel staining from donor mouse endothelial cells β-Galactosidase (β-Gel, encoded by the LacZ gene), and migrates from the arterial surgical interface to the central site over time (Hagensen et al. 2012). It is clearly pointed out that the re-endothelialization of blood vessels mainly comes from the migration of adjacent VECs, while EPCs do not contribute to the repair and regeneration of the endothelial layer. Douglas et al. implanted DES in a GCH-Tg transgenic mouse model overexpressing cyclohydrolase (specifically enhancing endothelial cell function and NO content) and found a significant increase in endothelial cell coverage (Douglas et al. 2013). Further observation revealed that the migration of VECs and the homing of EPCs contribute to stent segment vascular repair. The above contradictory research results have once again sparked discussions on the source of vascular endothelial repair cells. Van der Heiden speculated in subsequent review articles that the differences in the contribution of VECs and EPCs to re-endothelial repair observed in various studies may be due to the different animal pathological models, as well as different mechanical damage patterns and degrees (Van der Heiden et al. 2013). From this, it can be seen that vascular injury can be repaired through the migration of adjacent VECs and the recruitment of EPCs in the circulating blood. However, their exact contributions and molecular mechanisms are still unclear. The relatively simple vascular injury model has such great controversy, and for vascular stents with complex geometric configurations, the process of re-endothelialization after implantation will be more complex due to various factors such as shear stress, different degrees of vascular damage caused by dilation, and the physicochemical properties of the stent surface in vivo.

It is worth noting that due to the pressure applied to the balloon during the surgical process having a certain range (8–10 atm) and individual differences in the diameter of the patient's blood vessels, stent expansion has varying degrees of damage to the intima and mesomembrane of the blood vessels. Tahir et al. predicted the intimal thickness after vascular healing by numerically simulating the degree of damage to the internal elastic lamina (IEL) caused by stent implantation depth (90, 110, 130 μm) and the damage to intermediate smooth muscle cells. The degree of vascular damage will result in significant differences in the mechanical microenvironment in which the stent is located in the body (Tahir et al. 2013). The different exposure modes and hemodynamic changes of stents in blood vessels will determine the migration of VECs and the adhesion rate of EPCs, and ultimately determine their contribution to re endothelial repair. In addition, on the scale of vascular injury, the filling of the balloon and the expansion of the scaffold disrupt the integrity of the endothelial layer, leading to extensive erosion of endothelial cells, which also has an important impact on the process of re-endothelialization (Morlacchi et al. 2013). Immediately after the implantation of the stent, the animals were euthanized. The results showed that the filling of the balloon and the expansion of the stent would create an almost completely exposed endothelial layer throughout the entire stent segment, with only a small amount of detached endothelial cells remaining at the edge of the stent. Therefore, it is necessary to complete the repair of a single layer of endothelium throughout the entire vascular segment of the mesh stent.

Research has found that VECs and fluid shear stress can induce EPCs to differentiate into mature VECs. After co culturing with VECs for 2 weeks, the angiogenic ability of EPCs was enhanced, and the mRNA levels of endothelial nitric oxide synthase (eNOS) and vWF were significantly increased. The expression levels of markers such as CD31, ICAM, and Flt were not significantly different from mature VECs, except for KDR, in peripheral blood isolated $CD34^+$ cells induced by endothelial cell conditioned medium for 40 days. In addition, researchers also found that laminar shear stress can upregulate the expression of endothelial growth factor receptors (VEGFR, Flt-1/Flk-1), VE-cadherin and other endothelial markers and adhesion molecules in EPCs, and promote their homing at the site of injury to participate in vascular repair. One day after the implantation of a stent into the porcine carotid artery, it was found that the surface of the stent achieved approximately 30% endothelial coverage, and the endothelial coverage at the proximal end was significantly higher than that at the distal end. Endothelial cells are mainly concentrated at the edge of the scaffold, and there are also individual cells adhering to the exposed central area of the scaffold. By analyzing the above results, it is suggested that VECs have an inducing effect on the homing and differentiation of EPCs, and their synergistic involvement in the re-endothelialization repair of stent segment blood vessels is regulated by blood flow shear stress.

During the process of vascular remodeling caused by stress injury, the activation of coagulation and inflammatory systems may lead to delayed endothelialization. Vascular injury leads to rapid platelet aggregation, platelets and activated endothelial cells secrete SDF-1 and VEGF. At the same time, changes in shear stress can also stimulate endothelial cells to release a large amount of VEGF, PDGF, FGF, TGF-β. The activation of multiple signaling pathways, including PI3K and MAPKs, regulates cytoskeletal rearrangement, mobilizes EPCs to homing and differentiate into mature endothelial cells. Although there are many influencing factors that regulate the homing of EPCs involved in vascular repair in the complex mechanochemical microenvironment of the body, the increase in VEGF concentration in the blood is considered the most critical influencing factor. VEGF activates the PI3K/AKT/eNOS signaling pathway by binding to the VEGFR receptor (Flk/Flt) on the cell membrane surface. The upregulation of eNOS by shear stress does not entirely depend on the increase of intracellular Ca^{2+} concentration. The activation of AKT mediates the phosphorylation of eNOS in Ser1177, while eNOS dephosphorylates at the Thr495 site and binds to calmodulin in this region to increase NO production. The increase in NO concentration in the blood will initiate the mobilization, homing, and differentiation processes of EPCs, participating in vascular repair. Research has shown that the process of NO mobilization of EPCs is mediated by the activation of matrix metalloproteinases (MMPs) (Aicher et al. 2003). Under the action of activated MMP-9, KitL (kit ligand) transforms from membrane bound molecules to soluble molecules, namely sKitL (double Kit ligand), enters the bloodstream and binds to EPCs with cKit receptors, promoting their differentiation and mobilization to the peripheral circulation, reaching the site of action. In endothelial eNOS gene knockout mouse, the expression and activation of MMP-9 were sharply reduced, and significant EPCs mobilization defects were observed during neovascularization (Heieeig et al. 2002). The migration of VECs also depends on the activation of the PI3K signaling pathway. PI3K, the subtype of P110α, activates eNOS and releases NO by phosphorylating downstream molecules AKT. NO promotes the extracellular release of MMP by nitrating its tyrosine residues, degrading the extracellular matrix and promoting endothelial cell migration.

3.4 Surface Modification of Stents for Promoting Vascular Endothelial Repair

Percutaneous transluminal coronary angioplasty (PTCA) is a commonly used method for the treatment of coronary heart disease, but the postoperative injury response—restenosis is a difficult problem that troubles interventional therapy. The widespread application of coronary stent implantation significantly reduces the rate of restenosis after PTCA, but stent implantation also causes ISR, with an incidence rate of 15% to 20%. After implantation of coronary artery stent materials into the body, their surface has thrombogenic, allergenic, and inflammatory properties due to direct contact with blood and vascular walls, which leads to acute thrombosis and ISR. At present, research on the prevention and treatment of ISR mainly focuses on drug coated stents and intravascular radiation therapy, both of which can inhibit or disrupt DNA synthesis, thereby significantly inhibiting neointimal hyperplasia (Holme et al. 2012). However, they also affect endothelial cell regeneration, posing a risk of delayed thrombosis and stent exposure (Khan et al. 2012). At present, it is believed that rapid endothelialization of the material surface after stentimplantation is the most ideal way to improve stent-blood compatibility, inhibit ISR and thrombosis. VECs migration plays an important role in vascular physiology and pathological activities.

Clinical observations have shown that wound healing and re endothelialization of endothelial injury areas caused by vascular surgery are accompanied by the migration of VECs towards the wound site. For intravascular stents, after stent implantation and expansion, the migration of adjacent VECs is the main source of cells required for re-endothelial repair on the stent surface due to damage to the vascular intima.

3.4.1 Research on Surface Modification of Vascular Stents

The in vitro endothelialization of intravascular stents is considered an ideal way to improve the biocompatibility of biomaterials and reduce ISR. However, the in vitro endothelialization of scaffolds involves issues such as the source of seed cells, cell adhesion, biological function, transportation and storage of endothelial scaffolds. More researchers are focusing on the development of new materials and the design of stents, hoping to find a material with better biocompatibility, excellent corrosion resistance and antithrombotic performance (Shen et al. 2006). At the same time, they are seeking better preparation techniques in the design of stents, making the radial compression resistance and flexibility of the stents better, allowing them to pass through curved blood vessels with minimal damage to the vessel wall, thereby reducing stent restenosis caused by endometrial hyperplasia. Cobalt based alloy brackets (Co–Cr–Ni–Mo–Mn) are considered better scaffold materials due to their superior key performance compared to 316L stainless steel, but their cost will far exceed that of stainless steel brackets. Due to the respective advantages and disadvantages of drug eluting stents, cell endothelial stents, or new material stents, there is currently no consensus on which stent will ultimately solve the occurrence of ISR.

The surfaces of microporous and microgrooves were prepared by mechanical etching, chemical etching, and SiO_x:H ($0 < x < 2$) Nano-coating (Fig. 3.4), to study the effects of material surface morphology, nano-scale roughness and hydrophilicity on the adhesion of endothelial cells and platelet, as well as in vitro biocompatibility of plasma SiO_x:H nano-coated scaffold materials and in vivo animal experiments. Different micro morphologies such as micropores and micro grooves can be prepared on the surface of NiTi alloy substrate using chemical etching and mechanical etching methods; The average surface roughness of the micro roughened NiTi alloy prepared is within the range of nanometers; The thickness and hydrophilicity (hydrophobicity) of plasma SiO_x:H nanocoating can be controlled through parameters such as monomer selection, monomer flow rate ratio, and deposition time; The optimal conditions for plasma deposition of SiO_x:H nanocoates are TMS:O_2 = 1:4 sccm, 4 min, D C. 5W, system pressure 25 mTorr; Under these conditions, the thickness of the prepared coating is about 45 nm, and the water contact angle of the ionomer on the Si sheet is between 30 and 40°. Combining two surface modification techniques, micro roughening and plasma deposition of SiO_x:H nanocoates, can significantly promote the adhesion of VECs; On the order of nanoroughness, materials with microporous structures have the highest number of VECs adhered to their surfaces and exhibit good growth status; The micro groove structure on the

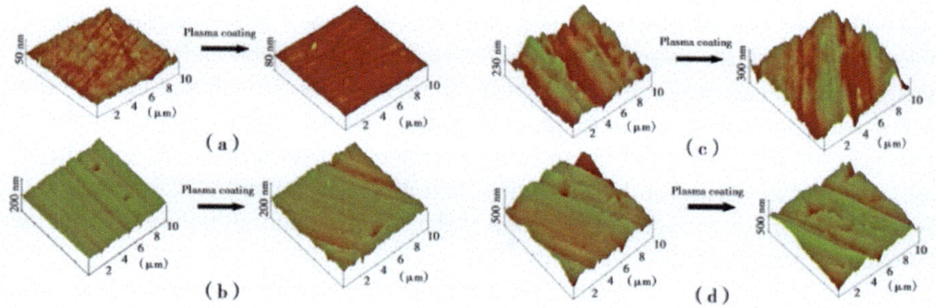

Fig. 3.4 Atomic Force Microscope (AFM) measurement of different microstructures combined with plasma deposition to prepare SiO_x:H nano-coated NiTi alloy surface. Adapted from Shen et al. (2009)

material surface has a contact induced effect on early cell adhesion. After plasma deposition of SiO_x:H nanocoating, the number of platelet adhesion on the surface of the material decreased; And it can significantly increase the corrosion resistance of NiTi matrix, reduce the dissolution rate of Ni^{2+}, improve the anticoagulant properties of the material, significantly reduce the hemolysis rate, cytotoxicity, and heat generation, which meets the safety requirements of biological materials measurement standards (Shen et al. 2009); Animal experiments have shown that plasma deposition of SiO_x:H nanocoats accelerates the surface endothelialization process of the scaffold compared to unmodified bare scaffolds, significantly inhibits intimal hyperplasia and reduces in-stent restenosis (Zhang et al. 2015). It has been confirmed that the combination of micro roughening material surface with micropores and micro grooves and surface modification with low-temperature plasma deposition SiO_x:H nanocoating not only improves the biocompatibility of the material matrix, but also accelerates the endothelialization process after stent implantation, which is expected to solve the problem of narrowing within the stent (Shen et al. 2009).

3.4.2 *The Modified Intravascular Stents of Transgenic Endothelial Cell*

The role of VECs in the formation of restenosis is receiving increasing attention. It is difficult to see spontaneous endothelialization when implanting intravascular stents into the human body. Simple endothelial cell implantation of intravascular stents results in a very slow rate of endothelial cell division and proliferation, which prevents rapid re-endothelialization at the site of endothelial cell detachment on the scaffold. Prolonged contact with blood flow poses a risk of thrombosis, and does not promote repair of damaged intima, making it difficult to prevent and treat stent restenosis. In addition, there is currently no consensus on the effects of transfection gene types, transfection methods, and cells from different species on cell function, especially adhesion retention function, after transfection. VECs transplantation can replace and repair the damaged endothelial function and inhibit endometrial hyperplasia. Endothelial cells cultured in vitro experience slow growth and loss of other biological characteristics due to changes in their growth environment. Combined with gene therapy, it is possible to stably transfect growth factors into endothelial cells cultured in vitro, thereby enhancing their proliferation ability. Combined with dynamic culture, it is expected to promote the adhesion and growth of endothelial cells on the surface of vascular stents.

We stably transfected VEGF121 into human umbilical vein endothelial cells (HUVECs); A rotating culture device for implanting cells on the surface of intravascular stents was designed and manufactured (Tang 2008). By analyzing the influencing factors such as rotation speed, rotation time, cell implantation density, and number of repeated rotations, the optimal combination of adhesion conditions was found; Preparation of protein coatings for cell adhesion using ultrasonic atomization spray method, and detection of the prepared protein coatings using scanning electron microscopy and energy dispersive spectroscopy; Based on the optimal rotational culture conditions obtained, a transgenic endothelial cell coated scaffold was prepared. Scanning electron microscopy was used to detect the number and spread of cells adhering to the surface of the stent after rotational culture. Immunofluorescence was used to detect the expression of VEGF factor in the cells adhering to the surface of the stent; In vitro flow chamber experiments were conducted to detect the adhesion and growth of HUVECs transfected with VEGF gene on the surface of intravascular stents after transmission loss and flow shear stress; Finally, the anti-restenosis and anti-thrombotic effects of transgenic endothelial cell coated stents were tested using an animal model of rabbit abdominal aortic stent implantation. By analyzing the influencing factors such as rotation speed, rotation time, cell seeding density, and number of repeated rotations, the optimal combination of adhesion conditions was found when the cell seeding density was 1×10^5 cells/mL, a rotation time of 6 h, and a rotation speed of 0.4 rpm, the number of cells adhering to the rotation culture was the highest; Prepare a mixed solution with a concentration of 2–6 μg/mL of gelatin and 10 μg/mL of polylysine. The protein coating obtained by ultrasonic atomization spraying method is tested to be tightly and uniformly coated on the surface of the scaffold; Transgenic endothelial cell covered intravascular scaffolds were prepared through optimized rotational culture conditions. Under optical microscopy and scanning electron microscopy, the cells adhered and grew well on the surface of the scaffold. Fluorescence microscopy was used to detect cells with high expression of VEGF protein. Through in vitro flow chamber experiments, it was found that some cells of transgenic endothelial cells will be lost during the simulated in vivo transport process, but under different shear stresses, the cells can quickly resume growth. The results of animal experiments showed that compared with the two control groups, the cell coated scaffold significantly inhibited endometrial hyperplasia and reduced in stent restenosis (Tang et al. 2011). This study preliminarily solved the cell implantation method for intravascular stents and proposed a laboratory preparation and detection method for tissue engineering intravascular stents. The relevant experimental results

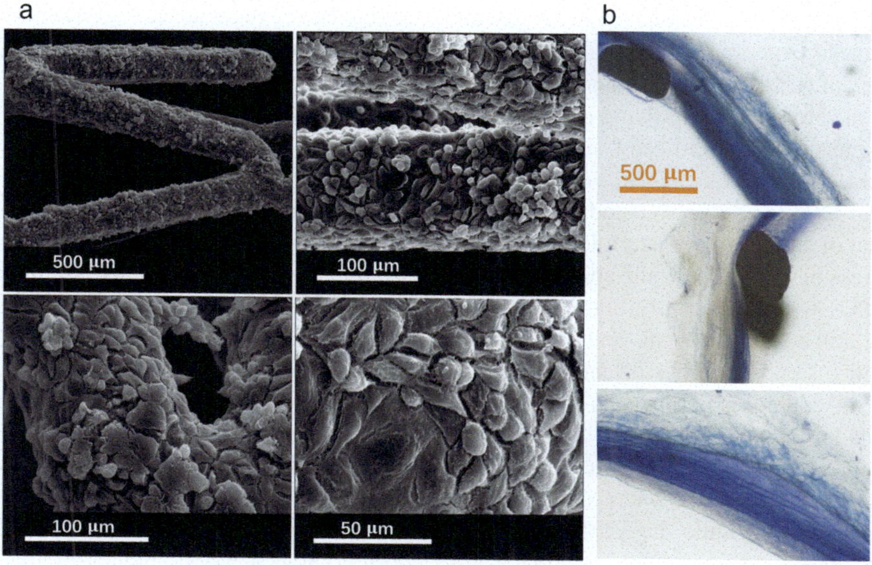

Fig. 3.5 Preparation of transgenic endothelial cells modified intravascular scaffolds and animal experimental results. **a** Observation of cell adhesion and growth on the surface of the scaffold in vitro by scanning electron microscopy; **b** Toluidine blue staining results of hard tissue sections at 1, 4 and 12 weeks after implantation of transgene endothelial cell modified scaffolds. Adapted from Tang et al. (2011)

obtained provide experimental basis for the clinical application of cell implantation in the prevention and treatment of intravascular stent restenosis (Fig. 3.5).

3.4.3 CD133 Antibody Capture EPCs Scaffold

At present, there are two main types of research on promoting stent re endothelialization repair: in vivo endothelialization (Wu et al. 2011) and in vitro endothelialization (Wu et al. 2013). Endothelialization in vivo involves antibody modification of the surface of scaffold materials, inducing the proliferation and differentiation of endothelial cells on the material surface after implantation, and exercising normal functions to form an intimal layer similar to healthy blood vessels. Antibodies capture EPCs in the circulating blood, inducing them to differentiate into mature endothelial cells and secreting growth factors to promote rapid repair of the endothelial layer on the scaffold. Currently, this study has been applied to clinical trials and has achieved promising therapeutic effects.

We prepared anti-CD133 antibody coated scaffolds and conducted in vitro evaluations. In previous studies, it has been proven that anti CD133 antibody coating is beneficial for EPC adhesion (Wu et al. 2015). EPCs were isolated from the bone marrow of New Zealand white rabbits, and confirmed through rat tail collagen angiogenesis experiments that anti-CD133 antibody can effectively promote the angiogenesis of EPCs; A uniformly coated anti-CD133 scaffold was prepared using an ultrasonic atomization spraying device, and the coating remained intact and stable after expansion; The ability of anti CD133 antibody coated scaffolds to capture EPCs was evaluated through in vitro experiments. In static environments, the ability of anti-CD133 antibody coated scaffolds to capture EPCs was significantly higher than that of gelatin coated scaffolds and bare scaffolds. In flow environments with a shear stress of 2.5 dyn/cm^2, the ability of anti-CD133 antibody coated scaffolds to capture EPCs was significantly higher than that of anti CD34 antibody coated scaffolds, gelatin coated scaffolds, and bare scaffolds. To test the endothelialization effect and anti restenosis ability of anti CD133 antibody coated scaffolds and VEGF transgenic cell implantation scaffolds. The scanning electron microscopy results indicate that both types of scaffolds can effectively accelerate the process of endothelialization. The endothelial layer on the surface of the anti CD133 antibody coated stent appeared within one week and may have been earlier than the anti CD34 antibody coated stent, while the VEGF interference group had a slightly weaker effect on the stent, and no normal endothelial layer appeared after 4 weeks, indicating that VEGF plays an important role in the process of endothelialization; The results of hard tissue sectioning after 12 weeks showed that both antibody coated scaffolds and cell implantation scaffolds achieved significant anti restenosis effects. Among them, the intimal hyperplasia and lumen loss of anti-CD133 antibody coated stents were significantly lower than those of gelatin coated stents and bare stents, and there was no significant difference compared to anti-CD34 antibody coated stents. The endometrial hyperplasia and lumen loss of VEGF overexpressing cell seeding scaffolds were significantly lower than those of VEGF interfering cell seeding scaffolds and bare scaffolds, and the restenosis rate was relatively lower compared to anti CD133 antibody coated scaffolds; In order to analyze the source of endothelial cells on the scaffold, a small number of implanted cells were observed in the 0 h sample using scanning electron microscopy. However, no fluorescence expression of transplanted cells was found in the samples at this time and thereafter. Therefore, it is speculated that the endothelial cells on the scaffold almost entirely originate from host autologous cells. In order to demonstrate the safety of VEGF transgenic cell implantation scaffolds, frozen sections of organ tissues were further performed, and no transgenic cells expressing fluorescence were observed. HE staining showed that the morphology of cells and tissues was consistent with that of normal tissues.

3.4.4 Molecular Mechanism Study on Surface Materials-Mediated Re-Endothelial Repair of Scaffold

The adhesion and migration of vascular endothelial cells are important processes for endothelial repair after implantation of cardiovascular materials, and the surface/

interface interactions between cells and materials are the key factors determining their adhesion/migration (Shen et al. 2015b). The design of material surface physicochemical properties to regulate biological behaviors such as cell adhesion and migration, thereby accelerating the process of endothelialization, is currently the focus of cellular biomechanics research, but its molecular mechanism is not yet clear (Bazellières et al. 2015). Materials factors (including composition, structure, surface chemistry, microstructure, etc.), like chemical signals and mechanical signals generated by mechanical stimuli, can participate in the cascade activation of regulating protein or gene expression, thereby regulating biological behaviors such as cell adhesion, migration, proliferation, and differentiation. The interaction between the surface/interface of a material and proteins, cells after implantation depends on the structural characteristics of the material itself and surface, as well as the microenvironment in which it is implanted. Numerous studies have shown that the physical and chemical properties of material surfaces, such as surface morphology and roughness, chemical functional groups, energy state and hydrophilicity/hydrophobicity, charge types and distribution, all affect the number and state of cell adhesion on their surfaces. However, there are few studies on the surface properties of materials that regulate cell migration to repair the endothelial layer.

(1) Molecular Mechanism of Cell Adhesion and Migration Mediated by the Surface of Hydrophilic Materials

The effect of hydrophilic and hydrophobic material surfaces on the adhesion and migration of vascular endothelial cells. The hydrophilic material surface (1:2–1:5 sample group) is more conducive to cell adhesion, but cells exhibit obvious filamentous and plate-like pseudopodia on the hydrophobic material surface, showing a migration trend. The statistical results of scratch experiments, cell migration distance, number of migrated cells, and final healing time all confirmed that the hydrophobic plasma SiO_x:H surface (1:0 and 1:1 samples) is more conducive to the migration of vascular endothelial cells. The continuous observation results for 30 min also proved that the migration rate of cells on hydrophobic materials is higher than that on hydrophilic surfaces. The adhesion and migration of cells on surfaces of materials with different wettability exhibit opposite behaviors.

The role of focal adhesion kinase (FAK) in the adhesion and migration of vascular endothelial cells. By measuring the adhesion and migration of VECs and the expression of downstream small G proteins with different concentrations of FAK inhibitors, it was found that inhibiting FAK caused pseudopodial contraction of VECs, nuclear protrusion, disruption of intercellular connections, and easy detachment of cells. The addition of FAK inhibitors significantly inhibits cell migration behavior. The higher the concentration of FAK inhibitors, the smaller the cell migration distance, in a concentration dependent manner. As the concentration of FAK inhibition increases, the expression of Rac1 protein gradually weakens, further confirming that the adhesion and migration of vascular endothelial cells are related to the FAK-Rho GTPases signal axis. The migration of VECs on SiO_x:H nanofilms with different wettability was measured using a concentration of 50 nm/mL FAK inhibitor. The results showed that the migration of vascular endothelial cells on the surfaces of different wettability materials depended on the tyrosine phosphorylation of FAK at Y397 and its downstream signaling pathway (Shen et al. 2012).

The effect of material surface hydrophobicity on the expression and distribution of FAK-Rho GTPase signaling pathway related proteins. Western blot combined with immunofluorescence was used to investigate the effect of material surface hydrophobicity on the distribution and expression of key signaling proteins in the FAK-Rho GTPases signaling pathway. The results showed that the surface of hydrophilic and hydrophobic materials caused different expressions of Integrinα and Integrinβsubunits of membrane surface receptors, which affected the different responses of Rac1, RhoA and Cdc42 in downstream Talin, pFAK, small G proteins to the surface of hydrophilic and hydrophobic materials (Fig. 3.6). The statistical results of several protein expressions confirm that the adhesion and migration of vascular endothelial cells on the surfaces of different hydrophilic and hydrophobic materials depend on the phosphorylation of the key enzyme FAK tyrosine at site 397 in the cells. The migration of cells on the surface of hydrophobic materials is related to the FAK-Rho GTPases signaling pathway, while on the surface of hydrophilic materials, it is uncertain that it is completely related to this pathway. Among them, 1:2 samples ($\theta = 47.5 \pm 3.0°$) is the turning point (Shen et al. 2013).

(2) Preliminary Study on Signal Pathways of Cell Adhesion and Migration Mediated by Different Functional Groups on the Surface of Materials

Cells are involved in a series of complex signaling pathways during migration. Research has shown that extracellular matrix, integrin, and FAK signaling molecules play important roles in cell migration. The differences in surface physicochemical properties of materials, such as the conformational changes of adhesive components

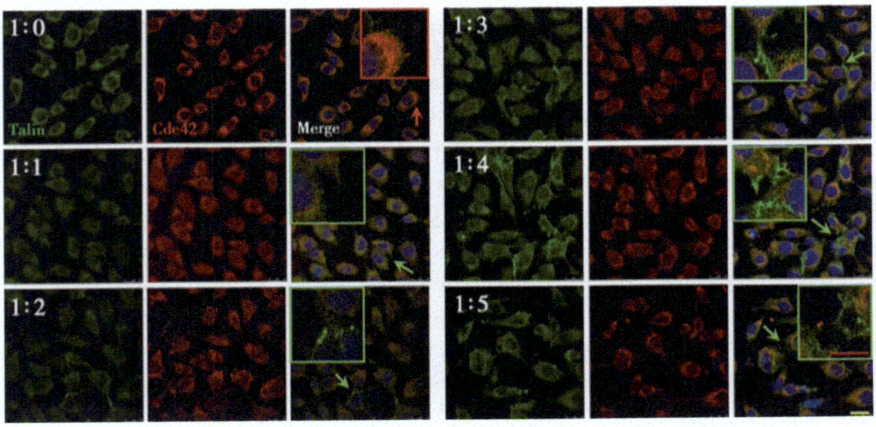

Fig. 3.6 Immunofluorescence double staining analysis of the effects of SiO_x:H coating surface deposited by different hydrophilic (1:3 ~ 1:5) hydrophobic (1:0 ~ 1:2) on the expression and distribution of Cdc42 protein (red fluorescence) and Talin (green fluorescence) in VECs cells. DAPI nuclear staining is shown in blue, the ruler in Figure B = 25 μm. Adapted from Shen et al. (2013)

in the extracellular matrix (ECM) caused by different surface chemical functional groups, may stimulate cells to have different adhesion states and migration rates on the material surface. ECM includes components such as laminin (LN), fibronectin (FN), and collagen, which are specifically adsorbed on the surface of the material, undergo conformational changes, expose active sites, and specifically bind to the cell membrane adhesion receptor integrin ($\alpha2\beta1$, $\alpha3\beta1$, $\alpha5\beta1$ and $\alpha V\beta3$, etc.), activating integrin and recruiting and phosphorylating a series of intracellular proteins (Talin, Paxillin, Vinculin, etc.), thereby transmitting signals into the cell. Subsequently, focal adhesion (FA) is formed locally on the cell membrane, and FAK is also activated and phosphorylated downstream proteins, such as the small G protein Rho GTPases, mediating a series of signal transduction, thereby regulating cell migration behavior.

Self-assembly monolayers (SAMs) prepared by strong adsorption between Au–S are considered a good model material for studying the interactions between cells, proteins, and material surfaces, due to their ability to effectively control the chemical composition of the material surface. A 20 nm Au coating deposited on the surface of the material binds to the S bond of n-dodecyl mercaptan, while chemical functional groups such as CH_3, NH_2, OH, COOH are exposed on the surface of the material. Using this model material surface, the effects of material surface chemistry on the adhesion and migration of vascular endothelial cells and the expression and distribution of key proteins in Integrin Rho GTPases signaling pathway were investigated. The research results indicate that the surfaces of CH_3 and NH_2 are more conducive to cell migration, while the COOH surface has the smallest cell migration distance. The migration of vascular endothelial cells on the surface of materials with different chemical functional groups depends on the tyrosine phosphorylation of FAK at the Y397 site, and the activation of FAK is highest on the CH_3 surface. The CH_3, NH_2, and OH groups showed higher expression of Rac1, RhoA, and Cdc42, consistent with migration results at the cellular level. Confirming that the adhesion and migration behavior of vascular endothelial cells on the surface of materials with different chemical functional groups depend on the Integrin-FAK-Rho GTPases signaling pathway (Figs. 3.7 and 3.8) (Shen et al. 2015a).

3.5 Research Progress and Prospects

In addition, bioabsorbable scaffolds can temporarily support blood vessels, maintain lumen patency, inhibit the occurrence of early thrombosis and late neointimal hyperplasia (Garcia-Garcia et al. 2014). With the development of vascular repair processes, scaffold materials are gradually absorbed and decomposed by vascular tissue, making them a hot topic in vascular stent research (Koppara et al. 2015). We recently analyzed the endothelial function, vascular remodeling, scaffold degradation, and vascular cell response after the implantation of poly L lactic acid (PLLA) biodegradable scaffolds, and explored the impact of the degradation product lactate on the phenotype of smooth muscle cells, confirming that:

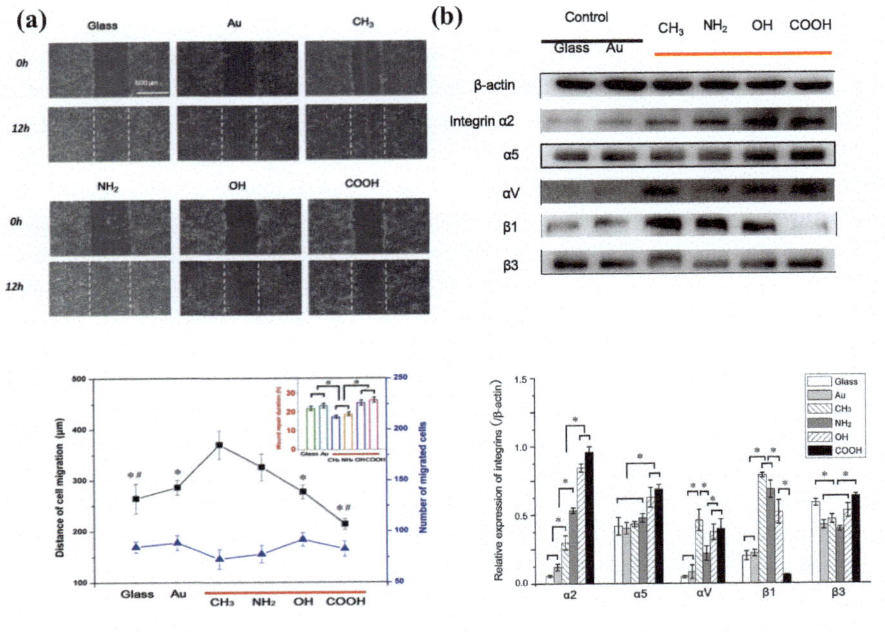

Fig. 3.7 Effects of material surfaces with different chemical functional groups on VEC migration and Integrin α and β subunit expression. **a** The influence of material surfaces with different chemical functional groups on VEC migration was measured by scratch method. **b** Western blot analysis of the effects of different chemical functional groups on the expression of Integrin subunits (α2, α5, αV, β1, β3) and the statistical results. Adapt from Shen et al. (2015a)

① After 4 weeks of implantation of PLLA intravascular stent, re-endothelialization is completed, which can obtain neointima with low inflammation, good endothelialization and excellent barrier function.

② Time-dependent intimal hyperplasia and vascular repair and reconstruction after PLLA intravascular stent implantation. After PLLA stent implantation, the degree of neointimal hyperplasia first increased and then decreased. The blood vessels began to reconstruct positively at 24 weeks, and gradually tended to return to normal after 48 weeks.

③ The degradation process of PLLA intravascular stent in vivo and the accompanying changes in vascular cell phenotype. After 24 weeks of stent implantation, significant degradation occurred, and vascular cells underwent a process of transformation from normal to osteogenic, followed by a decrease in the degree of transformation.

④ The degradation product of PLLA, lactate, promotes the osteogenic phenotype transformation of vascular smooth muscle cells (Du 2019).

This chapter provides a detailed evaluation of the vascular intimal repair and vascular reconstruction process after PLLA stent implantation, starting from animal models, providing more reference basis for clinical evaluation of bioabsorbable stents and providing more theoretical basis for drug coating design of stents.

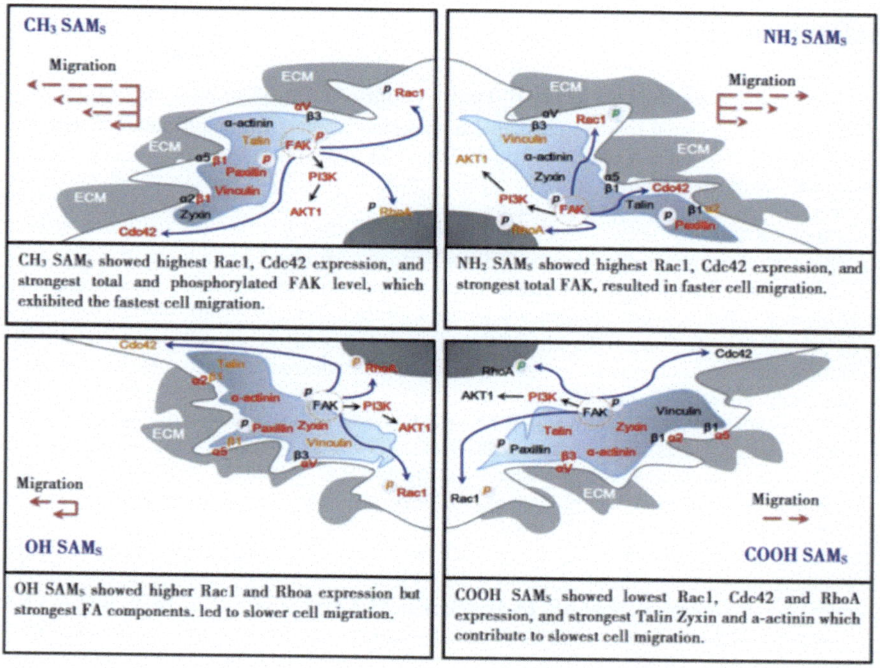

Fig. 3.8 The adhesion and migration behaviors of vascular endothelial cells on the surface of materials with different chemical functional groups depend on the Integrin-FAK-Rho GTPases signaling pathway. Adapt from Shen et al. (2015a)

Our research team has developed the hemodynamic theory of vascular injury repair and proposed a new perspective on repair kinetics that accelerates endothelialization, inhibits vascular restenosis, and relapses AS. It has revealed the mechanism of increasing endothelial cell adhesion to the substrate, inducing differentiation of circulating endothelial progenitor cells or mesenchymal stem cells, accelerating vascular endothelialization, inhibiting thrombosis, and promoting endothelial repair (Wang et al. 2019); Clarified the relationship between hemodynamic factors and damage repair regeneration in the process of vascular injury repair, proposed low-temperature plasma deposition technology combined with surface modification technology of scaffold materials on microstructure surface, and explored the molecular mechanism of VECs re endothelial repair induced by material surfaces with different wettability and chemical functional groups, On this basis, bioactive surface modification technologies such as CD133 antibody capture scaffolds and transgenic endothelial cell modified intravascular scaffolds were constructed, providing a theoretical basis for the clinical application of drug scaffolds such as arsenic trioxide.

References

Aicher A, Heeschen C, Mildner-Rihm C, Urbich C, Ihling C, Technau-Ihling K, Zeiher A M, Dimmeler S. Essential role of endothelial nitric oxide synthase for mobilization of stem and progenitor cells[J]. Nat Med, 2003, 9(11): 1370–1376.

Asahara T, Murohara T, Sullivan A, Silver M, van der Zee R, Li T, Witzenbichler B, Schatteman G, Isner JM. Isolation of putative progenitor endothelial cells for angiogenesis[J]. Science, 1997, 275(5302): 964–967.

Bazellières E, Conte V, Elosegui-Artola A, Serra-Picamal X, Bintanel-Morcillo M, Roca-Cusachs P, Muñoz JJ, Sales-Pardo M, Guimerà R, Trepat X. Control of cell-cell forces and collective cell dynamics by the intercellular adhesome[J]. Nat Cell Biol, 2015, 17(4): 409–420.

Chen Q, Li W, Quan Z, Sumpio BE. Modulation of vascular smooth muscle cell alignment by cyclic strain is dependent on reactive oxygen species and P38 mitogen-activated protein kinase[J]. J Vasc Surg, 2003, 37(3): 660–668.

Chiu JJ, Chien S. Effects of disturbed flow on vascular endothelium: pathophysiological basis and clinical perspectives[J]. Physiol Rev, 2011, 91(1): 327–387.

Citi S. The mechanobiology of tight junctions[J]. Biophys Rev, 201, 11(5): 783–793.

Cong X, Kong W. Endothelial tight junctions and their regulatory signaling pathways in vascular homeostasis and disease[J]. Cell Signal, 2020, 66: 109485.

Dotter C. Transluminally-placed Coilspring Endarterial Tube Grafts Long-term Patency in Canine Popliteal Artery. Invest Radiol, 1969, 4(5):329–332.

Douglas G, Van Kampen E, Hale AB, McNeill E, Patel J, Crabtree MJ, Ali Z, Hoerr RA, Alp NJ, Channon KM. Endothelial cell repopulation after stenting determines in-stent neointima formation: effects of bare-metal vs. drug-eluting stents and genetic endothelial cell modification[J]. Eur Heart J, 2013, 34(43): 3378–3388.

Du RL. Influence and mechanism of bioabsorbable PLLA stent on intimal repair and vascular remodeling[D]. Chongqing University, 2019.

Farhad R et al. Hemodynamics in coronary arteries with overlapping stents. J Biomech, 2014, 47 (2): 505–511

Fishman JA, Ryan GB, Karnovsky MJ. Endothelial regeneration in the rat carotid artery and the significance of endothelial denudation in the pathogenesis of myointimal thickening[J]. Lab Invest, 1975, 32(3): 339–351.

Foin N, Lee RD, Torii R, Guitierrez-Chico JL, Mattesini A, Nijjer S, Sen S, Petraco R, Davies JE, Di Mario C, Joner M, Virmani R, Wong P. Impact of stent strut design in metallic stents and biodegradable scaffolds[J]. Int J Cardiol, 2014, 177(3): 800–888.

Garcia-Garcia HM, Serruys PW, Campos CM, Muramatsu T, Nakatani S, Zhang YJ, Onuma Y, Stone GW. Assessing bioresorbable coronary devices: methods and parameters[J]. JACC Cardiovasc Imaging, 2014, 7(11): 1130–1148.

Gimbrone MA Jr, García-Cardeña G. Endothelial cell dysfunction and the pathobiology of atherosclerosis[J]. Circ Res, 2016, 118(4): 620–636.

Gordon PR, Leimig T, Chen X, Buhring HJ, Niethammer D, Kuci S. Biology and plasticity of CD133+hematopoietic stem cells[J]. Ann N&Y Acad Sci, 2003, 996: 141–151.

Hagensen MK, Raarup MK, Mortensen MB, Thim T, Nyengaard JR, Falk E, Bentzon JF. Circulating endothelial progenitor cells do not contribute to regeneration of endothelium after murine arterial injury[J]. Cardiovasc Res, 2012, 93 (2): 223–231.

Heissig B, Hattori K, Dias S, Friedrich M, Ferris B, Hackett NR, Crystal RG, Besmer P, Lyden D, Moore MA, Werb Z, Rafii S. Recruitment of stem and progenitor cells from the bone marrow niche requires MMP-9 mediated release of kit-ligand[J]. Cell. 2002, 109 (5): 625–637.

Holme MN, Fedotenko IA, Abegg D, Althaus J, Babel L, Favarger F, Reiter R, Tanasescu R, Zaffalon PL, Ziegler A, Müller B, Saxer T, Zumbuehl A. Shear-stress sensitive lenticular vesicles for targeted drug delivery[J]. Nat Nanotechnol, 2012, 7(8): 536–543.

Hu Y, Davison F, Zhang Z, Xu Q. Endothelial replacement and angiogenesis in arteriosclerotic lesions of allografts are contributed by circulating progenitor cells[J]. Circulation. 2003, 108 (25): 3122–3127.

Khan W, Farah S, Domb AJ. Drug eluting stents: developments and current status[J]. J Control Release, 2012, 161(2): 703–712.

Klomp M, Beijk MA, de Winter RJ. Genous endothelial progenitor cell-capturing stent system: a novel stent technology[J]. Expert Rev Med Devices, 2009, 6 (4): 365–375.

Kolluru GK, Sinha S, Majumder S, Muley A, Siamwala JH, Gupta R, Chatterjee S. Shear stress promotes nitric oxide production in endothelial cells by sub-cellular delocalization of eNOS: A basis for shear stress mediated angiogenesis[J]. Nitric Oxide-Biol Ch. 2010, 22(4): 304–315.

Koppara T, Cheng Q, Yahagi K, Mori H, Sanchez OD, Feygin J, Wittchow E, Kolodgie FD, Virmani R, Joner M. Thrombogenicity and early vascular healing response in metallic biodegradable polymer-based and fully bioabsorbable drug-eluting stents[J]. Circ Cardiovasc Interv, 2015, 8(6): e002427.

Koskinas KC, Chatzizisis YS, Antoniadis AP, Giannoglou GD. Role of endothelial shear stress in stent restenosis and thrombosis: pathophysiologic mechanisms and implications for clinical translation[J]. J Am Coll Cardiol, 2012, 59(15): 1337–1349.

Li X, Yang Q, Wang Z, Wei D. Shear stress in atherosclerotic plaque determination[J]. DNA Cell Biol, 2014b, 33(12): 830–838.

Morlacchi S, Migliavacca F. Modeling stented coronary arteries: where we are, where to go[J]. Ann Biomed Eng, 2013, 41 (7): 1428–1444.

Moss SC, Lightell DJ Jr, Marx SO, Marks AR, Woods TC. Rapamycin regulates endothelial cell migration through regulation of the cyclin-dependent kinase inhibitor p27Kip1[J]. J Biol Chem, 2010, 285(16): 11991–11997.

Padfield GJ, Newby DE, Mills NL. Understanding the role of endothelial progenitor cells in percutaneous coronary intervention[J]. J Am Coll Cardiol, 2010;55(15):1553–1565.

Pelliccia F, Cianfrocca C, Rosano G, Mercuro G, Speciale G, Pasceri V. Role of endothelial progenitor cells in restenosis and progression of coronary atherosclerosis after percutaneous coronary intervention: a prospective study[J]. JACC Cardiovasc Interv, 2010, 3(1): 78–86.

Qu MJ. Effect and its mechanism of cyclicmechanical strain on phenotypemodulation of vascular smooth muscle cells[D]. Shanghai Jiao Tong University, 2007.

Rehman J, Li J, Orschell CM, March KL. Peripheral Blood "Endothelial progenitor cells" are derived from monocyte/macrophages and secrete angiogenic growth factors[J]. Circulation. 2003, 107: 1164–1169.

Reyes M, Dudek A, Jahagirdar B, Koodie L, Marker PH, Verfaillie CM. Origin of endothelial progenitors in human postnatal bone marrow[J]. J Clin Invest, 2002, 109: 337–346.

Rikhtegar F, Wyss C , Stok KS, Poulikakos D, Müller R, Kurtcuoglu V. Hemodynamics in coronary arteries with overlapping stents[J]. J Biomech. 2014, 47(2):505–511.

Shen Y, Wang GX, Quan XJ. Study on surface modification and biocompatibility of NiTi alloy intravascular stents[J]. Chinese Journal of Medical Instrumentation, 2006(01): 3–6+38.

Shen Y. Study on plasma deposition of SiOx:H nanocoating combined with surface microroughness modified intravascular stents [D]. Chongqing University, 2009.

Shen Y, Wang GX, Chen L, Li H, Yu P, Bai MJ, Zhang Q, Lee J, Yu QS. Investigation of surface endothelialization on biomedical nitinol (NiTi) alloy: effects of surface micropatterning combined with plasma nanocoatings[J]. Acta Biomater, 2009, 5 (9): 3593–3604.

Shen Y, Wang G, Huang X, Zhang Q, Wu J, Tang C, Yu Q, Liu X. Surface wettability of plasma SiOx:H nanocoating-induced endothelial cells' migration and the associated FAK-Rho GTPases signalling pathways[J]. J R Soc Interface, 2012, 9 (67): 313–327.

Shen Y, Ma YH, Gao M, Lai Y, Wang GX, Yu QS, Cui FZ, Liu XH. Integrins-FAK-Rho GTPases pathway in endothelial cells sense and response to surface wettability of plasma nanocoatings[J]. ACS Appl Mater Inter. 2013, 5(11):5112–5121.

Shen Y, Gao M, Ma YL, Yu HC, Cui FZ, Liu XH, Gregersen H, Yu QS, Wang GX, Liu XH. Effect of surface chemistry on the integrin induced pathway in regulating vascular endothelial cells migration[J]. Colloid Surface B, 2015a, 126: 188–197.

Shen Y, Leng MT, Yu HC, Zhang Q, Luo XL, Gregersen H, Wang GX, Liu XH. Effect of amphiphilic PCL–PEG nano-micelles on hepg2 cell migration macromol[J]. *Biosci.* 2015b, 15: 372–384.

Shmelkov SV, St Clair R, Lyden D, Rafii S. AC133/CD133/Prominin-1[J]. Int J Biochem Cell Biol, 2005, 37(4): 715–719.

Shyy YJ, Zhen C, Wei W, Wei S. Shear-stress activation of AMP-activated protein kinase in endothelial homeostasis[J]. Cell Mol Bioeng. 2011, 4(4): 538–546.

Sigwart U, Puel J, Mirkovitch V, Joffre F, Kappenberger L. Intravascular stents to prevent occlusion and restenosis after transluminal angioplasty. N Engl J Med, 1987, 316(12):701–706.

Tahir H, Bona-Casas C, Hoekstra AG. Modelling the effect of a functional endothelium on the development of in-stent restenosis. PLoS One, 2013, 13;8(6):e66138.

Tang CJ. The study of endovascular stent seeded with gene-transfection endothelial cells[D]. Chongqing University, 2008.

Tang C, Wang G, Wu X, Li Z, Shen Y, Lee JC, Yu Q. The impact of vascular endothelial growth factor-transfected human endothelial cells on endothelialization and restenosis of stainless steel stents[J]. J Vasc Surg, 2011, 53 (2): 461–471.

Tesfamariam B. Endothelial repair and regeneration following intimal injury[J]. J Cardiovasc Transl Res, 2016, 9(2): 91–101.

Tsuzuki M. Bone marrow-derived cells are not involved in reendothelialized endothelium aAs endothelial cells after simple endothelial denudation in mice[J]. Basic Res Cardiol, 2009, 104 (5): 601–611.

Van der Heiden K, Gijsen FJ, Narracott A, Hsiao S, Halliday I, Gunn J, Wentzel JJ, Evans PC. The effects of stenting on shear stress: relevance to endothelial injury and repair[J]. Cardiovasc Res, 2013, 99 (2): 269–275.

Vyalov S, Langille BL, Gotlieb AI. Decreased blood flow rate disrupts endothelial repair in vivo[J]. Am J Pathol, 1996, 149(6): 2107–2118.

Wang Y, Zhang K, Qin X, Li T, Qiu J, Yin T, Huang J, McGinty S, Pontrelli G, Ren J, Wang Q, Wu W, Wang G. Biomimetic Nanotherapies: Red Blood Cell Based Core-Shell Structured Nanocomplexes for Atherosclerosis Management[J]. Adv Sci (Weinh), 2019, 6(12): 1900172.

Wu X, Wang G, Tang C, Zhang D, Li Z, Du D, Zhang Z. Mesenchymal stem cell seeding promotes reendothelialization of the endovascular stent[J]. J Biomed Mater Res A, 2011, 98(3): 442–449.

Wu X. Re-endothelialization study on endovascular stents coated by antibody and endovascular stents seeded by endothelial cells through VEGF up/down- regulation[D]. Chongqing University, 2013.

Wu X, Yin T, Tian J, Tang C, Huang J, Zhao Y, Zhang X, Deng X, Fan Y, Yu D, Wang G. Distinctive effects of CD34- and CD133-specific antibody-coated stents on re-endothelialization and in-stent restenosis at the early phase of vascular injury[J]. Regen Biomater, 2015, 2(2): 87–96.

Xiao L. The screening of specific substrates for endothelial progenitor cell capturing stent and effect of sophorcarpidine on number and function of cells[D]. Chongqing University, 2007.

Xiao L, Wang G, Jiang T, Tang C, Wu X, Sun T. Effects of shear stress on the number and function of endothelial progenitor cells adhered to specific matrices[J]. J Appl Biomater Biomech, 2011, 9(3): 193–198.

Yin AH, Miraglia S, Zanjani ED, Almeida-Porada G, Ogawa M, Leary AG, Olweus J, Kearney J, Buck DW. AC133, a novel marker for human hematopoietic stem and progenitor cells[J]. Blood. 1997, 90: 5002–5012.

Zhang Q, Shen Y, Tang C, Wu X, Yu Q, Wang G. Surface modification of coronary stents with SiCOH plasma nanocoatings for improving endothelialization and anticoagulation[J]. J Biomed Mater Res B Appl Biomater, 2015, 103 (2): 464–472.

Zhou J, Li YS, Nguyen P, Wang KC, Weiss A, Kuo YC, Chiu JJ, Shyy JY, Chien S. Regulation of vascular smooth muscle cell turnover by endothelial cell-secreted microRNA-126: role of shear stress[J]. Circ Res, 2013, 113(1): 40–51.

Chapter 4
Mechanobiological Mechanism of Cardiovascular and Cerebrovascular Development

The cardiovascular and cerebrovascular system is the first functional system established during embryonic development in vertebrates, which ensures material exchange and energy supply for the growth and development of other tissues and organs. Vascular cells are derived from the mesoderm of the embryo. The development of blood vessels generally involves a series of processes, such as primitive blood vessel formation, vascular stabilization, blood vessel branch formation, vessel remodeling and vascular function specialization. There are two forms of new blood vessel formation during blood vessel development. One is vasculogenesis, in which vascular stem cells differentiate into endothelial cells (ECs) in situ and form primitive new blood vessels. The other is angiogenesis, which is the sprouting and invagination of the preexisting vascular endothelium to form the capillary plexus. In the early embryonic period, primitive blood vessels are formed through the process of angiogenesis. Vascular stem cells derived from the mesoderm gradually differentiate and migrate, eventually differentiating into ECs and migrating to specific sites to assemble into primitive blood vessels, forming major arterial and venous vessels, heart primordia and extraembryonic capillary networks. The structure of these primitive blood vessels is extremely unstable and is composed of only a layer of ECs. Vascular smooth muscle cells (SMAs) migrate to the surface of primitive blood vessels to form a protective layer and stabilize the structure of the blood vessels. After the establishment of blood circulation in the embryo, these immature vascular systems also need to be reconstructed to form vascular branches to establish a capillary network of blood circulation. Vascular development also requires the formation of specialized blood vessels and a complete vascular network system through the process of arteriovenous differentiation. After the above series of processes, a mature vascular network system is established, and life activities rely on the blood circulation system to complete the exchange of material and energy.

The development of the embryonic cardiovascular and cerebrovascular system is a complex process that is regulated by multiple stimulatory and inhibitory signals.

G. Wang, *The Mechanobiology in Atherosclerosis*,
https://doi.org/10.1007/978-981-97-6512-6_4

These signals must act in coordination to ensure that each stage of vascular development proceeds normally. The process of vascular development is to some extent controlled by genes, and research on this process has been carried out over a wide range. However, recent studies have found that biomechanics is a necessary factor for embryonic vascular development, and different cellular biomechanical mechanisms are involved in the process of embryonic vascular development. From the establishment of blood circulation, hemodynamics plays a very important role in the development of cardiovascular and cerebrovascular diseases. Blood flow dynamics are particularly important for the differentiation of arteries and veins and the formation and sprouting of new blood vessels. The mechanical interaction between blood and blood vessels also plays a key role in the regulation of vessel remodeling and homeostasis (Xie et al. 2018). Recent studies have shown that microgravity can lead to cardiovascular dysfunction and local vascular malformations in astronauts. In addition, microgravity can also affect early vascular development and lead to vessel remodeling (Xie et al. 2017, 2019).

This chapter starts with the influence of biomechanical factors on the cardiovascular and cerebrovascular development process, briefly describes the influence of hemodynamic changes on cardiovascular development, analyzes the relationship between several hemodynamic factors and vascular development in depth, and elucidates the relationship between force and vascular development. This chapter focuses on the molecular mechanism of hemodynamic regulation of vascular development and the mechanical signal transduction mechanism of hemodynamic regulation of vascular development, introduces the potential effect of hemodynamics on cerebrovascular development and its related molecular mechanism, and discusses the effect of microgravity on vascular development and its related molecular mechanism.

4.1 The Effect of Hemodynamics on Cardiovascular and Cerebrovascular Development and Its Mechanical-Molecular Mechanism

The biological organism itself is a mechanical structure that contains rich mechanical laws not only from the overall level of the organism but also from the level of organs, tissues and cells. Mechanical action accompanies us all the time from the beginning of the cell division of the biological organism to the end of life. From the establishment of blood circulation, the vascular system is a mechanical system centered on the heart. There are a variety of mechanical interactions in the process of blood circulation, such as blood flow dynamics, blood cell deformation and tumbling migration, axial stress and transverse pressure of blood vessels, and the interaction between blood and the blood vessel wall. Therefore, hemodynamic factors play a particularly important role in the development of cardiovascular and cerebrovascular diseases and their physiological and pathological processes. Flow shear stress (FSS) and vascular stretch stress can regulate changes in cardiovascular cellular components, including membrane receptors and integrins, signal protein kinases, growth factors and cytokines, and thus participate in the development and pathophysiological process of cardiovascular and cerebrovascular vessels. There are many basic biomechanical and biological issues involved in the process.

4.1.1 The Influence of Hemodynamics on Vascular Development

4.1.1.1 Hemodynamics Regulates Vascular Development

During vascular development, the primitive vascular plexus exhibits great plasticity since the establishment of blood circulation. The increase in blood flow is accompanied by the widening and thickening of blood vessels; blood vessels with small blood flow simply disappear, and fragments of blood vessels begin to break off and form new fragments. Moreover, in areas with high blood flow, there is a tendency of strong blood vessel branching. On the basis of the initial primitive blood vessels, the branch formation of arterioles and venules and the establishment of capillary networks are related to hemodynamic factors. Although the aorta and cardinal vein can be formed in zebrafish without initial blood circulation during vascular development, recent studies have found that the development of aortic arch vessels is still dependent on hemodynamic factors (Nocoli et al. 2010). In addition, hemodynamics has recently been shown to influence the development of collateral circulation in zebrafish (Gray et al. 2007).

In our previous study, we found that hemodynamic changes significantly affected the vascular development of the caudal vein plexus (CVP) in zebrafish (Xie et al. 2018). In the experimental control group, the lumen of the ventral vein (VV, yellow arrow) was smooth and full, and the CVP network structure was clear. However, in the inhibitor treatment group, the CVP did not form, the vessel shape was irregular, and no complete network structure was formed (Fig. 4.1a). Quantitative analysis of CVP showed that angiogenesis was significantly inhibited after inhibitor treatment (Fig. 4.1c). As shown in Fig. 4.1b, d, inhibition of heart beating by injection of *tnnt2a*-morpholino oligonucleotides (MO) resulted in anthemic blood flow in zebrafish embryos and a significant inhibition of CVP development.

Due to the limitations of current experimental conditions and methods, the study of the effect of hemodynamics on vascular development in vivo still has certain limitations. Many studies have used tnnt2a-MO-injected zebrafish (without heartbeat) as the research object since there is no initial blood circulation due to the absence of heartbeat to investigate the effect of the absence of blood circulation on vascular development. In addition, some studies have observed the effect of hemodynamics on angiogenesis by laser cutting or ligation of blood vessels (Lucitti et al. 2007; Nocoli et al. 2010), but these studies cannot regulate the magnitude of blood

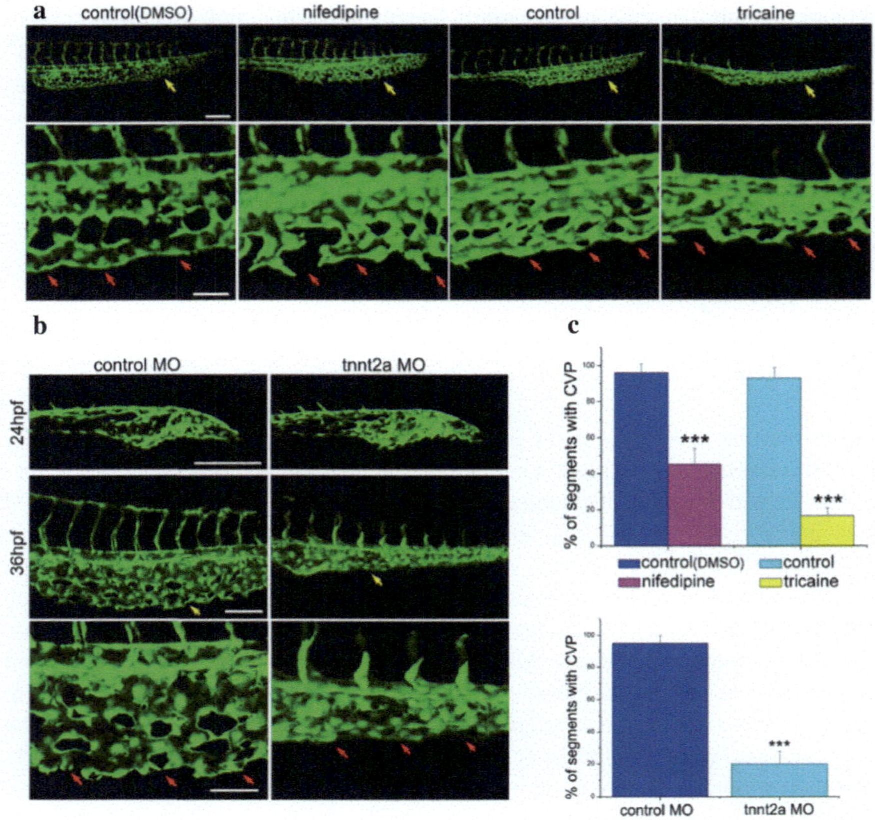

Fig. 4.1 Hemodynamics regulates zebrafish embryos CVP angiogenesis. **a** Treatment with a blood flow regulator (nifedipine: 1 mM, tricaine: 2 mM) treatment inhibits CVP angiogenesis. Transgenic Tg (flk1:GFP) embryos were treated with chemicals 24 h after fertilization (hour post fertilization, hpf) and imaged at 36 hpf by in vivo confocal microscope. A representative example after drug treatment is shown. Side view, front left, back up. CVP is marked with a yellow arrow. The scale bar is 100 μm. The image below shows an enlarged view of CVP. VV is shown by the red arrow. The scale bar is 50 μm. **b** The effect of tnnt2a MO on the formation of CVP at 24 and 36 hpf. Zebrafish embryos were injected with tnnt2a MO and imaged at 24 and 36 hpf using a confocal microscope. Side view, front left and back up. CVP is marked with a yellow arrow. Scale bar, 100 μm. The figure below shows an enlarged view of the CVP area at 36 hpf. VV is shown with a red arrow. The scale bar is 50 μm. **c** Percentage of body segments containing CVP. In each treatment, thirty embryos were used for quantification. Nifedipine or tricaine blocked the formation of CVP and the fusion of VV (Mean ± SD, t test, *** $P < 0.001$). **d** The percentage of somites containing CVP, quantified in control MO ($n = 56$) and tnnt2a MO ($n = 49$) injected embryos. tnnt2a MOs prevented CVP angiogenesis to a large extent (Mean ± SD, t test, *** $P < 0.001$). Adapted from Xie et al. (2018)

flow and only observe the effect of no blood flow on blood vessel development, which to further study the relationship between blood flow and blood vessel development. At present, there is a plan to use chemical inhibitors that can regulate the magnitude of blood flow and find that blood flow can regulate the process of blood vessel development (Xie et al. 2018). Therefore, could the drug inhibitors themselves cause malformations in blood vessel development? The latest study found that at the same inhibitor concentration, vascular malformations of zebrafish were observed, but there was no significant change in the appearance and survival rate of the treated and control zebrafish. Therefore, although chemical inhibitors may have some effects on the development of zebrafish, they are not significant, so severe malformations of vascular development are largely caused by changes in hemodynamics (Xie et al. 2013).

4.1.1.2 Relationship Between Various Hemodynamic Factors and Vascular Development

With decreases in blood flow, blood flow velocity and FSS, CVP angiogenesis is also reduced. However, there was no linear relationship between the changes in blood flow, blood flow velocity and CVP angiogenesis. However, there is a linear relationship between FSS and CVP angiogenesis, which indicates that these three hemodynamic factors are all involved in the mechanical regulation of CVP vascular development, and the change in FSS is the hemodynamic factor that can most directly reflect the changes in CVP vascular development (Fig. 4.2). When blood flow, blood flow velocity and FSS were changed within a certain range (blood flow reduced to 65 pl/s, blood flow velocity reduced to 220 μm/s, and FSS reduced to 0.86 dynes/cm^2), there was no significant change in CVP vascular development. However, when these factors are lower than this value, the vascular development of CVP will be greatly affected. Blood flow, blood flow velocity and FSS have no significant effect on vascular development as long as they are maintained at the amount needed for the basic development of the body. However, if the blood flow is below a certain value, even if the basic nutrients and oxygen content cannot be met, vascular development will be greatly affected. Therefore, whether there is such a threshold in the development of blood vessels in different organisms, as long as the threshold is above it, there is no great effect on the development of blood vessels, but once below this threshold, the development of blood vessels will produce malformations. Further studies are needed to confirm the important role of this threshold in vascular development and the occurrence and development of vascular diseases, which may also provide new clues and targets for the prevention and treatment of vascular development, including diseases.

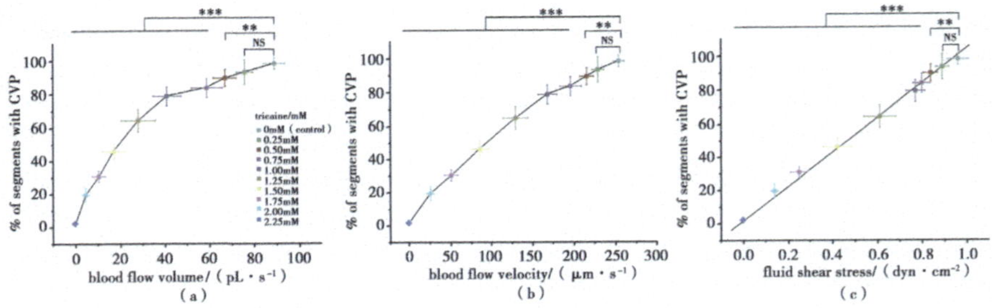

Fig. 4.2 The relationship between the three hemodynamic factors and the percentage of CVP. 26–36 hpf embryos (n = 30) were treated with a gradient concentration of tricaine (from 0 mM to 2.25 mM), and different colors indicate the concentration (mean ± standard deviation, significant and control, analysis of variance, *** $P < 0.001$ and ** $P < 0.01$, n = 3). **a** There is a nonlinear correlation between blood flow and the percentage of CVP. **b** The blood flow rate and the percentage of CVP have a nonlinear correlation. **c** The FSS has a linear relationship with the percentage of CVP. Adapted from Xie et al. (2018)

4.1.2 Molecular Mechanisms of Hemodynamic Regulation of Vascular Development

In recent years, a large number of studies on vascular development have been carried out around mechanics, and the relationship between vascular development and mechanics has been explored from the genetic level to the overall biological level. The focus of research is on how mechanics causes abnormal vascular development, as well as the related mechanical signaling and regulatory pathways. Previous studies have found that the expression levels of hypoxia-inducible factors hif-1α and hif-2α in zebrafish without blood flow caused by tnnt2a-MO treatment or slowed blood flow by chemical inhibitor treatment were not significantly different from those in the control group, suggesting that hypoxia was not involved in the development of CVP malformations caused by lack of blood flow. It was found that there was obvious apoptosis in the tail of zebrafish with no blood flow or slowed blood flow. However, further studies showed that injection of p53 MO could rescue the expression of flk1 but not the angiogenesis of zebrafish CVP, indicating that the apoptosis pathway is not involved in the hemodynamic regulation of CVP angiogenesis (Xie et al. 2018).

4.1.2.1 ERK5-Klf2a Signaling is Involved in the Process of Angiogenesis Regulated by Hemodynamics

The Klf family is a recently widely studied and reported class of zinc-finger structural transcription factors involved in vascular system activities. Kruppel-like factor is a general term for a class of transcription factors that contain three closely linked acetyl-like zinc finger structures at the carboxyl terminus. Klf2, klf4, klf5 and klf6 in this family are several factors that have received great attention and are related to vascular development and vascular diseases. Klf2(KLF2/LKLF) was first found to be mainly expressed in the lung, so its full name is lung Kruppel-like transcription factor. A series of studies have shown that klf2 is the most critical mechanical response factor of hemodynamics in the vascular circulation system (Boon et al. 2009; Lee et al. 2006a; Wang et al. 2006). FSS induces the expression of klf2 transcription factor in mouse microvascular ECs. By transfection of the fluorescent gene into the $5'$ flanking region of the klf2 gene, the gene response region of klf2 expression induced by FSS is from -157 to -95 bp at the transcription start site. Blood flow can regulate klf2 expression through ERK5/MEF2 signaling pathway-induced AMPK activity (Young et al. 2009). Klf2 is an important transcriptional regulator in vascular ECs. Klf2 changes the expression levels of factors such as eNOS, VEGFR2 and TNF-α through JNK, MAPK and other signaling pathways. It is involved in the regulation of physiological processes related to endothelial cell migration, vasodilation function, inflammation, cell homeostasis, and stress fiber-related cell morphological changes (Boon et al. 2010; Dekker et al. 2006; Lin et al. 2005; SenBanerjee et al. 2004). Hemodynamically induced klf2 in specific areas of the artery has an anti-atherosclerotic effect.

Many studies have shown that klf2 participates in different physiological and pathological processes in cells. Recently, it has been found that klf2, as a hemodynamic response factor, plays an important role in vascular development, especially angiogenesis, and can regulate the expression of a variety of enzymes and transcription factors to regulate vascular development. Studies have shown that klf2a, as a hemodynamic response factor, regulates angiogenesis through the miR-126 signaling pathway and VEGF signaling pathway (Lucitti et al. 2007). Overexpression of klf2 in the process of angiogenesis can effectively inhibit the expression of KDR and VEGFR2 and regulate angiogenesis through the VEGFR2/KDR signaling pathway. The klf2 and ETS-family protein ERG synergistically activate the expression of flk1 in vascular ECs to regulate vascular development. Furthermore, klf2 plays an important role in erythropoiesis during the early development of the cardiovascular system, regulating the expression of human and mouse embryo-like globin genes. Recently, tnnt2a-MO injection resulted in a significant decrease in klf2a expression, suggesting hemodynamic regulation of klf2a expression (Fig. 4.3e), and injection of KLf2a-Mo into embryos at the single-cell stage resulted in not only downregulation of flk1 expression (Fig. 4.3d) but also a significant CVP defect (Fig. 4.3f–h). These results suggest that klf2a plays a critical role in CVP angiogenesis in zebrafish (Xie et al. 2018).

Extracellular signal-regulated kinase (ERK) is an important signal transduction protein. It normally localizes in the cytoplasm and translocates to the nucleus when activated to regulate the activity of transcription factors and produce cellular effects. The ERK signaling pathway is one of the most important signal transduction pathways and can regulate a variety of important biochemical reactions, such as cell proliferation and differentiation. Recent studies have shown that hemodynamics cause morphological changes and cytoskeletal rearrangement in ECs. The expression of Klf2 changes with changes in the cytoskeletal microtubule state and is affected by the ERK5 signaling pathway (Parmar et al. 2006; Vartanian et al. 2010). It has

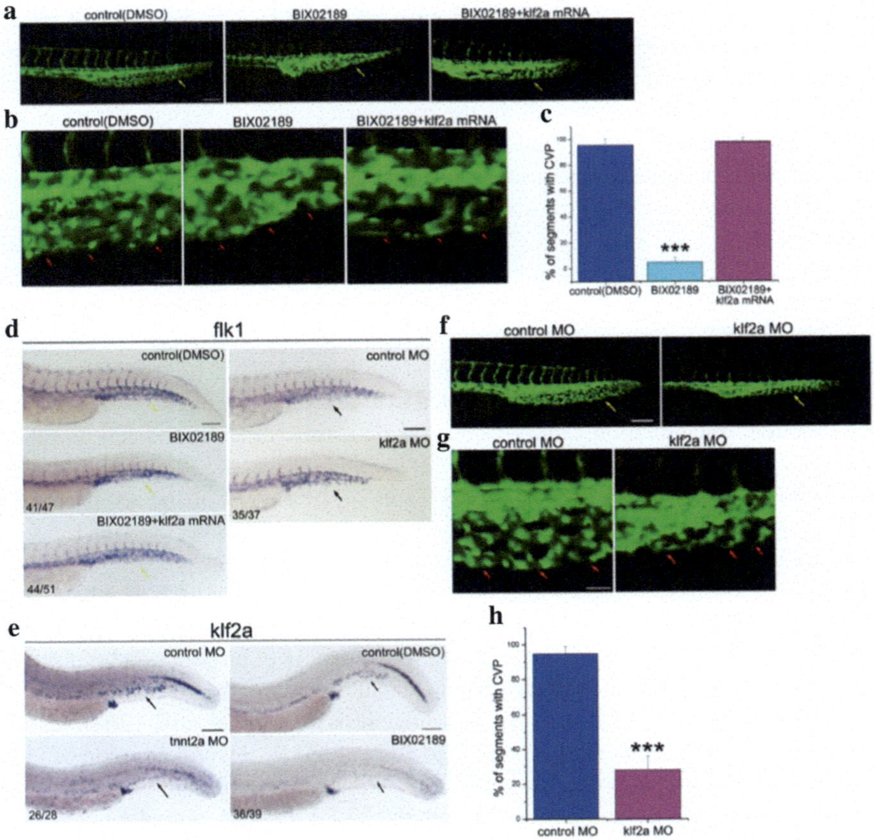

Fig. 4.3 Blood flow-mediated ERK5 regulates zebrafish CVP angiogenesis through Klf2a. **a** At 36 hpf, the formation of zebrafish CVP by BIX02189 and klf2a mRNA was evaluated by fluorescence microscopy. Seen from the front left side, the back side is upward. CVP is marked with a yellow arrow. The scale bar is 100 μm. **b** The higher magnification of the CVP area obtained from A. VV is marked with a red arrow. The scale bar is 50 μm. **c** Embryos treated with DMSO (n = 56), BIX02189 (n = 50), BIX02189 + klf2a mRNA (n = 47). CVP deficiency in embryos treated with BIX02189 can be rescued by klf2a mRNA in zebrafish angiogenesis (mean ± SD, significance compared with control, ANOVA, *** $P < 0.001$). **d** In situ hybridization of the tail region at 36 hpf flk1 whole embryo. The left panel shows the control (DMSO), BIX02189 and BIX02189 + klf2a mRNA. CVP is marked with a yellow arrow. The right image shows embryos treated with control MO, klf2a MO. CVP is marked with a black arrow. Seen from the front left side, the back side is upward. Scale bar, 100 μm. **e** After the embryo is injected with tnnt2a MO, in situ hybridization of klf2a whole embryos in the CVP region and surrounding tissues at 36 hpf. The left image shows embryos treated with control MO, tnnt2a MO. The right panel shows embryos treated with control (DMSO) BIX02189. Side view, front left, back side up. The scale bar is 100 μm. **f** Fluorescence microscope image of flk1:GFP transgenic embryo at 36 dpf. Seen from the front left side, the back side is upward. CVP is marked with a yellow arrow. The scale bar is 100 μm. **g** An enlarged view of Figure F. The CVP area in the VV is marked with a red arrow. The scale bar is 50 μm. **h** The percentage of body segments containing CVP. Control MO (n = 58), klf2a MO (n = 52) (Mean ± SD, t test, *** $P < 0.001$). Adapted from Xie et al. (2018)

also been reported that the changes in cytoskeletal intermediate filament, vimentin and desmin gene expression induced by FSS can cause arterial dilatation and remodeling, and changes in the cytoskeletal microtubule network structure can affect the expression of the eNOS gene (Loufrani et al. 2008).

Recent studies have shown that treatment of zebrafish embryos with the ERK5 inhibitor BIX02189 not only downregulated flk1 expression at the CVP site (Fig. 4.3d) but also caused a significant CVP defect (Fig. 4.3a, b). Further studies showed that overexpression of klf2a mRNA could rescue the defect in CVP angiogenesis induced by the ERK5 inhibitor BIX02189 (Fig. 4.3a, b), suggesting that ERK5 participated in zebrafish CVP angiogenesis by regulating klf2a.

4.1.2.2 ENOS is Involved in Hemodynamically Regulated Angiogenesis

Under the condition of normal endothelial function, the synthesis of NO to maintain the normal physiological operation of the vascular system is mainly guided by endothelial nitric oxide synthase (eNOS). FSS not only affects the intracellular distribution of eNOS but also regulates the expression of eNOS by regulating cytokines such as VEGFR2, c-Src tyrosine kinase, and AMP protein kinase. The cytoskeleton regulates eNOS expression through G-actin and F-actin, and this process is cell growth- and density dependent (Kondrikov et al. 2006; Su et al. 2003). The regulation of endothelial nitric oxide kinase is mostly at the transcriptional level, but there are a variety of modifications and posttranscriptional regulation, such as its acylation and phosphorylation through intracellular localization and protein interaction. As a key mechanical response factor of hemodynamics in the vascular circulation system, eNOS not only has anti-atherosclerotic effects but also participates in vascular development (Kubes et al. 1991; Moncada et al. 1992).

The expression pattern of eNOS is closely related to the development of collateral vessels during arteriogenesis, suggesting that eNOS plays an important role in the process of arteriogenesis (Cai et al. 2004). We found that eNOS knockout (KO) mice exhibited loose and transparent arteries both in natural development and disease models, and this process was proven to be regulated by cell proliferation in the disease model (Dai et al. 2010). eNOS not only regulates vitelline sac angiogenesis in mice but also has a rescue effect on hyperglycemia-induced vasculopathy.

NO regulates early development of the cardiovascular system in zebrafish and plays an important role in the maintenance of the vascular system during ontogeny (Pelster et al. 2005). A digital motion analysis video technique was used to study the effects of NO and epinephrine on the arteries and veins of developing zebrafish vasculature. The results showed that vascular lumen development in larval zebrafish tissues was regulated by endothelial cell secretions such as NO and epinephrine (Fritsche et al. 2000). No homologous gene of eNOS has been found in fish, but NOS2b in zebrafish has been found to contain a special functional region that is common with mammalian eNOS: the N-terminal tetradecanoylation sequence (Lepiller et al. 2009). In mammals, the N-terminal tetradecanoylation (Ntetradecanoylation) and hexadecanoylation (hexadecanoylation) sequences of eNOS are targeted in tiny regions of the plasma membrane (generally recognized as plasma membrane microvesicles), which facilitate interactions with other cellular proteins and lipids to initiate signal transduction pathways (Dudzinski et al. 2006). Not only do zebrafish NOS2b and mammalian eNOS share common functional domains, but their expression in the heart region and the stress response to lipopolysaccharides (LPS) are also found to be the same (Forstermann et al. 1998). Recent studies found that injection of NOS2b-MO not only downregulated flk1 expression (Fig. 4.4b) but also caused significant CVP defects (Fig. 4.4a, c), indicating that zebrafish NOS2b is involved in hemodynamically regulated vascular development.

Recent studies have shown that klf2 can regulate the expression of eNOS, indicating that klf2 acts as an important molecular switch for eNOS in response to FSS (Dekker et al. 2005; Groenendijk et al. 2004, 2005). Studies have shown that statins (statins) exert antiatherosclerotic and antithrombotic effects through klf2, and this process is achieved by klf2 regulating the activity of endothelial nitric oxide kinase and thrombomodulin. It has been found that continuous shear stress is better than statins (statins) in maintaining the sustained expression of klf2, which in turn maintains the expression of eNOS and thromtomodulin, which in turn has better anti-atherosclerotic and anti-thrombotic effects. Recent studies found that overexpression of NOS2b rescued the CVP angiogenesis defects induced by tnnt2a-MO or KLf2a-MO injection (Fig. 4.4a, c), and NOS2b was located downstream of klf2a to regulate zebrafish CVP angiogenesis.

4.1.2.3 Klf6a-Tagln2 is Involved in the Process of Vessel Remodeling Regulated by Hemodynamics

During zebrafish embryonic development, the ventral capillary of the CVP develops into the caudal vein (CV) (Xie et al. 2010). However, the details of this developmental process and the underlying mechanisms remain elusive. To address this question, we used confocal microscopy to visualize CV development in wild-type (WT) Tg (flk1:EGFP) transgenic embryos from 32 to 72 hpf after fertilization (Fig. 4.5a). We show that CV formation is a dynamic process and that CV is remodeled by the ventral capillaries of the CVP in zebrafish embryos. As shown in Fig. 4.5b, the ventral capillaries of the CVP migrate horizontally and connect with each other at 32 hpf. At 36 hpf, they develop into luminal vessels with vascular loops. These luminal vessels undergo a dramatic remodeling process due to blood flow and develop CV at 60 hpf (Campinho et al. 2020; Gebala et al. 2016). Interestingly, there is a negative correlation between the number of vascular loops (white arrows) and CV formation. As shown in Fig. 4.5c, as the number of vascular loops decreases from 8.47 at 36 hpf to 2 at 54 hpf, the ventral capillaries of the CVP eventually realign to establish homogeneous vessels called CV at 60 hpf. The remodeling of the ventral capillaries of the CVP to establish the CV is accompanied by a reduction in vascular loops in the zebrafish embryo.

The Kruppel-like factor (Klf) transcription factor family plays an important role in regulating cardiovascular homeostasis mediated by hemodynamic forces (Xie et al.

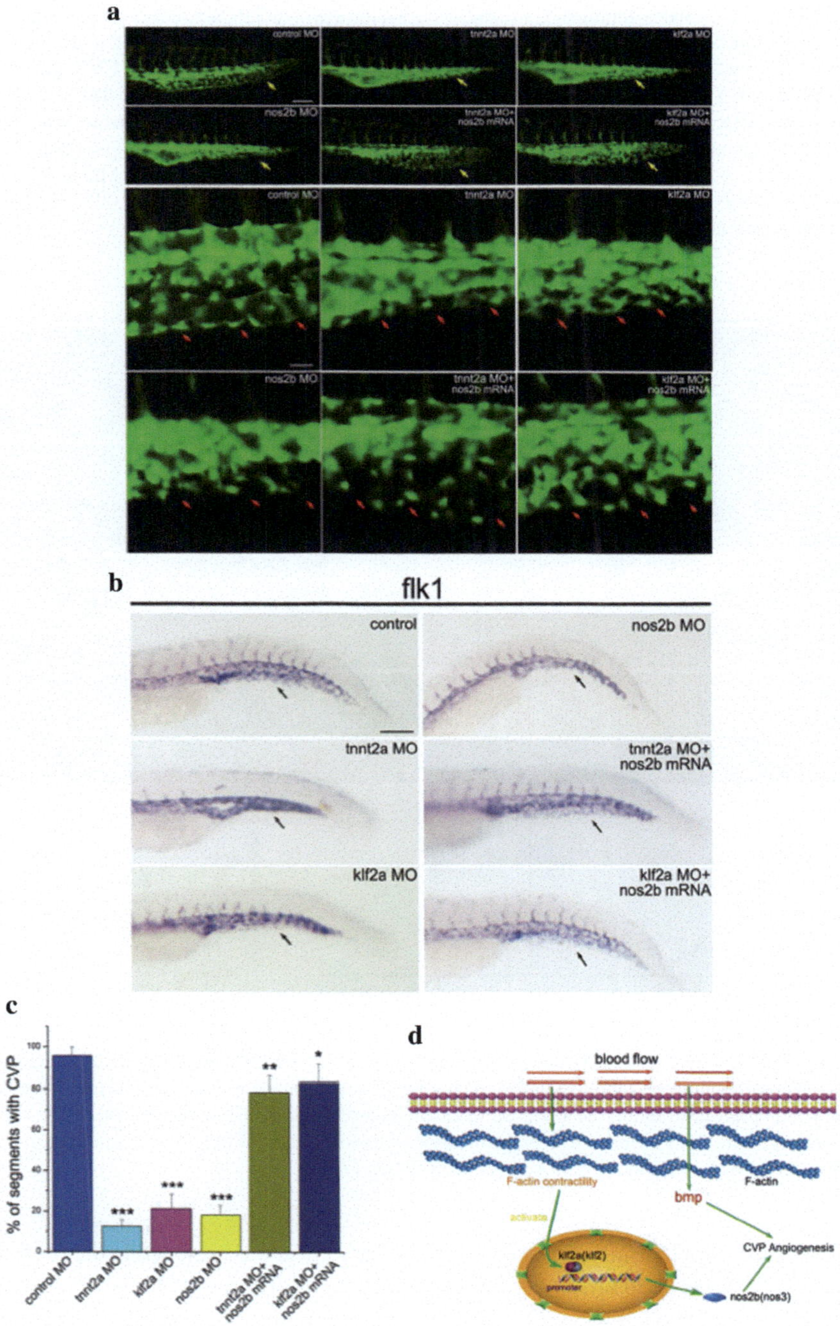
a
control MO
tnnt2a MO
klf2a MO
nos2b MO
tnnt2a MO+
nos2b mRNA
klf2a MO+
nos2b mRNA
b
flk1
control
nos2b MO
tnnt2a MO
tnnt2a MO+
nos2b mRNA
klf2a MO
klf2a MO+
nos2b mRNA
c
% of segments with CVP
100
80
60
40
20
0
***
***
***
**
*
control MO
tnnt2a MO
klf2a MO
nos2b MO
tnnt2a MO+
nos2b mRNA
klf2a MO+
nos2b mRNA
d
blood flow
F-actin contractility
F-actin
bmp
activate
klf2a(klf2)
promoter
CVP Angiogenesis
nos2b(nos3)

◂**Fig. 4.4** Klf2a regulates zebrafish CVP angiogenesis through NOS2b. **a** Confocal laser image of embryo, 36 hpf. Seen from the front left side, the back side is upward. CVP is marked with a yellow arrow. The scale bar is 100 μm. The image below shows an enlarged view of the CVP area. VV is marked with a red arrow. The scale bar is 50 μm. **b** In situ hybridization map of flk1 at 36 hpf. Seen from the front left side, the back side is upward. CVP is marked with a black arrow. Scale bar, 100 μm. **c** Percentage of somites containing CVP: control MO (n = 67), tnnt2a MO (n = 61), klf2a MO (n = 54), NOS2b MO (n = 42), tnnt2a MO + NOS2b mRNA (n = 46), klf2a MO + NOS2b mRNA (n = 39) (Mean ± SD, significant relative to control, ANOVA, *** $P < 0.001$). **d** Signal path regulation diagram. Adapted from Xie et al. (2018)

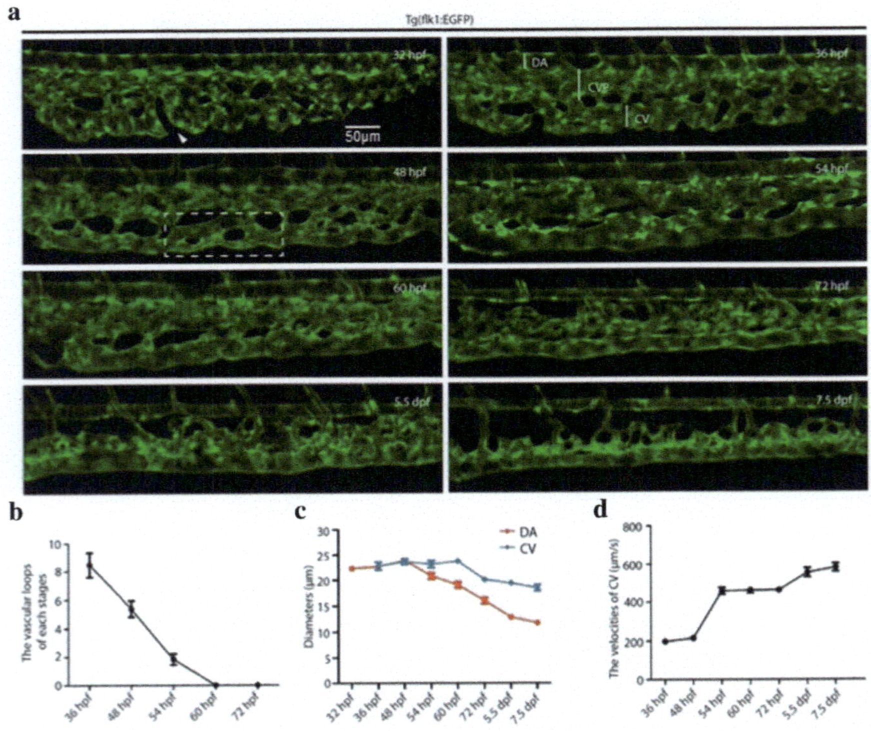

Fig. 4.5 The zebrafish caudal vein formation involves vessel remodeling. **a** Confocal images of the caudal vasculature of zebrafish embryos. **b** Quantification of the number of vascular loops in the CV: 36 hpf, n = 17 embryos; 48 hpf, n = 23 embryos; 54 hpf, n = 19 embryos; 60 hpf, n = 22 embryos; 72 hpf, n = 13 embryos. **c** Quantification of the diameters of the CV. **d** Quantification of the blood flow velocities in the CV. Adapted from Wang (2019)

2018). Of the 24 Klf family genes in zebrafish, only klf6a was enriched in the CVP of zebrafish embryos at 36 hpf by whole body in situ hybridization (WISH). Therefore, we hypothesized that k lf6a might be involved in the pruning of CV zebrafish embryos (Jin et al. 2018). To determine whether klf6a responds to blood flow, we performed a flow chamber experiment in which we exposed human umbilical vein ECs (HUVECs) to static (0 dyn/cm^2) or physiological shear stress (12 dyn/cm^2) for 12 h. Our results show that FSS induces KLF6 expression at both the mRNA and protein levels in HUVECs in vitro. In addition, WISH showed that klf6a mRNA levels in the CVP region were significantly disrupted at 36 hpf due to decreased blood flow velocity caused by treatment with tricocane and tnnt2a MO. These results indicate that klf6a responds to hemodynamic forces both in vitro and in vivo. To further verify whether klf6a is expressed in the CV of zebrafish embryos, we generated KI(klf6a-ha-p2a-GAL4) using CRISPR/Cas9 technology; Tg(UAS:EGFP) KI fish. The HA P2A-gal4 sequence is inserted into the last exon of the klf6a gene before the stop codon and fused into klf6A-HA-P2a-Gal4, as shown in Fig. 4.6a. The adult F0 KI(klf6a-HA-P2A gal4) line was then crossed with Tg(UAS:EGFP) fish. We identified the progeny by PCR analysis of their genomic DNA using target-site-specific and donor-specific primers and subsequent sequencing analysis. Consistent with our WISH results and the expression of klf6a in zebrafish in other reports, KI (klf6a-ha-p2a-Gal4) embryos showed weak but clear EGFP expression in the vasculature of the CV (Fig. 4.6b). These results indicate that klf6a is expressed in the CV of zebrafish embryos and responds to hemodynamics both in vitro and in zebrafish embryos (Xie et al. 2012).

To confirm the role of klf6a in zebrafish CV pruning, we injected klf6a MO into Tg(flk1:EGFP; gata1a:dsRed) embryos. We observed that knockdown of klf6a significantly increased the number of vascular rings in the CV at 72 hpf (mean rings: control = 0.6667, klf6a MO = 1.7037), although embryos showed normal morphology. To further confirm the effect of klf6a on CV trimming in zebrafish

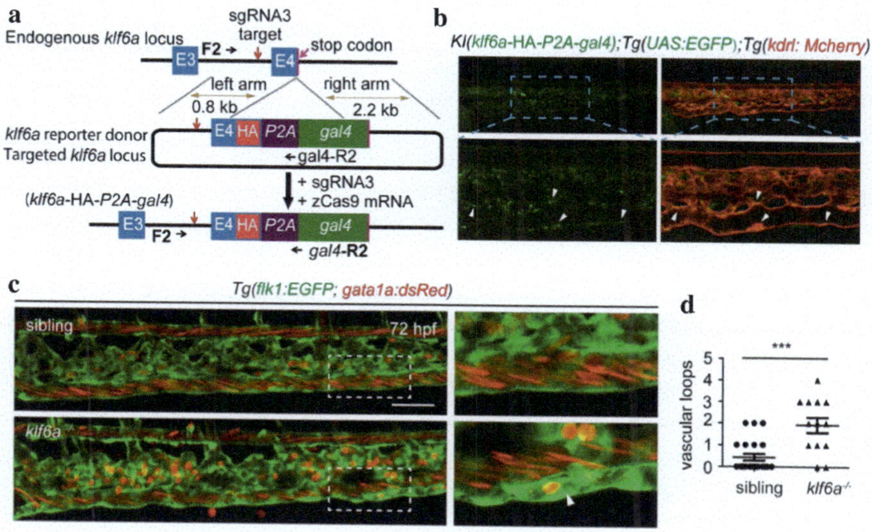

Fig. 4.6 Transcription factor klf6a regulates CV pruning in zebrafish. **a** Generation of KI (klf6a -HA-P2A-gal4) fish. **b** Expression pattern of KI (klf6a-HA-P2A-gal4) fish. Arrowheads indicate the location of klf6a at the vasculature. **c** Role of klf6a in CV pruning in zebrafish embryos. Boxes show enlarged images of the CV. The arrowhead indicates the unpruned vessel. Scale bar: 50 μm. **d** Quantification of vascular loops in the sibling and klf6a$^{-/-}$: sibling, n = 23 embryos; klf6a$^{-/-}$, n = 13 embryos. P < 0.001. Student's unpaired two-tailed t test. Adapted from Wang (2019)

embryos, we generated a zebrafish mutant of klf6a by CRISPR/Cas9, in which a 7 bp DNA fragment was deleted from exon 2. Deletion of klf6a increased the number of vascular rings in the CV at 72 hpf compared with the control group (average rings: sibling = 0.43478, klf6$^{-/-}$ = 1.92308; Fig. 4.6c, d). These results indicate that klf6a is needed for CV pruning in zebrafish embryos.

The effect of the transcription factor klf6a on actin cytoskeleton arrangement raises the possibility that cytoskeleton-related genes may be involved in KLF6A-mediated CV trimming in zebrafish embryos (Searles et al. 2004). After analyzing the available RNAseq profiles of KLF6-related transcripts in non-EC types reported by Laitman et al., we found that two actin-related genes are downregulated by Klf6: activity-regulated cytoskeleton-associated protein (Arc) and transgelin 2 (Tagln2) (Silvia et al. 2007). Tagln2 expression is enriched in the vasculature of zebrafish embryos. Therefore, we hypothesized that tagln2 might be a downstream target of klf6a deletion to regulate CV pruning in zebrafish embryos. In support of this hypothesis, we found that knockdown of KLF6 with siKLF6 reduced TAGLN2 expression at the mRNA and protein levels in vitro (Fig. 4.7a, b). Moreover, the klf6a homozygous mutant showed reduced tagln2 mRNA in the CVP region at 36 hpf, as determined by WISH (Fig. 4.7c). To confirm whether tagln2 is a direct target of klf6a in zebrafish, we performed bioinformatics analysis to predict the potential sequence within the zebrafish tagln2 promoter to which the transcription factor Klf6a could bind. As shown in Fig. 4.7d, the three CACCC sequences that Klf6a could bind to were located in the −1977 to −1643 bp region of the tagln2 promoter. We then performed chromatin immunoprecipitation (ChIP) analysis on zebrafish embryos to verify this. The results showed that Klf6a could bind to this promoter region within the tagln2 gene but not to exon 1 (Fig. 4.7e). These results indicate that tagln2 is a direct downstream target of klf6a in zebrafish, suggesting that tagln2 may be involved in KLF6A-mediated CV pruning.

To determine the role of tagln2 in zebrafish CV trimming, we first verified the efficiency of tagln2 MO using the tagLN2-EGFP fusion protein, whose expression was clearly inhibited by tagln2 MO. This MO was then used to knock down tagln2 in embryos to assess the effect on CV pruning at 72 hpf, where disruption of tagln2 by MO increased the number of vascular rings in CV (average rings: control = 0.6, tagln2 MO = 1.4615). To further validate the role of tagln2 in CV trimming, we subsequently generated tagln2 mutants in zebrafish using CRISPR/cas9 technology, in which a 7-bp fragment was deleted from exon 1. Compared with WT, tagln2 homozygous mutants showed a significantly increased number of vascular rings at CV at 72 hpf (average rings: WT = 0.44, tagln2$^{-/-}$ = 1.81818; Fig. 4.8a, b).

Next, we used Tg(fli1a:nEGFP; time-lapse real-time imaging of kdrl:mCherry) embryos to analyze the effect of tagln2 on EC behavior. In MO-injected control embryos, retrograde flow migration of EC nuclei (white circles) was readily observed within the regressive vascular branches at 48–54 hpf (Fig. 4.8c). However, EC nuclei (white circles) did not move significantly from 48 to 58 hpf in tagln2 morphants (Fig. 4.7c). Knockdown of tagln2 by MO resulted in a 38% decrease in EC nuclei (control: n = 14/24 ECs, tagln2 MO: n = 3/15 ECs) in retrograde flow migration and a 3% increase in EC nuclei (control: n = 4/24 ECs, tagln2 MO: n = 3/15

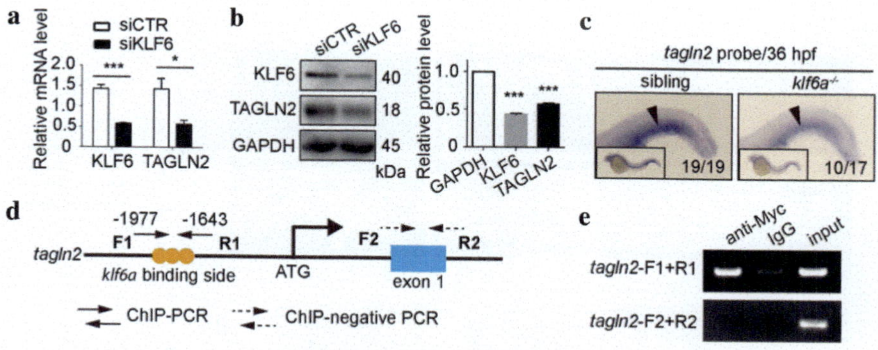

Fig. 4.7 Tagln2 is a direct downstream target of klf6a. **a** and **b** SiKLF6 in HUVECs efficiently decrease TAGLN2 expression at the mRNA (KLF6: P = 0.0009, TAGLN2: P = 0.0352) and protein (KLF6: P < 0.0001, TAGLN2: P < 0.0001) levels. Student's unpaired two-tailed t test. *P < 0.05, ***P < 0.001. **c** WISH of the tagln2 gene of sibling and klf6a$^{-/-}$ zebrafish at 36 hpf. Arrowheads indicate the CVP region expressing tagln2. **d** Structure of the tagln2 promoter. The primer tagln2-F1 + R1 contains the conserved Klf6a binding side-CACCC, whereas tagln2-F2 + R2 is a negative control located in the tagln2 exon 1. **e** ChIP-PCR of the tagln2 promoter shows that Klf6a can bind directly to the tagln2 promoter. Adapted from Wang (2019)

ECs) migrating with blood flow, but EC nuclei increased by 35% (control: n = 6/ 24 ECs, tagln2 MO: n = 9/15 ECs) and did not migrate (Fig. 4.8d). These results strongly support our claim that tagln2 is needed for CV pruning by regulating EC nuclear migration (Wang 2019). To assess the effect of TAGLN2 on EC migration in vitro, we silenced TAGLN2 in cultured HUVECs using siTAGLN2 and found that TAGLN2 expression was significantly inhibited at both the mRNA (Fig. 4.8e) and protein (Fig. 4.8f) levels. The results of wound healing assays performed to assess the effect of TAGLN2 on EC migration in HUVECs showed that knockdown of TAGLN2 by siRNA significantly inhibited EC migration in HUVECs (Fig. 4.8g). Our results above suggest that tagln2, as a downstream target of klf6a, regulates CV pruning in zebrafish by promoting EC nuclear migration (Wen et al. 2021).

4.1.3 Hemodynamic Regulation of the Mechanical Conduction Mechanism of Vascular Development

In the process of vascular development and pathological angiogenesis, ECs regulate the integrity of vascular structure by maintaining a dynamic balance between their survival and apoptosis. Friction-shear stress generated by blood flow has a variety of effects on the structure and function of ECs. ECs respond to changes in mechanical forces, causing the modulation of signaling networks and cellular functions. Accumulating evidence suggests that shear stress exerts physiological effects by regulating gene expression in ECs. In vitro studies have demonstrated that the hemodynamic environment has a decisive effect on the phenotypic changes of heart valve ECs, and

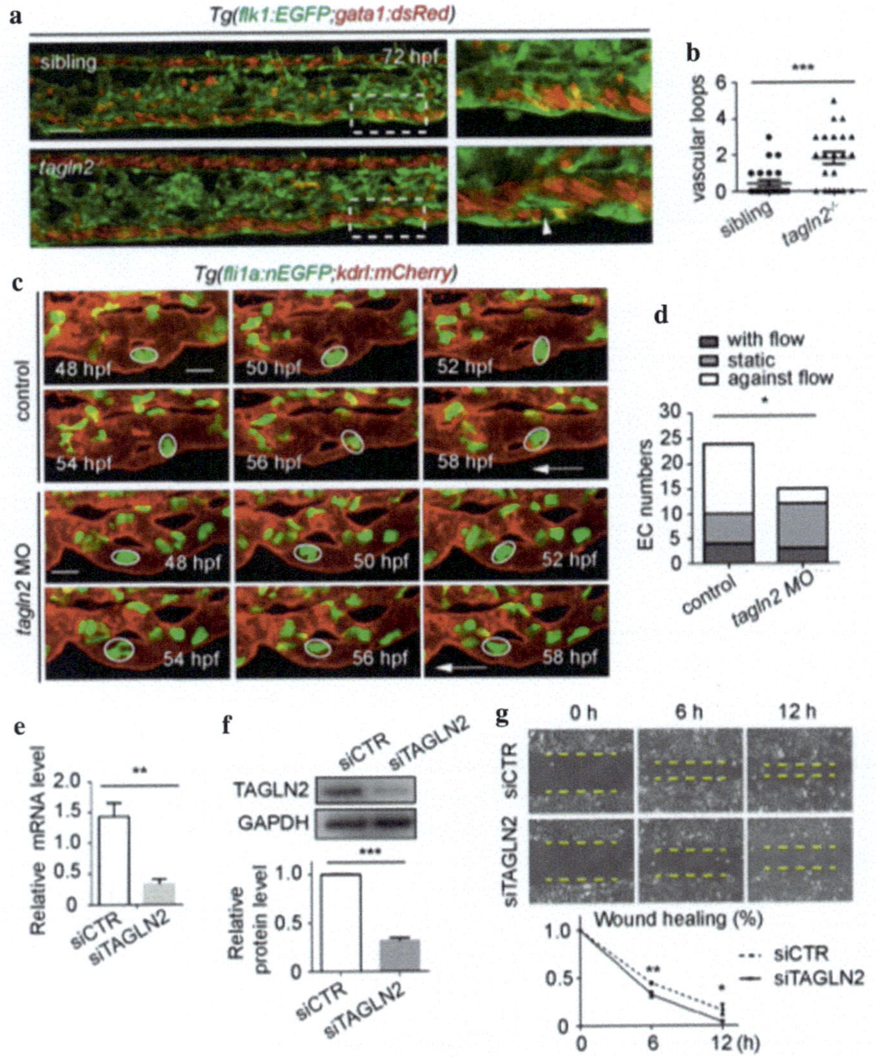

Fig. 4.8 Tagln2 regulates CV pruning by promoting EC nucleus migration. **a** Role of tagln2 in CV pruning in zebrafish embryos. Boxes show enlarged images of the CV. The arrowhead indicates unpruned vessel. Scale bar: 50 μm. **b** Quantification of vascular loops in sibling and tagln2$^{-/-}$ embryos: sibling, n = 25 embryos; tagln2$^{-/-}$, n = 23 embryos. P = 0.0004. Student's unpaired two-tailed t test. **c** Time-lapse live imaging of Tg (fli1a:EGFP; kdrl:mCherry) embryos shows EC nucleus migration in CV pruning in control or tagln2 morphant. Arrows indicate the direction of the blood flow. Colored circles indicate the EC nuclei. Scale bar: 10 μm. **d** Direction of EC nucleus migration in the regressing vessel. The direction is classified into three types: with the flow, static, and against the flow. A total of 12 vascular loops with 24 EC nuclei and 8 vascular loops with 15 EC nuclei were calculated in the control group and tagln2 morphant, respectively. Unpaired two-tailed chi-square test. P = 0.0471. **e** and **f** Efficiency of siTAGLN2 in HUVECs at the mRNA (P = 0.0094) and protein (P < 0.0001) levels. **g** Wound healing of siCTR-transfected and siTAGLN2-transfected HUVECs. Statistics for wound healing after siCTR and siTAGLN2 transfection: 6 h, P = 0.0037; 12 h, P = 0.0252. E–G Student's unpaired two-tailed t test. *P < 0.05, *P < 0.01, ***P < 0.001. Adapted from Wang (2019)

this process is achieved through the klf2 and MCP-1 signaling pathways (Weinberg et al. 2010).

As the mechanical support structure of cells, the cytoskeleton plays an important role in mechanical signal transduction. Under static culture, the cytoskeleton of ECs is in a dispersed state. Under shear stress, the number of cytoskeleton components and the direction of fiber arrangement will change significantly. Under flow conditions, the cytoskeleton was rearranged, and the most prominent change was observed in the distribution of the filamentous motor protein F-actin. Under stress, F-actin rearrangement generates stress fiber bundles. Both in vitro and in vivo studies have shown that the changes in cell morphology are the result of the changes in stress fiber bundles (McCue et al. 2004). Vascular ECs cultured in vitro are subjected to FSS, which can cause cytoskeletal rearrangement and change the location of cell gene expression. Studies have shown that FSS regulates the cytoskeleton rearrangement of ECs by regulating the expression of small gtpase, which in turn alters gene expression. Hemodynamics affects endothelial function and gene expression by affecting the morphology and structure of ECs, especially cytoskeletal changes, and thus affects embryonic vascular development (Poelmann et al. 2008).

Recent studies have shown that FSS can not only induce the expression of klf2 and eNOS but also induce the translocation of klf2 into the nucleus (Fig. 4.9). Hemodynamics induces klf2 rearrangement in ECs, but whether the cytoskeleton in turn influences klf2 expression is not known. Further studies showed that cytochalasin D, but not colchicine, inhibited FSS-induced KLF2 expression induction and nuclear translocation by inhibiting F-actin polymerization (cytochalasin D specifically inhibited F-actin polymerization; colchicine inhibited microtubules) (Fig. 4.9). The above studies suggest that the cytoskeleton has a feedback regulatory effect on klf2.

4.1.4 The Effect of Hemodynamics on Cerebrovascular Development and Its Mechanism

Cerebral small vessel disease (CSVD) and stroke are currently the leading causes of death and disability worldwide. CSVD and stroke are associated with developmental defects, and impaired integrity of the blood brain barrier (BBB) is often one of the important signs of the development of CSVD and stroke. The BBB is generally composed of ECs, pericytes, astrocytes, and extracellular matrix. Endothelial junctions and pericyte coverage are essential for the integrity of the BBB.

The importance of normal blood flow for the development of a healthy brain microvasculature has been demonstrated since infancy. The maintenance of a healthy brain in adults is also closely linked to normal hemodynamics, and people with heart disease will also suffer from brain aging (Farzam et al. 2017). One of the regulators between neural activity and human brain microvascular ECs (HBMECs) is pericytes, which control the diameter of capillaries within the central nervous system according to neurotransmitters. Pericytes have also been detected to act as regulators of blood

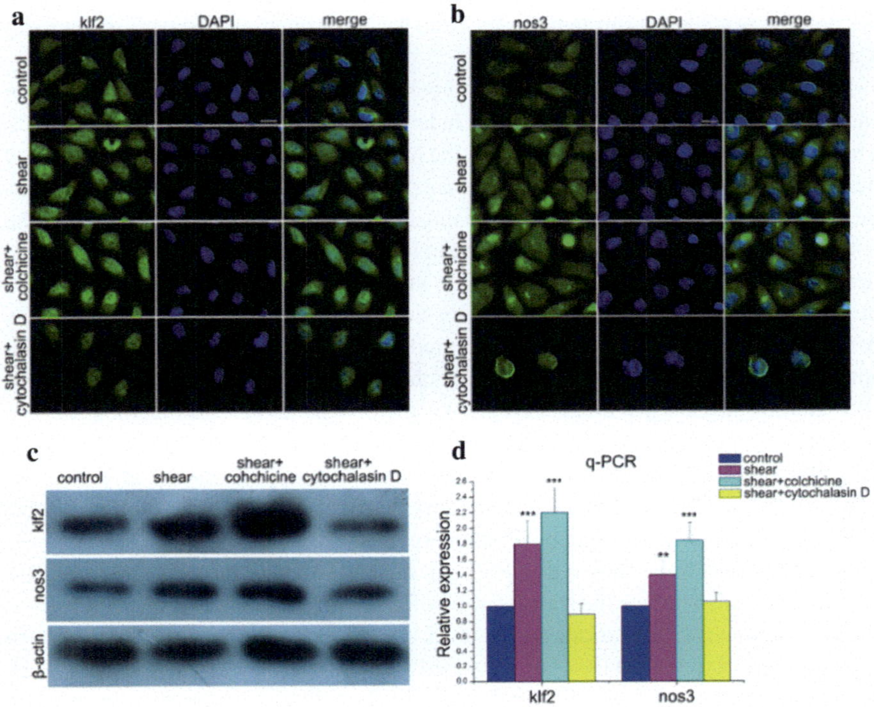

Fig. 4.9 FSS regulates the expression of klf2 and eNOS in HUVECs through F-actin polymerization. **a** Immunofluorescence image. The fluid shear force can induce klf2 to enter the nucleus, the cytochalasin D treatment can inhibit the fluid shear-induced klf2 entry into the nucleus, and the colchicine treatment cannot inhibit the fluid shear force-induced klf2 entry into the nucleus. **b** Immunofluorescence image. The fluid shear force can induce the expression of eNOS, the cytochalasin D treatment can inhibit the eNOS expression induced by the fluid shear force, and the colchicine treatment cannot inhibit the eNOS expression induced by the fluid shear force. **c** Western blotting diagram. Fluid shear force can promote the expression of klf2 protein, and cytochalasin D treatment can inhibit the expression of klf2 protein induced by fluid shear force. **d** Figure c shows the relative quantification of protein. Adapted from Xie (2018)

flow in the neurovascular unit during the aging of the adult brain (Lei et al. 2017). Phenotypes such as BBB breakdown, neurodegeneration and neuroinflammation were observed in pericyte-deficient model mice. The ability of pericytes to express a-SMA indicates the contractile capacity of the cells to regulate blood vessel diameter and blood flow (Alarcon-martinez et al. 2018).

In vitro studies on bovine brain microvascular ECs showed that some proteins involved in the BBB, Occludin and ZO-1 tight junctions were regulated by blood flow. Reduced blood flow alters ion homeostasis and insulin-mediated endocytosis at the BBB but not paracellular trafficking. In particular, HBMECs can maintain a cobblestone-like appearance under high shear stress and minimize paracellular transport, most likely by minimizing the length of tight junctions.

Several components of the TGFβ/BMP superfamily, such as Alk1, Alk5, TGFR2, Bmpr2, Endoglin and smads, are involved in angiogenesis and have been implicated in vascular diseases. In addition, in the pathological process of CSVD, the destruction of extracellular matrix is related to the dysfunction of TGFβ/BMP signaling, suggesting that the precise spatiotemporal regulation of TGFβ/BMP signaling is crucial for cerebral vascular development. Bone morphogenetic protein-3 (BMP3) is a member of the TGFβ superfamily. It has been reported that BMP3 regulates TGFβ/BMP signaling in a downstream signal-dependent manner. Recent studies have shown that BMP3 can regulate the integrity of the blood–brain barrier in the zebrafish brain by promoting pericyte development. Knockdown of BMP3 by MO resulted in cerebral hemorrhage and BBB leakage in zebrafish embryos (Fig. 4.10), as well as reduced pericyte coverage in zebrafish embryos (Fig. 4.11). Mechanistically, knockdown of BMP3 disrupts the pattern and activity of BMP signaling in the zebrafish brain, thereby disrupting the balance of TGFβ/BMP signaling in the zebrafish embryo.

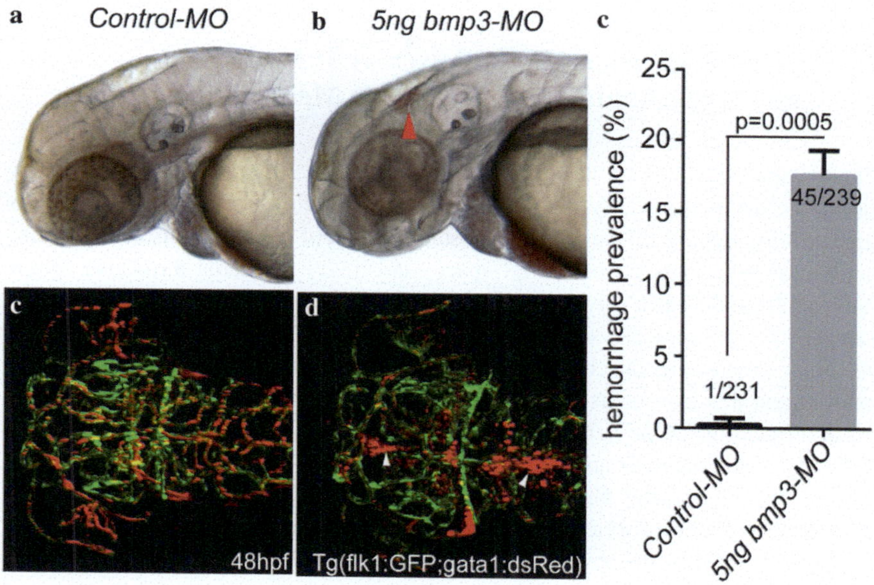

Fig. 4.10 Zebrafish cerebral hemorrhage (48 hpf). The cerebral hemorrhage of Tg(fli1:GFP; gata1:dsRed) zebrafish was observed by confocal microscope at 48 hpf. **a**, **c** Zebrafish embryos injected with control MO. **b**, **d** Zebrafish embryos injected with bmp3-MO, the arrow shows the location of zebrafish cerebral hemorrhage. **e** Statistical results of zebrafish cerebral hemorrhage. Adapted from Lei et al. (2017)

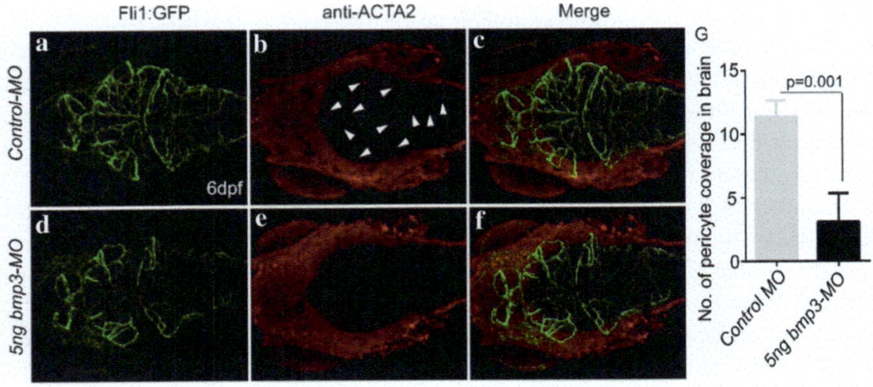

Fig. 4.11 The coverage of cerebral perivascular cells in zebrafish (6 dpf). Tg(fli1:GFP) zebrafish embryo pericytes were stained by immunofluorescence technique. **a–c** Zebrafish embryos injected with control MO, the arrow shows marked pericytes in the zebrafish brain; **d–f** zebrafish embryos injected with bmp3-MO; **g** statistical results of zebrafish brain cell coverage. Lei et al. (2017)

4.2 Effect of Microgravity on Vascular Development and Its Mechanical and Molecular Mechanisms

Manned space technology has become the most competitive field among all countries in the world, especially among the great powers. Now, it has entered a period of rapid development. With the successive establishment of space stations, an increasing number of astronauts have entered space, and many problems have also arisen. Different from the earth's living environment in space, astronauts will be affected by many space environments, such as high vacuum, a variety of radiation, microgravity and other environmental characteristics, among which microgravity is the space environment that astronauts must face and cannot avoid. The physiological indicators and functions of the human body have been adapted to the gravity environment of the earth, and the microgravity environment is a special mechanical environment different from the Earth. The astronauts who face this environment for a long time are bound to produce a series of unadapted symptoms. For example, cardiovascular dysfunction, muscle atrophy, decreased bone density, decreased immune function, etc., among which astronauts suffer from changes in cardiovascular function caused by microgravity, is a more serious problem. Therefore, the study of the biological effects of microgravity on organisms has become one of the most urgent space problems for scientists.

4.2.1 *Effects of Microgravity on Blood Vessel Development*

In space flight, astronauts are exposed to a microgravity environment, especially when G = 0, the blood is redistributed, the hydrostatic pressure gradient disappears, the blood pressure changes, and cardiovascular function is disordered. The recovery of astronauts back to the ground is uneven; some recover for a period of time after return, and some do not. Researchers have performed much research on the cardiovascular effects of microgravity and the mechanisms involved. When the researchers examined the X-ray exposure of the astronauts' chest in microgravity, they found that the volume and mass of the astronauts' hearts were smaller than those before the flight, and the mass of the heart was reduced by 8%. The results showed that the geometry and arrangement of vascular ECs changed when the blood vessels responded to internal shear stress, which indicated that the blood flow of the rats would be greatly changed under simulated microgravity. After two weeks of tail suspension under simulated weightlessness, the arterial structure of rats can be changed, and the changes in blood vessels in different parts of the body are different. After four weeks of weightlessness, it was found that the blood vessels of the predecessor of the rat (such as the brain and neck) had hypertrophic changes, while the blood vessels of the muscle arterioles of the hind limbs had atrophic changes (Mao et al. 1999a, b).

ECs are one of the most important components of the blood vessel wall and are located in a very thin layer between the circulating blood and the subcutaneous tissue within the blood vessel wall. They regulate the contractile and diastolic functions of blood vessels mainly by producing vasoactive substances. ECs are important "regulatory tissues" that maintain the homeostasis of the cardiovascular system. They are sensitive to changes in gravity, and almost all known cardiovascular diseases are related to endothelial dysfunction. ECs can be used as a selective permeability barrier for the exchange of substances inside and outside the blood vessels. Some gases, liquids and macromolecules can be selectively absorbed. In addition, ECs can sense a variety of signal stimuli in the intravascular environment, so they are special receptors, such as the interaction between some growth factors and vasoactive substances, signal transduction, and the synthesis and release of regulatory factors. In addition, researchers have found that vascular tension, permeability, platelet, leukocyte adhesion are also related to ECs and can also regulate vessel remodeling. Therefore, ECs are of great significance for homeostasis of the human body environment and physiological life activities (Kharbanda et al. 2001; Veerasamy et al. 2015).

In addition, the growth, morphology and cytoskeleton of ECs respond to microgravity. Versari et al. (2007) used two simulated microgravity devices, an random positioning machine (RPM) and an RWV (rotating wall vessel), to simulate the effect of microgravity on the growth of human umbilical vein ECs. The experimental results showed that simulated microgravity for 48 h could promote the growth of human umbilical vein ECs for a short time, and the effect time was as low as 8 min and was blocked immediately after overweight. They further showed that the actin cytoskeleton was reconstructed and the total amount of actin was reduced when the ECs sensed microgravity. Spaceflight can cause significant changes in the morphology of human umbilical vein ECs, such as enlarged cells, eccentric nuclei, and a large amount of secretions. ECs are very sensitive to gravity. It has been found that weightlessness can lead to the reconstruction of the cytoskeleton of ECs, and the expression of adhesion molecule-1 (ICAM-1), E-selectin and vascular endothelial cell adhesion molecule-1 (VCAM-1) on the cell surface changes after long-term weightlessness treatment of ECs (Buravkova et al. 2005).

In previous studies, most of the ECs were tail-suspended rats or rotated ECs to simulate microgravity, which were greatly limited in sample size and treatment (Sun et al. 2013). Moreover, rotated ECs were subjected to in vitro experiments, and there were few studies on the in vivo simulation of microgravity. Therefore, it is necessary to find a model organism that can be used for in vivo experiments and accurately study the effect of microgravity on vascular development. At present, zebrafish have been used to study the biological effects of microgravity in vivo. Zebrafish is a kind of in vivo research material that can accurately control microgravity and is usually not limited by the sample size, which can well observe and analyze the effects of microgravity on vascular development.

Recent studies have shown that microgravity can accelerate the heart rate of zebrafish but has little effect on the survival rate of zebrafish, indicating that microgravity does not kill zebrafish but has certain effects on the cardiovascular system (Xie et al. 2017, 2019). Further studies showed that the tail blood vessels of zebrafish treated with microgravity from 12 hpf developed normally at 24 hpf in the control and simulated microgravity (SM) groups, suggesting that the tail blood vessels developed normally before blood circulation (Fig. 4.12a). At 36 hpf, the CVP of control embryos showed a clear reticular structure with capillary spacing and fused VV (Fig. 4.12a, b). However, in the SM group, CVP width, CVP intercapillary number, and CVP intercapillary area were significantly increased (Fig. 4.12a–e). Moreover, flk1 expression was normal in SM-treated embryos at 24 h (Fig. 4.12f, g), while in the SM-treated group, flk1 expression in CVP and its surrounding area was significantly increased at 36 hpf (Fig. 4.12f, g). These findings suggest that SM is an important regulator in the formation of the CVP network in zebrafish.

4.2.2 Molecular Mechanisms of Microgravity Affecting Vascular Development

During the development of the vascular system in zebrafish, there are two stages: vasculogenesis and angiogenesis. The stage of angiogenesis is from 12 to 24 hpf. Angiogenesis occurs after 24 hpf, and the vasculature is basically formed by 36 hpf. From 24 to 36 hpf, blood circulation is established, which is an important stage of angiogenesis. Due to microgravity stimulation, the heart rate changes, suggesting that the angiogenesis stage may be affected. It can be concluded that simulated microgravity affects the vascular development of zebrafish embryos.

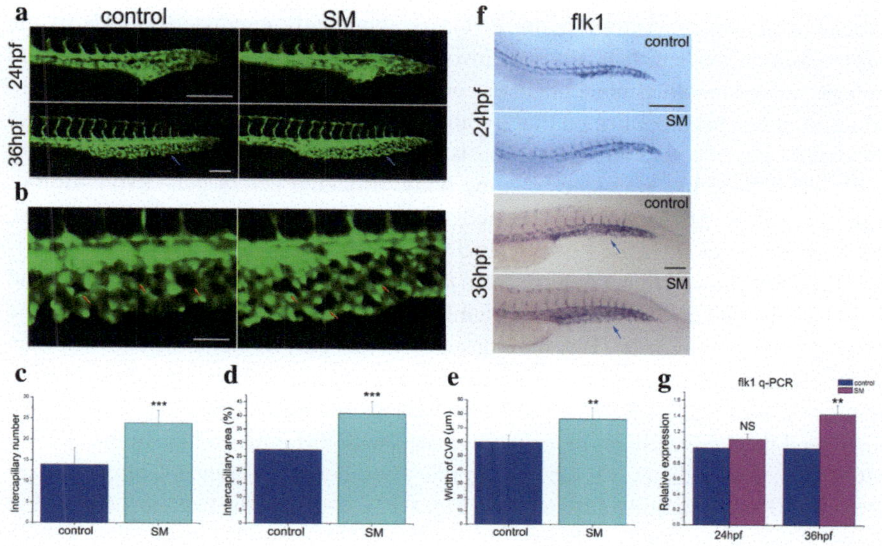

Fig. 4.12 SM regulates the formation of zebrafish CVP network. **a** Transgenic flk1: GFP embryos are SM treated from 12, and 24 and 36 hpf are observed under an in vivo fluorescence microscope. Side view, front left, back side up. CVP is indicated by a blue arrow. Scale bar, 100 μm. **b** An enlarged view of the CVP area A. The red arrow indicates the vascular network structure. Scale bar, 50 μm. **c** Calculate the number of vascular network structures. A total of 30 embryos in each case were counted (Mean ± standard deviation, t test, ***$p < 0.001$). **d** Quantitative vascular network structure area percentage. A total of 30 embryos were used for measurement in each treatment (Mean ± standard deviation, t test, ***$P < 0.001$). **e** Measure the width of CVP. A total of 30 embryos per treatment were used for measurement (Mean ± standard deviation, t test, **$P < 0.01$). **f** SM treatment can increase the expression of flk1 in the CVP region. Scale bar, 100 μm. **g** The qPCR results of relative expression of 24 hpf and 36 hpf zebrafish flk1 (Mean ± standard deviation, t test, **$P < 0.01$). Adapted from Xie et al. (2019)

In the space environment, astronauts are affected by microgravity, the hydrostatic pressure difference in blood disappears, body fluids are diverted to the head, and blood is redistributed. This change will cause the upregulation of eNOS expression in vascular ECs, an increase in NO release, and the dilation of blood vessels. This is actually an adaptive response, but if this adaptive response is returned to Earth, however, if this adaptive response does not return to normal quickly, it may lead to poor orthostatic tolerance of astronauts. Cytological experiments have shown that simulated microgravity has multiple effects on endothelial cell function, including stimulating endothelial cell proliferation, apoptosis, migration and cytoskeleton remodeling, thereby affecting cell movement (Ferranti et al. 2014). In his studies, Chen Ronggui showed that simulated microgravity could not only promote the apoptosis of ECs but also cause vessel remodeling in the arteriovenous region of rats, which further caused changes in the expression of some genes in ECs and increased NO secretion (Chen 2014). In addition, simulated microgravity can significantly stimulate the synthesis of NO, which has a significant effect on the activity of NOS

(Vaziri et al. 2004). Among them, eNOS is abundantly expressed in ECs of the cardiovascular system. It is not only involved in the production of NO but also an indispensable signaling molecule in various endothelial cell-dependent reactions. In addition, it has been confirmed that gyration-simulated weightlessness can increase the expression of eNOS and the synthesis of NO in pulmonary microvessels. eNOS is closely related to angiogenesis in the body. Blocking eNOS gene expression or inhibiting eNOS production reduces neovascularization in animal models (Cai et al. 2004; Momken et al. 2004; Zhao et al. 2002).

The study showed that zebrafish were treated with microgravity from 12 hpf, and at 36 hpf, the NO concentration was significantly higher in the SM group than in the control group (Fig. 4.13b). This finding suggests that the NO signaling pathway may be involved in the regulation of vascular development by SM in zebrafish. Further studies showed that NOS2b MO was injected into zebrafish embryos at the 1–4 cell stage, and then zebrafish embryos were treated with SM until 36 hpf. In the NOS2b MO group, the width of CVP, the intercapillary number of CVP, and the intercapillary area of CVP were significantly reduced (Fig. 4.13a, c–e), and the expression of flk1 in embryos was significantly reduced (Fig. 4.13f, g). These results indicate that NOS2b is a downstream factor of SM in regulating CVP network formation in zebrafish.

The phosphoinositide 3-kinase (PI3K)-Akt signaling pathway is one of the most important classical signal transduction pathways in cells and plays an important role in cell survival. PI3K is an intracellular phosphatidylinositol 3-kinase (PI3K) that specifically phosphorylates the 3′-hydroxyl group of the inositol ring. The activation of PI3K can activate Akt and its downstream signaling cascades through a series of reactions, which are involved in growth, development, differentiation and cell survival (Zhao et al. 2002). Studies have found that the PI3K/Akt pathway is inhibited after cerebral hypoxic-ischemic injury, and the expression of vascular endothelial growth factor (VEGF) is significantly downregulated, indicating that the PI3K/Akt signaling pathway plays an important role in the regulation of the brain internal environment and tissue repair after cerebral hypoxic-ischemic injury (Li et al. 2008). The PI3K/Akt signaling pathway plays an important role in the proliferation, differentiation, adhesion, apoptosis and angiogenesis of ECs. Recent studies have shown that after zebrafish were treated with a PI3K inhibitor (LY294002), the structure of the CVP capillary network could not be formed (Fig. 4.14a), and the width, number of capillaries and area between capillaries of CVP were significantly reduced (Fig. 4.14a–d). At 36 hpf, flk1 expression in the CVP region was significantly reduced in LY294002-treated embryos (Fig. 4.14e, f). Co-treatment with SM and LY294002 partially rescued the LY294002-induced loss of CVP network structure (Fig. 4.14a–d) and flk1 expression (Fig. 4.14e, f). Sm-induced CVP network structure generation was mediated by PI3K signaling.

For ECs, a variety of exogenous stimuli can induce the phosphorylation of eNOS through the PI3K-Akt signaling pathway, leading to its activation and subsequent increase in NO production, thus playing physiological and pathological roles (Uruno et al. 2004). The PI3K-Akt signaling pathway plays a key role in regulating the expression and activity of HUVECs in the process of enhanced angiogenesis under simulated weightlessness. Recent studies have shown a significant reduction in

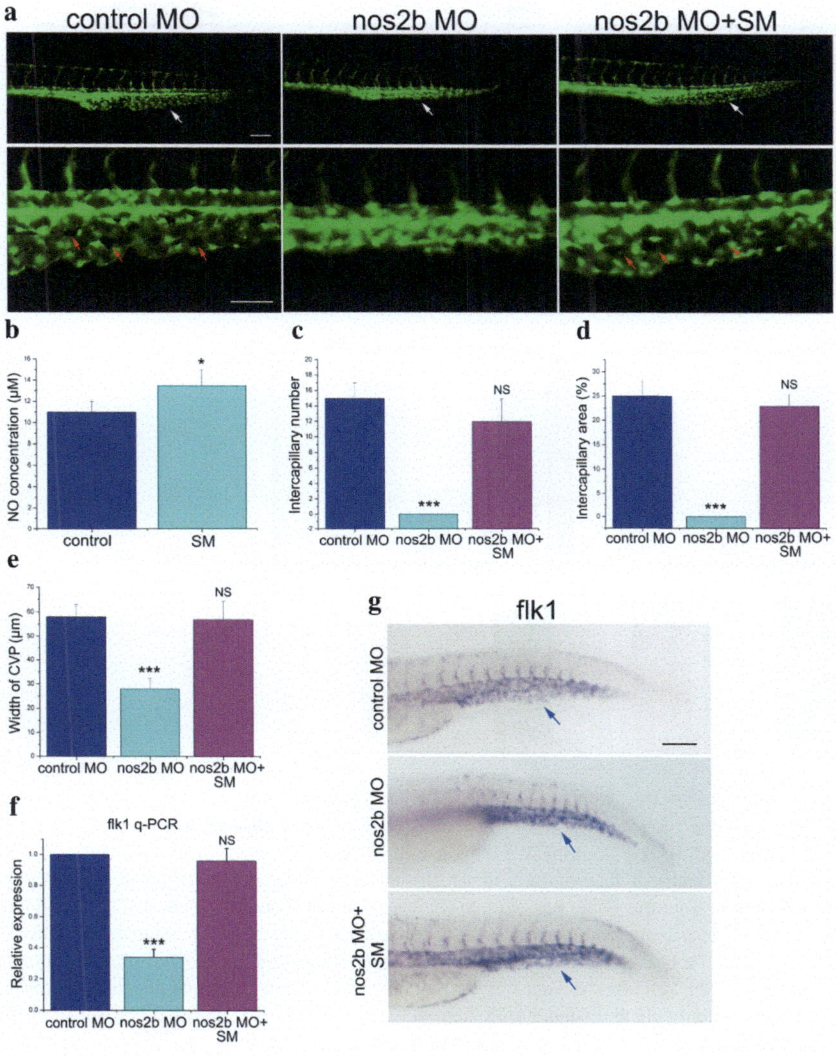

Fig. 4.13 NOS2b is the key factor that SM regulates the formation of CVP network. **a** Fluorescence microscope image of embryo at 36 hpf. Side view, front left, back side up. CVP is indicated by a white arrow. Scale bar, 100 μm. The image below shows an enlarged view of the CVP area of the upper image. The red arrow indicates the capillary network structure. Scale bar, 50 μm. **b** NO concentration of 36hpf zebrafish under SM and control conditions (Mean ± standard deviation, t test, *P < 0.05). **c** Calculate the number of capillary network. Thirty embryos were counted for each treatment (Mean ± standard deviation, t test, ***P < 0.001). **d** Percentage of area between quantitative capillaries. 30 embryos were used for the measurement per treatment (Mean ± standard deviation, t test, ***P < 0.001). **e** Measure the width of CVP. Thirty embryos per treatment were used for measurement (Mean ± standard deviation, t test, **P < 0.01). **f** The qPCR result of the relative expression of flk1 in zebrafish at 36hpf (Mean ± standard deviation, t test, ***P < 0.001). G. SM can rescue the expression of flk1 injected with NOS2b MO. Scale bar, 100 μm. Adapted from Xie et al. (2019)

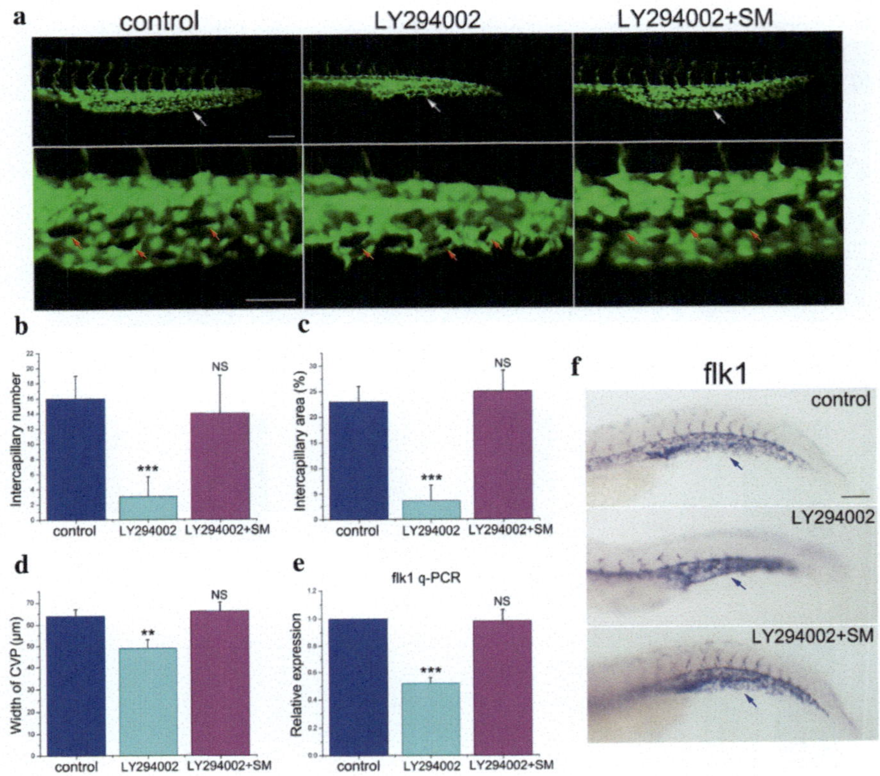

Fig. 4.14 SM regulates the formation of CVP network through PI3K signal. **a** At 36 hpf, the effect of LY294002 on the formation of CVP network was observed with a fluorescence microscope. Side view, front left, back side up. CVP is indicated by a white arrow. Scale bar, 100 μm. The image below is an enlarged view of the CVP area of the upper image. The red arrow indicates the capillary network structure. Scale bar, 50 μm. **b**. The capillary network of the LY294002 group is significantly reduced, and SM can be partially rescued ($n = 30$) (mean ± standard deviation, t test, ***$P < 0.001$). **c** Quantitative capillary network area percentage ($n = 30$) (mean ± standard deviation, t test, ***$P < 0.001$). **d** Measure the width of CVP. Thirty embryos per treatment were used for measurement (Mean ± standard deviation, t test, **$P < 0.01$). E. The qPCR result of the relative expression of zebrafish flk1 at 36 hpf (Mean ± standard deviation, t test, ***$P < 0.001$). F. At 36 hpf, the tail region flk1 whole in situ hybridization. Scale bar, 100 μm. Adapted from Xie et al. (2019)

NO concentration after PI3K inhibition (Fig. 4.15a). It is hypothesized that PI3K signaling could regulate NOS2b activity, thereby mediating the formation of the CVP network. Further studies showed that after PI3K inhibitor treatment, CVP was inhibited and unable to connect with adjacent vessels, while there was no obvious vascular abnormality in CVP in the LY294002 + NOS2b mRNA group (4–15 days). NOS2b mRNA partially rescued the decreased CVP width, intercapillary number, intercapillary area and flk1 expression caused by LY294002 treatment (Fig. 4.15b, c, e, f). In conclusion, NOS2b mRNA partially rescued the LY294002-induced loss of

CVP network formation in zebrafish embryos. Whole embryo in situ hybridization also showed the same results (Fig. 4.15g). These findings suggest that SM regulates CVP network formation through PI3K-NOS2b signaling (Zhang et al. 2016).

4.3 Research Progress and Prospects

Vascular ECs are a single layer of cells present in the innermost part of the vascular lumen, which separates the vascular lumen from the surrounding tissue. During embryonic development, ECs not only need to be tightly packed to form a barrier but also need to adapt to growing blood vessels to undergo changes and remodeling. ECs become polarized, migrate and change shape, and even proliferate or leave the tissue to become other cell types under complex mechanical conditions. Differences in endothelial cell behavior can lead to differences in vascular morphogenesis. Currently, the known types of vascular morphogenesis include angiogenesis, vessel pruning, lumen formation, changes in vessel caliber, changes in vessel stability, and changes in vascular endothelial cell fate.

4.3.1 Morphogenesis of the Vascular System

ECs (ECs) are a single layer of squamous cells that line the lumen of all blood vessels. ECs are connected to each other through cell junctions and aggregate into a giant tree-shaped tubular network, the vasculature, which allows nutrients, gases, cells, hormones and other factors to reach the entire body by circulating through the lumen of the blood vessels. The vascular system is formed through a complex series of steps. First, the primary vascular network is formed through vasculogenesis, which is mainly composed of the main trunk blood vessels, the aortic arch and the umbilical blood vessels. Subsequently, the rest of the vasculature is continuously generated and modified under the action of blood flow to adapt to embryonic growth and new physiological requirements, a process called angiogenesis. New branches originate from preexisting blood vessels in a process called vascular sprouting (nonangiogenesis). The newly grown blood vessels fuse with each other or with preexisting blood vessels to form new connections. The new connections will allow a lumen to form and blood to circulate. This process is called anastomosis. To meet the needs of the body, the primary vascular network will be optimized or remodeled; that is, redundant vessels will be removed to further form mature functional vessels. This selective degradation process is called vessel pruning. In addition, another type of angiogenesis, intussusceptive angiogenesis, is involved in the remodeling and optimization of the vascular network.

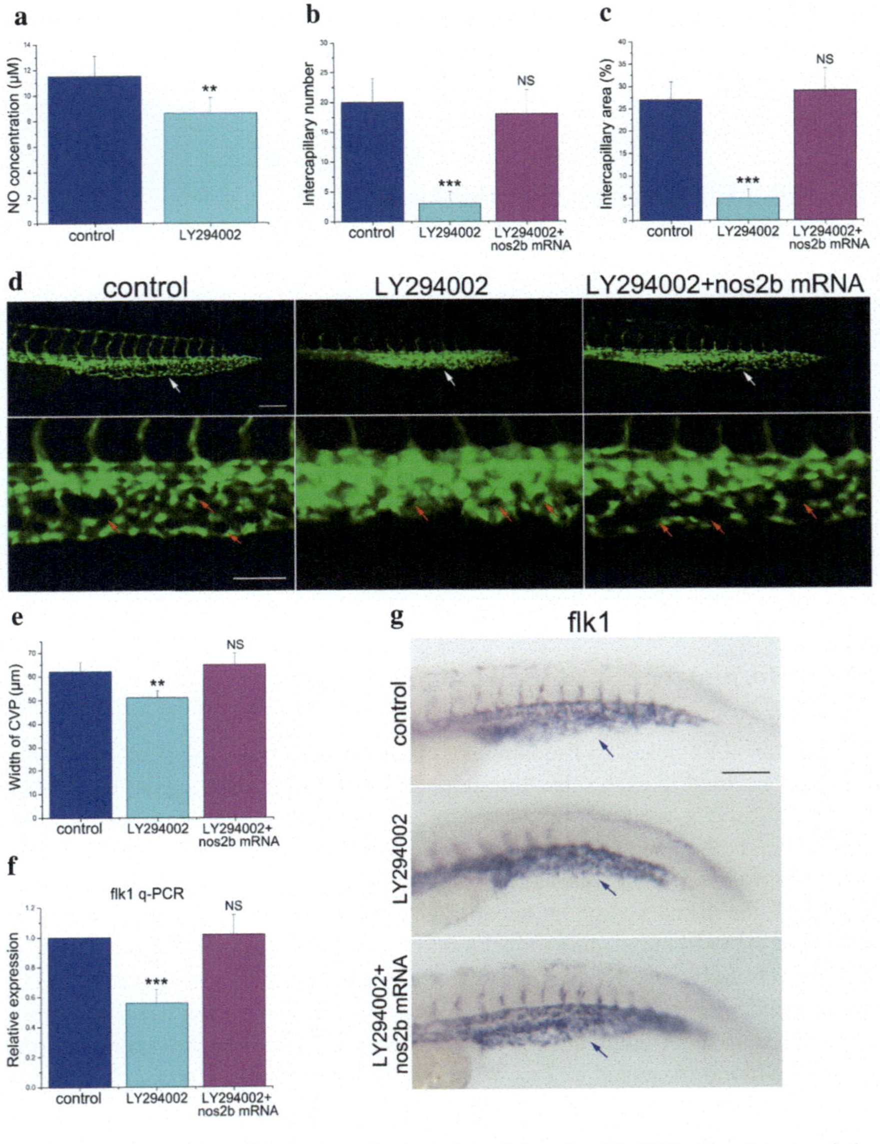

Fig. 4.15 SM-mediated PI3K regulates the formation of the zebrafish CVP network by regulating NOS2b. **a** 36 hpf detects NO concentration in zebrafish (Mean ± standard deviation, t test, $^{**}P < 0.01$). **b** NOS2b mRNA partially rescues the number of capillary networks of embryos treated with LY294002 (n = 30) (mean ± standard deviation, t test, $^{***}P < 0.001$). **c** NOS2b mRNA can also partially rescue the percentage of capillary area (n = 30) (mean ± standard deviation, t test, $^{***}P < 0.001$). **d** Fluorescence microscope image of embryo at 36hpf. Side view, front left, back side up. CVP is indicated by a white arrow. Scale bar, 100 μm. The image below shows a magnified view of the CVP area above the image. The red arrow indicates the capillary network. Scale bar, 50 μm. **e** The width of CVP can be restored by NOS2b mRNA (n = 30) (mean ± standard deviation, t test, $^{***}P < 0.001$). **f** The qPCR result of the relative expression of flk1 in zebrafish at 36hpf (Mean ± standard deviation, t test, $^{***}P < 0.001$). **g** whole in situ hybridization of flk1 in the tail region at 36 hpf. Scale bar, 100 μm. Adapted from Xie et al. (2019)

4.3.2 Stress on Vascular ECs

Most vascular growth and remodeling occur after the initiation of blood circulation. Vascular ECs are exposed to the fluid microenvironment to generate mechanical forces, such as shear stress, circumferential stress, and axial stress. Shear stress is a force parallel to the tissue surface due to the shear flow of a viscous fluid. It depends on the blood flow rate, viscosity, and geometry of the tube. The other two forces are controlled by intraluminal pressure, where circumferential stress refers to the force tangential to the vessel wall in the azimutrical direction (around the circumference); axial stress refers to the force along the longitudinal axis (long axis) of the vessel. These 3 stresses determine the mechanical properties of blood vessels and affect the geometric parameters of blood vessels, such as radius, wall thickness, and length.

4.3.3 The Formation of the BBB

The basis for the formation of the BBB is that ECs in the brain interact with other cells present in the brain microenvironment, such as pericytes, astrocytes, neurons, and microglia, to form an organized functional structure, the neurovascular unit (NVU). In the blood vessels that form the BBB, capillary ECs specialize and increase barrier tightness by forming additional cell–cell junctions. In addition to adherens junctions, the brain endothelium is sealed by tight junction proteins that prevent paracellular diffusion of substances and generate high transendothelial electrical resistance. At the same time, the transcellular transport capacity of brain ECs is significantly decreased. Thus, in mammals and zebrafish, ECs with a mature BBB lose the expression of plasma membrane vesicle-associated protein (plvap). In addition, mural cells, such as pericytes and SMAs, tightly encase the vessel wall to maintain the integrity of barrier function.

4.3.4 Hemodynamics and the BBB

4.3.4.1 Mechanical Microenvironment of Brain Tissue

As the softest tissue in the human body, the brain has a special biomechanical microenvironment. Under physiological conditions, the intracranial pressure is maintained between 0.05 and 0.1 kPa, and it rises sharply to 0.3–0.5 kPa during coughing or sneezing. Circulating blood flow in the brain generates FSS, which acts on the vascular endothelium and is transmitted to neurons and glial cells surrounding the brain capillaries. FSS ranges from 10 to 70 dyn/cm^2 in the arterial circulation, 2.8–95.5 dyn/cm^2 in the capillary circulation, and 5–23 dyn/cm^2 in the capillaries of the cerebral vascular network. Evidence has shown that FSS in capillaries promotes the differentiation of vascular ECs into the BBB phenotype, which is beneficial to the maintenance of the integrity of BBB structure and function. High shear stress caused by systemic vascular diseases or brain bypass grafting will lead to barrier destruction. Other studies have shown that physiological FSS inhibits inflammatory signaling and activates vascular stabilization pathways, while FSS levels that are too low or too high to activate inflammatory pathways and inhibit stabilizing pathways.

4.3.4.2 The Protective Effect of FSS on the BBB Under Physiological Conditions

Physiological FSS can enhance the barrier function of BMECs by improving the tightness of brain microvascular ECs and strictly controlling the exchange of substances into the cerebral microcirculation. The destructive effects of tumor necrosis factor-α (TNF-α) and interleukin-6 (IL-6) in human brain microvascular ECs (HBMECs) can be alleviated by physiological shear stress (8 dyn/cm^2). The significantly upregulated capillaroid shear stress level (6.2 dyn/cm^2) was associated with the gene expression of tight junction proteins, adhesion molecules, drug transporters, and integrins. In addition, capillary shear stress (10–20 dyn/cm^2) upregulated the expression of tight junction proteins (ZO-1, Claudin-5) and efflux transporter (P-gp) in primary HBMECs.

4.3.4.3 Damage of FSS to the BBB Under Pathological Conditions

Vascular bypass surgery can effectively restore the blood supply to the ischemic brain and increase the blood flow in the cortical hypoperfusion area. However, the sharp increase in FSS and perfusion pressure after bypass surgery and the process of cerebral ischemia/reperfusion may lead to the destruction of the BBB and then cause vasoedema and intracranial hemorrhage. Elevated FSS impairs BBB barrier function, leading to structural and functional abnormalities of the cerebral microvascular endothelium. Studies have shown that high FSS (40 dyn/cm^2) and pulsatile flow downregulate the expression of tight junction protein markers, transfer ZO-1 to the cytoplasm and nucleus, and reduce the efflux activity of P-gp. ZO-1 protein, which acts as a rapid trigger for the loss of barrier phenotype, is downregulated immediately after exposure to nonphysiological flow, followed by claudin-5 downregulation. In addition, recent findings have proposed that ischemia/reperfusion induces early BBB damage within 30–60 min, which is caused by subtle cytoskeletal rearrangements in BMECs.

4.3.5 *Prospects of Mechanical Developmental Biology Research*

The role of hemodynamic factors in ontogeny has attracted extensive attention in recent years. In the past, due to the limitations of research methods, it was difficult to track the dynamic role of mechanical factors in ontogeny in vivo. The vascular system is a complex system. With the development of the vascular system, the types and modes of force involved are more complex. With the development and utilization of high-definition microscopic imaging systems (such as two-photon laser confocal microscopy) and related software, as well as the wide application of atomic force microscopy (AFM), which can measure the forces between cells and molecules, it is not only possible to perform accurate and quantitative measurements of mechanical interactions in vitro but also to track cell movement and observe cell fate in vivo. Based on the development and application of zebrafish as a model animal in the field of cardiovascular development, we used the advantage of embryonic transparency to observe and calculate mechanical movement in vivo and established the in vivo vascular fluid dynamics network structure. At the same time, a variety of cardiovascular-related transgenic strains can be used for mechanobiological research. For example, the use of transgenic zebrafish without heartbeat can observe the effect of hemodynamic changes on vascular development in vivo. Developmental biology will move toward this new research field in the future. Meanwhile, the acquisition of more transgenic animal lines and the establishment of artificial mutant genetic engineering animal models will provide valuable resources for the study of mechanodevelopmental biology. Therefore, it is necessary to seize this opportunity to carry out a systematic study of developmental biology from the perspective of mechanics to lay a solid foundation for the establishment of a new interdisciplinary discipline, mechano-developmental biology.

In this regard, domestic institutions such as Chongqing University, the Institute of Zoology, the Chinese Academy of Sciences, the Institute of Neuroscience, the Shanghai Academy of Biological Sciences, and the Chinese Academy of Sciences have carried out fruitful research. For example, Chongqing University relied on the Key Laboratory of Biomechanics and Tissue Engineering of the Ministry of Education (later the Key Laboratory of Bio-Rheology Science and Technology of the Ministry of Education) and the Public Experimental Center of Chongqing National Biological Industry Base. In 2009, the Laboratory of Mechanical Developmental Biology was established to carry out a series of studies related to mechanical developmental biology. For example, in our previous study, we used zebrafish as a model to test the hypothesis of lipid concentration intensification as a focal hemodynamic cause of atherosclerosis in vivo for the first time and initially elucidated the molecular mechanism of low-density lipoprotein (LDL) polarization in atherogenesis. Subsequently, we used double transgenic zebrafish (flk1: GFP $\times$ gata1: DsRed) as the research subject and systematically studied the whole process of zebrafish tail venous vascular plexus (CVP) development by laser confocal microscopy combined with in vivo imaging technology, gene knockdown, gene editing, and molecular cell biology technology. The role and mechanism of hemodynamics in CVP angiogenesis and tail vein formation were quantitatively analyzed. We found that the development of zebrafish tail vein vessels can be divided into three distinct stages: the CVP angiogenesis stage, tail vein formation stage, and CVP regression stage. Changes in hemodynamics perturb CVP angiogenesis in zebrafish. Quantitative analysis showed a linear relationship between the change in FSS and the change in CVP development. Mechanistically, hemodynamics promote CVP angiogenesis by activating the ERK5-klf2a-NOS2b signaling pathway. At the same time, studies have found that tail vein formation mainly includes two ways: vessel fusion and vessel trimming. Subsequently, laser confocal imaging combined with in vivo zebrafish imaging technology found that when the blood flow rate in the two bifurcating vessels is the same, the two bifurcating vessels fuse and finally form the tail vein. When the blood flow velocity of the two bifurcated vessels is different, the vessel with higher blood flow velocity will be stably retained and eventually form the tail vein, while the vessel with lower blood flow velocity will undergo endothelial cell rearrangement and be pruned off. Mechanistically, hemodynamic KLf6A-Tagln2 signaling was found to regulate zebrafish tail vein formation by promoting endothelial cell migration and cytoskeletal polymerization. This study has established a method for introducing mechanical factors into the study of developmental biology, which validates the hypothesis proposed earlier by the authors of this monograph (i.e., FSS is involved in the regulation of vascular development by changing the cytoskeletal state of vascular ECs and mediating the ERK5 signaling pathway to regulate the expression of klf2a and enos genes. This study provides a new basis for further understanding and elucidating the mechanobiological mechanism of FSS in regulating vascular development.

References

Alarcon-martinez L, Yilmaz-ozcan S, Yemisci M, Schallek J, Kılç K, Can A. Capillary pericytes express a-smooth muscle actin, which requires prevention of filamentous-actin depolymerization for detection[J]. eLife, 2018, 7: e34861.

Boon RA, Horrevoets AJ. Key transcriptional regulators of the vasoprotective effects of shear stress[J]. Hamostaseologie, 2009, 29: 39–43.

Boon RA, Leyen TA, Fontijn RD, Fledderus JO, Baggen JM, Volger OL, van Nieuw Amerongen GP, Horrevoets AJ. KLF2-induced actin shear fibers control both alignment to flow and JNK signaling in vascular endothelium[J]. Blood, 2010, 115: 2533–2542.

Buravkova L, Romanor Y, Rykova M. Cell to cell interactions in changed gravity: ground-based and flight experiments[J]. Acta Astronaut, 2005, 57: 67–74.

Cai WJ, Kocsis E, Luo X, Schaper W, Schaper J. Expression of endothelial nitric oxide synthase in the vascular wall during arteriogenesis[J]. Mol Cell Biochem, 2004, 264: 193–200.

Campinho P, Lamperti P, Boselli F, Vilfan A, Vermot J. Blood flow limits endothelial cell extrusion in the zebrafish dorsal aorta[J]. Cell Rep, 2020, 31(2): 107505.

Chen RG. Vascular autophagy changes and the role of autophagy in endothelial cells dysfunction after simulated microgravity[D]. Fourth Military Medical University, 2014.

Dai X, Faber JE. Eendothelial nitric oxide synthase deficiency causes collateral vessel rarefaction and impairs activation of a cell cycle gene network during arteriogenesis[J]. Circ Res, 2010, 106: 1870–1881.

Dekker RJ, van Thienen JV, Elderkamp YW, Seppen J, de Vries CJM, Biessen EA, van Berkel TJ, Pannekoek H, Horrevoets AJG. Endothelial KLF2 links local arterial shear stress levels to the expression of vascular-tone regulating genes[J]. Am J Pathol, 2005, 167: 609–618.

Dekker RJ, Boon RA, Rondaij MG, Kragt A, Volger OL, Elderkamp YW, Meijers JC, Voorberg J, Pannekoek H, Horrevoets AJ. KLF2 provokes a gene expression pattern that establishes functional quiescent differentiation of the endothelium[J]. Blood, 2006, 107: 4354–4363.

Dudzinski DM, Igarashi J, Greif D, Michel T. The regulation and pharma-cology of endothelial nitric oxide synthase[J]. Annu Rev Pharmacol Toxicol, 2006, 46: 235–276.

Farzam P, Buckley EM, Lin PY, Hagan K, Grant PE, Inder TE, Carp SA, Franceschini MA. Shedding light on the neonatal brain: probing cerebral hemodynamics by diffuse optical spectroscopic methods[J]. Sci Rep, 2017, 7(1): 15786.

Ferranti F, Caruso M, Cammarota M, Masiello MG, Scheri KC, Fabrizi C, Fumagalli L, Schiraldi C, Cucina A, Catizone A, Ricci G. Cytoskeleton modifications and autophagy induction in tcam-2 seminoma cells exposed to simulated microgravity[J]. Biomed Res Int, 2014, 2014(8): 66–70.

Forstermann U, Boissel JP, Kleinert H. Expressional control of the constitutive isoforms of nitric oxide synthase (NOS I and NOS III)[J]. FASEB J, 1998, 12: 773–790.

Fritsche R, Schwerte T, Pelster B. Nitric oxide and vascular reactivity in developing zebrafish, Danio rerio[J]. Am J Physiol Regulatory Integrative Comp Physiol, 2000, 279: R2200–R2207.

Gebala V, Collins R, Geudens I, Phng LK, Gerhardt H. Blood flow drives lumen formation by inverse membrane blebbing during angiogenesis in vivo[J]. Nat Cell Biol, 2016, 18(4): 443–50.

Gray C, Packham IM, Wurmser F, Eastley NC, Hellewell PG, Ingham PW, Crossman DC, Chico TJ. Ischemia is not required for arteriogenesis in zebrafish embryos[J]. Arterioscler Thromb Vasc Biol, 2007, 27(10): 2135–2141.

Groenendijk BCW, Hierck BP, Gittenberger-de Groot AC, Poelmann RE. Development-related changes in the expression of shear stress responsive genes KLF-2, ET-1, and NOS-3 in the developing cardiovascular system of chicken embryos[J]. Dev Dyn, 2004, 230: 57–68.

Groenendijk BCW, Hierck BP, Vrolijk J, Baiker M, Pourquie MJBM, Gittenberger-de-Groot AC, Poelmann RE. Changes in shear stress-related gene expression after experimentally altered venous return in the chicken embryo[J]. Circ Res, 2005, 96: 1291–1298.

Jin XP. The study of Klf6a involved in zebrafish caudal venous plexus pruning regulated by blood flow[D]. Chongqing University, 2018.

Kharbanda RK, Deanfield JE. Functions of the healthy endothelium[J]. Coronary Artery Dis. 2001, 12: 485–491.

Kondrikov D, Han HR, Block ER, Su Y. Growth and density-dependent regulation of NO synthase by the actin cytoskeleton in pulmonary artery endothelial cells[J]. Am J Physiol Lung Cell Mol Physiol, 2006, 290: L41–L50.

Kubes P, Suzuki M, Granger DN. Nitric oxide: an endogenous modulator of leukocyte adhesion[J]. Proc Natl Acad Sci USA, 1991, 88: 4651–4655.

Lee JS, Yu Q, Shin JT, Sebzda E, Bertozzi C, Chen M, Mericko P, Stadtfeld M, Zhou D, Cheng L, Graf T, MacRae CA, Lepore JJ, Lo CW, Kahn ML. Klf2 is an essential regulator of vascular hemodynamic forces in vivo[J]. Developmental Cell, 2006a, 11: 845–857.

Lee SH, Chun W, Kong PJ, Han JA, Cho BP, Kwon OY, Lee HJ, Kim SS. Sustained activation of akt by melatonin coniribútes to the protection against kainic acid-induced neuronal death in hippocampus[J]. J Pineal Res. 2006b, 40: 79–85.

Lei D, Jin X, Wen L, Dai H, Ye Z, Wang G. bmp3 is Required for Integrity of Blood Brain Barrier by Promoting Pericyte Coverage in Zebrafish Embryos[J]. Curr Mol Med. 2017, 17(4): 298–303.

Lepiller S, Franche N, Solary E, Chluba J, Laurens V. Comparative analysis of zebrafish nos2a and nos2b genes[J]. Gene, 2009, 15, 445(1–2): 58-–65

Li L, Qu Y, Mao M, Xiong Y, Mu D. The involvement of phosphoinositid 3-kinase/akt pathway in the activation of hypoxia-inducible factor-lalpha in the developing rat brain after hypoxia-ischemia[J]. Brain Res. 2008, 1197: 152–158.

Lin Z, Kumar A, SenBanerjee S, Staniszewski K, Parmar K, Vaughan DE, Gimbrone MA Jr, Balasubramanian V, García-Cardeña G, Jain MK. Kruppel-lLike fFactor 2 (KLF2) regulates endothelial thrombotic function[J]. Circ Res, 2005, 96: e48–e57.

Loufrani L, Henrion D. Role of the cytoskeleton in flow (shear stress)-induced dilation and remodeling in resistance arteries[J]. Med Biol Eng Comput, 2008, 46(5): 451–460.

Lucitti JL, Jones EA, Huang C, Chen J, Fraser SE, Dickinson ME. Vvascular remodeling of the mouse yolk sac requires hemodynamic force[J]. Development, 2007, 134(18): 3317–3326.

Mao QW, Zhang LF, Ma J. Differentiated remodeling changes of medium-sized arteries from different body parts in tail-suspended rats and their reversibility[J]. Space Med Med Eng, 1999a;12(2):92–96.

Mao QW, Zhang LF, Zhang LN. Ultrastructural changes of arterial wall from different body parts of rats during simulated weightlessness[J]. Space Med Med Eng, 1999b;12(4):252–253.

McCue S, Noria S, Langille BL. Shear-induced reorganization of endothelial cell cytoskeleton and adhesion complexes[J]. Trends Cardiovasc Med, 2004, 14: 143–151.

Moncada S. The 1991 uUlf von eEuler lLecture. The L-arginine: nitric oxide pathway[J]. Acta Physiol Scand, 1992, 145: 201–227.

Momken I, Lechene P, Ventura-Clapier R, Veksler V. Voluntary physical activity alterations in endothelial nitric oxide synthase knockout mice[J]. Am J Physiol Heart Circ Physiol, 2004, 287(2): H914–H920.

Nicoli S, Standley C, Walker P, Hurlstone A, Fogarty KE, Lawson ND. MicroRNA-mediated integration of haemodynamics and Vegf signalling during angiogenesis[J]. Nature, 2010, 464: 1196–1200.

Parmar KM, Larman HB, Dai G, Zhang Y, Wang ET, Moorthy SN, Kratz JR, Lin Z, Jain MK, Gimbrone MA Jr, García-Cardeña G. Integration of flow-dependent endothelial phenotypes by Kruppel-like factor 2[J]. J Clin Invest, 2006, 116: 49–58.

Pelster B, Grillitsch S, Schwerte T. NO As a mediator during the early development of the cardiovascular system in the zebrafish[J]. Comp Biochem Physiol B Biochem Mol Biol, 2005, 142: 215–220.

Poelmann RE, Gittenberger-de Groot AC, Hierck BP. The development of the heart and microcirculation: role of shear stress[J]. Med Biol Eng Comput, 2008, 46: 479–484.

Searles CD, Ide L, Davis ME, Cai H, Weber M. Actin Cytoskeleton Organization and Posttranscriptional Regulation of Endothelial Nitric Oxide Synthase During Cell Growth[J]. Circ Res, 2004, 95: 488–495.

SenBanerjee S, Lin Z, Atkins GB, Greif DM, Rao RM, Kumar A, Feinberg MW, Chen Z, Simon DI, Luscinskas FW, Michel TM, Gimbrone MA Jr, García-Cardeña G, Jain MK. KLF2 is a novel transcriptional regulator of endothelial proinflammatory activation[J]. J Exp Med, 2004, 199(10): 1305–1315.

Silvia V, Alessandro V, Silvia B, Maier JAM. Alterations of the actin cytoskelet on and increased nitric oxide synthesis are common features in human primary endothelial cell response to changes in gravity[J]. Biochim Biophys Acta, 2007, 1773(11): 1645–1652.

Su Y, Edwards-Bennett S, Bubb MR, Block ER. Regulation of endothelial nitric oxide synthase by the actin cytoskeleton[J]. Am J Physiol Cell Physiol, 2003, 284: C1542–C1549.

Sun T, Xie X, Wang GX. Effect of horizontal rotary culture on zebrafish vascular development[J]. Hereditas, 2013(4): 502–510.

Uruno A, Sugawara A, Kanatsuka H, Arima S, Taniyama Y, Kudo M, Takeuchi K, Ito S. Hepatocyte growth factor stimulates nitric oxide production through endothelialnitric oxide synthase activation by the phosphoinositide 3-kinase/Akt pathway and possibly by mitogen-activated protein kinase kinase in vascular endothelialcells[J]. Hypertens Res, 2004, 27(11): 887–895.

Vartanian KB, Berny MA, McCarty OJ, Hanson SR, Hinds MT. Cytoskeletal structure regulates endothelial cell immunogenicity independent of fluid shear stress[J]. Am J Physiol Cell Physiol, 2010, 298: C333–C341.

Vaziri ND, Ding Y, Sangha DS, Purdy RE. Upregulation of NOS by simulated microgravity, potential cause of orthostatic intolerance[J]. J Appl Physiol, 2004, 89(1): 338–344.

Veerasamy M, Bagnall A, Neely D, Allen J, Kunadian V. Endothelial dysfunction and coronary artery disease: a state of the art review[J]. Cardiol Rev, 2015, 23: 119–129.

Versari S, Villa A, Bradamante S, Maier JA. Alterations of the actin cytoskeleton and increased nitric oxide synthesis are common features in human primary endothelial cell response to changes in gravity[J]. Biochim Biophys Acta, 2007, 1773(11): 1645–52.

Wang N, Miao H, Li YS, Zhang P, Haga JH, Hu Y, Young A, Yuan S, Nguyen P, Wu CC, Chien S. Shear stress regulation of Krüppel-like factor 2 expression is flow pattern-specific[J]. Biochem Biophys Res Commun, 2006, 341: 1244–1251.

Wang G. Flow-dependent vessel pruning governs zebrafish CV formation by klf6a/tagln2 axis. The 7th Sino-American Workshop on Biomedical Engineering and Biomechanics. Chongqing, 2019.

Weinberg EJ, Mack PJ, Schoen FJ, García-Cardeña G, Kaazempur Mofrad MR. Hemodynamic environments from opposing sides of human aortic valve leaflets evoke distinct endothelial phenotypes in vitro[J]. Cardiovasc Eng, 2010, 10: 5–11.

Wen L, Zhang T, Wang GX. The blood flow-klf6a-tagln2 axis drives vessel pruning in zebrafish by regulating endothelial cell rearrangement and actin cytoskeleton dynamics[J]. PLoS Genet 2021, 17(7): e1009690.

Xie X, Wang G, Tan J, Tang C, Qiu J, Luo L. Angiogenesis during Embryonic Development is Dependent on Blood Flow in Zebrafish[J]. Circulation. 2010, 122: A18135 (Abstract).

Xie X, Hu JJ, Wang GX. Advance in biomechanical study of embryonic vascular system development[J]. Hereditas, 2012,34(9):1123–1132.

Xie X. The mechanisms study of hemodynamics on atherosclerosis and vascular development[D]. Chongqing University, 2013.

Xie X, Liu D, Lei DX, Liu YF, Wang Q, Wen ZE, Gregersen H, Qiu JH, Wang GX. Effects of simulated microgravity on vascular development in zebrafish[J]. Mol Cell Biomech, 2017, 14(3): 171–186.

Xie X. Study on the mechanobiological mechanisms of hemodynamic factors affecting vascular development in zebrafish. Chongqing: Report on Postdoctoral Research Work at Chongqing University, 2018

Xie X, Zhou T, Wang Y, Chen H, Lei DX, Huang L, Wang Y, Jin XP, Sun T, Tan J, Yin TY, Huang J, Gregersen H, Wang GX. Blood flow regulates zebrafish caudal vein plexus angiogenesis by ERK5-klf2a-nos2b signaling[J]. Curr Mol Med, 2018, 18 (1): 3–14.

Xie X, Lei DX, Zhang Q, Wang YQ, Wen L, Ye ZY, Ahmad UD, Jia DY, Apicella A, Wang GX. Effect of simulated microgravity induced PI3K-nos2b signalling on zebrafish cardiovascular plexus network formation[J]. J Biomech, 2019, 87: 83–92.

Young A, Wu W, Sun W, Benjamin Larman H, Wang N, Li YS, Shyy JY, Chien S, García-Cardeña G. Flow aActivation of AMP-Activated protein kinase in vascular endothelium leads to krüppel-like factor 2 expression[J]. Arterioscler Thromb Vasc Biol, 2009, 29(11): 1902–1908.

Zhao X, Lu X, Feng Q. Deficiency in endothelial nitric oxide synthase impairs myocardial angiogenesis[J]. Am J Physiol Heart Circ Physiol, 2002, 283(6): H2371–H2378.

Zhang Q. The study of PI3K-NOS2b Involved in zebrafish embryo vascular development regulated by simulated microgravity[D]. Chongqing University, 2016.

Chapter 5
Mechanobiological Mechanisms of Atherosclerotic Plaque Formation

This chapter begins by discussing the influence of hemodynamic factors, especially changes in shear stress, on vascular endothelial dysfunction, the inflammatory response and angiogenesis related to AS. It then goes to introduce the influence of shear stress on the formation of AS and its regulatory mechanism. This chapter focuses on the mechanical mechanism of hemodynamic changes that affect the stability and pathogenicity of AS plaques and the guiding significance of regulating blood flow shear stress for the clinical treatment of AS and briefly describes the potential application of other mechanical and biological factors in the treatment of AS plaques.

5.1 Shear Stress and Atherosclerosis

5.1.1 Shear Stress and Atherosclerosis-Associated Endothelial Dysfunction

Arterial endothelial dysfunction is an early event leading to vascular sclerotic disease. It can be observed from the tissue, cell and molecular levels that the functional inactivation of VECs affects the occurrence and development of AS. Shear stress regulates the expression of several AS-related genes and affects endothelial function to a large extent (Cunningham et al. 2005). For a long time, researchers have evaluated the function of the vascular endothelium by eNOS activity and NO production, which fundamentally regulate the physiological activity of endothelial cells. The release of NO in VECs can be regulated by enhancing the expression of eNOS at the transcriptional level or by eNOS protein phosphorylation at the posttranscriptional level. Laminar shear stress on VECs greatly acceleratd the release of NO. The results showed that after shear stress on VECs for 8–24 h, the secretion of NO in VECs was

G. Wang, *The Mechanobiology in Atherosclerosis*,
https://doi.org/10.1007/978-981-97-6512-6_5

significantly increased compared with that in the static culture group. In particular τ=15 dyn/cm^2, the secretion of NO was significantly increased compared with other shear stress levels. However, endothelial dysfunction often appears in the arterial region with disturbed flow, mainly due to the obvious limitation of NO production capacity, which further induces the formation of AS in combination with other local risk factors for endothelium (as shown in Fig. 5.1). For example, in vivo studies showed that in plaque rupture sites affected by turbulence may exhibit downregulated eNOS and activated AS factors. In addition, the change in shear stress at the AS site can promote the development of AS by downregulating endothelial vasodilator factors such as prostacyclin (PGI2) and upregulating vasoconstrictor factors such as endothelin-1 (ET-1). The results showed that compared with the static culture control, low fluid shear stress (5, 10 dyn/cm^2) significantly promoted the secretion of ET-1, while high fluid shear stress (15 dyn/cm^2) significantly inhibited the secretion of ET-1, and higher fluid shear stress had a greater impact on the accumulation of ET-1 in cells. In addition, the level of ET-1 secretion depends not only on the shear stress but also on the shear action time (Fig. 5.2). Therefore, a variety of AS-inducing factors and AS protective factors in vivo play an important role in balancing the response of endothelial function to shear stress by dynamically coordinating endothelial cell homeostasis. In general, laminar shear stress ($\geq$15 dyn/cm^2) produced by high-speed, unidirectional blood flow has antithrombotic, anti-proliferative and anti-inflammatory effects, while abnormal shear stress (<15 dyn/cm^2) produced by low-speed, oscillatory and whirling blood flow can promote inflammation and lead to AS.

Although the shear sensing mechanism on the surface of VECs has been widely studied, it is still not completely clear how a variety of receptors on the surface of VECs respond to hemodynamic signals, integrate mechanical signals into chemical signals to affect vascular endothelial function, and ultimately determine the occurrence and development of AS.

5.1.2 Shear Stress and Atherosclerosis-Related Inflammation

Atherosclerosis is considered a chronic inflammatory disease, and the level of inflammation plays a decisive role in the development of AS lesions. The aggregation and infiltration of inflammatory cells in the intima of the vascular wall is an important event in the formation of AS. Shear stress activates or inhibits the expression of some transcription factors and related genes, which is a key factor affecting the inflammatory state of AS plaques and ultimately mediates the homing, rolling and adhesion of leukocytes in the vascular endothelial layer in the blood circulation (Fisslthaler et al. 2003; Xu et al. 2019). In areas of arterial vessels with turbulent flow, shear stress upregulates the pro-inflammatory transcription factor NF-κB, triggering the progressive activation of inflammation in AS lesion areas. Low shear stress is thought to activate in part the expression of NF-κB-dependent proinflammatory genes, including adhesion molecules such as intercellular adhesion molecule-1

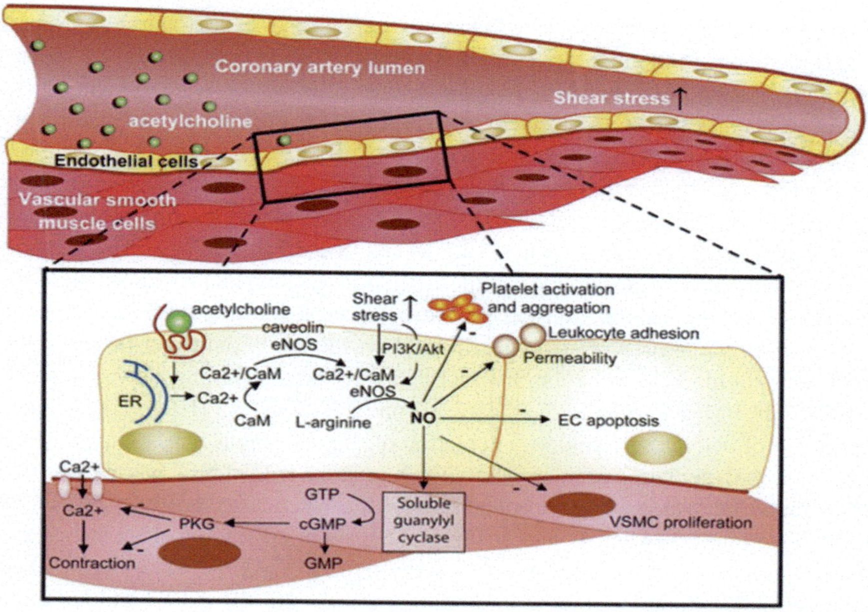

Fig. 5.1 Synthesis of NO. Illustration of the stimulation of endothelial NO synthase by acetylcholine and shear stress leading to increased nitric oxide (NO) production in endothelial cells by receptor and nonreceptor and calcium-dependent and non-calcium-dependent pathways. Adapted from Herrmann et al. (2010)

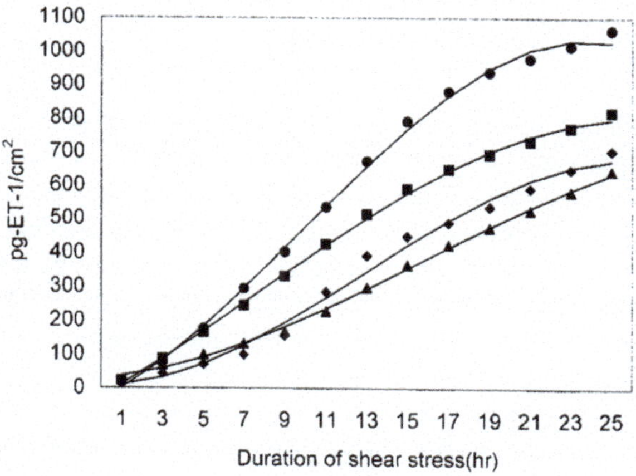

Fig. 5.2 Cumulative ET-1 secretion of ECs under static conditions (◆) or exposed to steady laminar shear stress of 5 dyn/cm^2 (■), 10 dyn/cm^2 (●), and 15 dyn/cm^2 (▲) at different time points. Adapted from Wang et al. (2002)

(ICAM-1), cellular chemokines such as monocyte chemoattractant protein (MCP-1), and proinflammatory cytokines such as tumor necrosis factor (TNF-α), which accelerate the formation of plaques in individual coronary arteries. The combined action of these proinflammatory factors increases the homing of leukocytes and monocytes, which penetrate deep into the endothelium to differentiate into macrophages and phagocytose lipids to transform into foam cells. On the other hand, foam cells produce a large number of inflammatory cytokines, growth factors, reactive oxygen species and matrix-degrading enzymes, thus further accelerating the development of AS plaques. In contrast, VECs exposed to laminar shear stress possess anti-inflammatory and antioxidative properties. Kruppel-like factor 2 (KLF2) and nuclear transcription-related factor 2 (Nrf2) are the key factors that constitute the transcription of shear stress-induced AS (Huang et al. 2021). They can upregulate the expression of anti-inflammatory and antioxidant genes, such as eNOS and heme oxygenase-1 (HO-1). In vivo studies have demonstrated that these two transcription factors contribute to the maintenance of arterial function and have revealed a role for their selective activation in protective regions of the vasculature. In addition, the study confirmed the molecular mechanism of blood flow regulation of KLF2 expression, including activation of the MEK5/ERK5/MEF2 signaling pathway, histone deacetylase 5 (HDAC5)-mediated regulation of active transcription, interaction of thioredoxin as a protein, and post-transcriptional regulation of microRNA. The regulatory role of Nrf2 has not yet been fully elucidated. Studies have revealed that Nrf2 can sequester KELCH-like ECH-associated protein 1 (Keap-1) in a resting state, respond to laminar shear stress, and translocate to the nucleus for accumulation through PI3K/Akt (Koskinas et al. 2013; Qiu et al. 2012; Wang et al. 2010). More importantly, NF-κB, KLF2 and Nrf2 can interact, but whether the mechanism of action is regulated by shear stress is still unclear.

5.1.3 Relationship Between Shear Stress and Angiogenesis in Atherosclerosis

Plaque rupture is the most common cause of arterial thrombosis and is responsible for approximately 75% of acute arterial thrombosis (Fang et al. 2020; Verhamme et al. 2021). Although plaque rupture is always accompanied by localized inflammation of the fibrous cap, the inflammatory infiltrate does not cover the entire plaque, most plaque components are characterized by few or no inflammatory cells, and inflammation is often present in stable plaques as well. Therefore, shear stress-induced inflammation alone cannot be fully identified as a cause of plaque rupture and acute coronary events.

Intraplaque hemorrhage has been identified as a key factor leading to plaque rupture during AS plaque progression. The traditional concept is that the adventitia nourishes blood vessels, breeds the outer layer of blood vessel walls, and plays a role in resisting AS. However, some scholars believe that the formation of new blood vessels makes the plaque expand beyond the critical thickness, which is also an inducer of AS. In addition, the density of nutrient vessels is considered to be the most directly related factor in the development of AS. Studies have found that the removal or inhibition of nutrient vessels can lead to intimal changes, similar to the occurrence of AS. Infiltration of medial necrotic cells, macrophages and VSMCs, even together with morphologically intact endothelial tissue, supports the adventitia vascular nutrient tube in the initial stage of AS. In addition, the study also found that atorvastatin inhibited the AS process while inhibiting vascular trophoblasts, which did not depend on the reduction in cholesterol levels. Therefore, it is gradually believed that vascular trophoblasts play an important role in the initiation and development of AS. With the progress of medical imaging technology, the use of new contrast-enhanced ultrasound has realized the visualization of vascular trophoblasts and plaque neovascularization processes and has developed into a diagnostic method to determine the location of early AS plaques.

Shear stress plays an important role in budding angiogenesis and the formation of the microvasculature network, affecting the development of AS plaques. Angiogenesis was obvious in the high shear stress area near the heart end of the stenotic vessel (Fig. 5.3a, c), while angiogenesis was extremely low in the low shear stress area at the distal end of the stenotic vessel (Fig. 5.3b, d). Low shear stress is likely to promote intimal vascularization by stimulating the secretion of pro-angiogenic factors, the most typical example of which is VEGF. In addition to promoting angiogenesis by increasing the viability of VEGF and subsequent activation of VEGF-dependent vascular endothelial growth factor receptor (VEGFR2) through shear stress (Fig. 5.4), shear stress-guided VEGFR2 ligand-independent activation is important for multiple cellular signaling responses. Shear stress-dependent activation of VEGFR2 causes p38 and ERK1/2 activation, which leads to shear stress-induced angiogenesis. In addition, angiopoietin-2 (ANG2) is another important proangiogenic factor for shear stress-mediated angiogenesis and intraplaque hemorrhage. More specifically, oscillatory shear stress upregulated ANG2 to promote angiogenesis, while laminar shear stress downregulated ANG2 and inhibited the formation of the vascular lumen. The specific mechanism by which shear stress regulates ANG2 expression can be attributed to the rapid induction of ANG2 expression by the transcription factor FOXO1 after inhibiting the PI3K/AKT pathway. Therefore, laminar shear stress activates the PI3K/AKT pathway, which may lead to the inhibition of ANG2 expression by phosphorylated FOXO1.

5.1.4 *The Effect of Blood Flow Shear Stress on the Formation of Atherosclerosis*

Influenced by the characteristics of blood flow velocity, the shear stresses in arterial and venous vascular trees are different. The arterial wall shear stress is 10–70 dyn/cm^2, while the normal value of the corresponding venous shear stress is significantly lower than that of the artery (1–6 dyn/cm^2). In the vasculature, regions of high shear stress also exist, especially in areas where turbulence or flow velocity increases due to vascular anatomy or geometry, such as arch artery-like bends, bifurcations, and vascular anastomoses. Moreover, due to the existence of a countercurrent, the direction of shear stress may also change, which depends on the hemodynamic conditions.

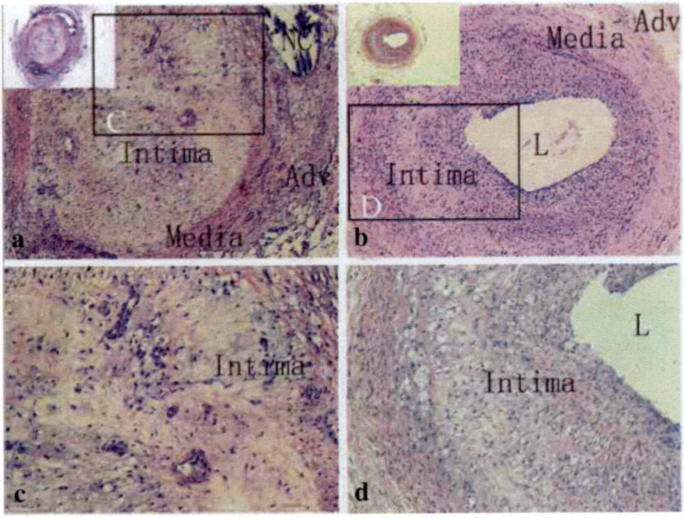

Fig. 5.3 MVD Increases with Atherosclerotic Progression in the upstream of Carotid Collar Placement. **a** Detail of hematoxylin and eosin-stained the upstream and the downstream of Carotid Collar Placement with thick cap fibroatheroma (necrotic core, NC); inset shows entire coronary artery. Boxed regions, **c** and **d** illustrate the regions of interest and correspond to **c** showing microvessels in the intraplaque region (**A**, **B** 100×; **C**, **D** 200×). Cited from Qiu (2011)

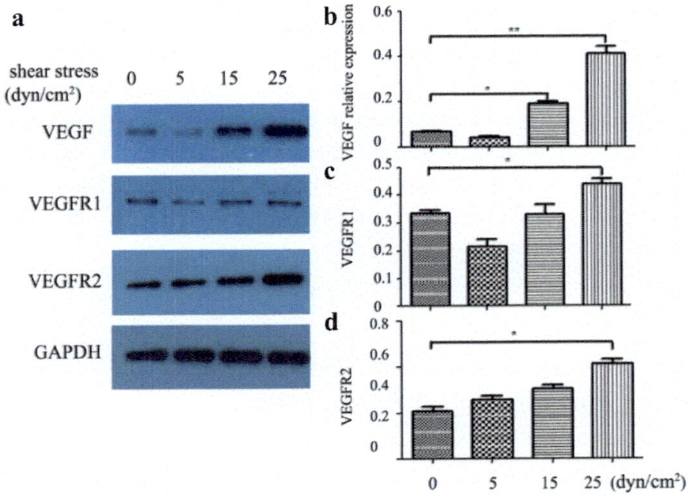

Fig. 5.4 Different shear stress regulates VEGF, VEGFR1, VEGFR2 protein expression in EC. A. Western blotting detects expression of the VEGF, VEGFR1, VEGFR2 protein in EC. B-D. The statistic of the expression of VEGF, VEGFR1, VEGFR2 protein expression in EC. Cited from Ling et al. (2017)

Low shear stress areas are often accompanied by the presence of unstable fluids, such as turbulence, blood reflux areas, and blood "stagnation" areas (congestion), which are often seen in large arteries or arterial bifurcations.

Atherosclerosis is closely related to many risk factors, such as hypertension, hypercholesterolemia, low HDL, diabetes and smoking, but the distribution of AS damage is nonrandom, showing a high degree of focal and no obvious individual difference. AS plaques mostly occur in complex blood flow areas, such as the coronary artery, carotid artery, cerebral artery, abdominal aorta, and femoral artery, and most of them are in places where the geometry and blood flow velocity of arteries have changed, such as bifurcation, stenosis or bending (Gijsen et al. 2007). The blood vessel diameter, blood flow velocity and blood viscosity determine the level of shear stress.

5.1.4.1 Effect of Normal Blood Flow Shear Stress

Under normal physiological conditions, laminar shear stress represents a pivotal factor in maintaining vascular health (Davies et al. 2019). It remains constant within a defined range while preserving its direction and magnitude. This steady-state laminar shear stress exerts preventive effects against AS, thrombus formation, leukocyte adhesion, VSMC proliferation, vascular remodeling, and endothelial cell apoptosis. It holds significant importance in regulating vascular diameter, cellular proliferation within the vascular wall, thrombogenesis inhibition, and inflammation suppression. Laminar shear stress stimulates vascular endothelial cells (VECs) to release a plethora of bioactive substances, including superoxide dismutase (SOD), glutathione, nitric oxide (NO), and endothelin-1 (ET-1). While NO induces vascular smooth muscle relaxation, ET-1 induces contraction, and their equilibrium maintains vascular tone within a defined range. NO also inhibits platelet adhesion, dampens excessive synthesis and release of growth factors by VECs, and restrains the hyperproliferation of VSMCs (Herrmann et al. 2010). Thus, stable laminar shear stress constitutes a crucial factor in countering the formation of AS (Yoshizumi et al. 2003).

Recent studies have indicated that stable laminar shear stress promotes the release of endothelial cell factors, inhibits coagulation, suppresses leukocyte migration, and leads to vasodilation by inducing VSMC activation. Laminar shear stress within the physiological range of 15–70 dyn/cm^2 induces endothelial cells to remain quiescent and maintains the normal expression of anti-atherosclerotic genes. Shear stress below 4 dyn/cm^2 stimulates the expression of genes associated with AS, while shear stress exceeding 70 dyn/cm^2 triggers early thrombus formation.

Relevant research suggests that stable laminar flow reduces VEC apoptosis and TNF-α-mediated VEC activation (Fig. 5.5), with effects proportional to the magnitude of shear stress. VEC apoptosis or activation is considered a major cause of plaque disruption and can further lead to platelet aggregation. Stable laminar shear stress at 12 dyn/cm^2 applied to human umbilical vein endothelial cells (HUVECs) activated three mitogen-activated protein kinases (MAPKs) in response to TNF-α: extracellular signal-regulated kinase (ERK 1/2), c-Jun N-terminal kinase (JNK), and

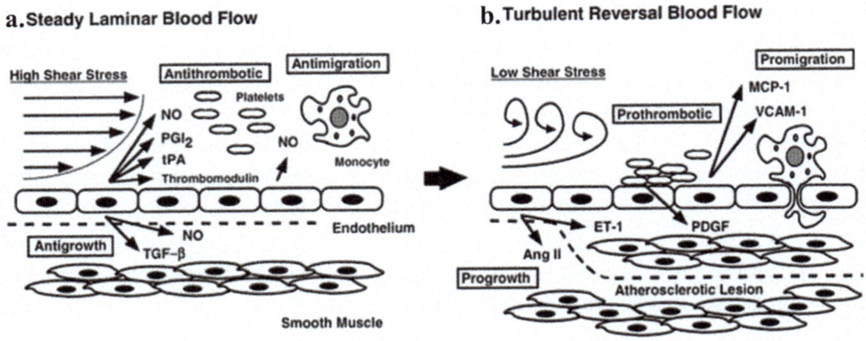

Fig. 5.5 Endothelial cell biology and shear stress. **a** Steady laminar shear stress promotes the release of factors from endothelial cells that inhibit coagulation, migration of leukocytes, and smooth muscle proliferation, while simultaneously promoting endothelial cell survival. **b** In contrast, low shear stress and turbulent flow shift the profile of secreted factors and expressed surface molecules to one that favors the opposite effects, thereby contributing to the development of atherosclerosis. PGI_2 indicates prostacyclin; tPA, tissue plasminogen activator; TGF-β; transforming growth factor-β; Ang II, angiotensin II; ET-1, endothelin-1; PDGF, platelet-derived growth factor; MCP-1, monocyte chemoattractant protein-1; VCAM-1, vascular cell adhesion molecule-1. Cited from Yoshizumi et al. (2003)

p38. In HUVECs, TNF-α activation of ERK1/2, JNK, and p38 takes as little as 15 min, but when HUVECs are pretreated with stable laminar flow for 10 min, JNK activation is inhibited, while ERK1/2 and p38 activation show no significant differences. Coincubation of HUVECs with PD98059 (a specific inhibitor of the protein kinase MEK1/2) blocks the shear stress-mediated inhibitory effect on JNK activation by TNF-α (Yoshizumi et al. 2003).

5.1.4.2 Effects of Abnormal Blood Flow Shear Stress

At bifurcations, branching outlets, and curvatures of blood vessels, nonlaminar flow (turbulent or disturbed flow) generates various vortices due to the chaotic alteration of fluid direction, leading to alterations in shear stress magnitudes (Fig. 5.6). These variations in blood flow patterns can induce diverse responses in the vascular endothelium at both the cellular and molecular levels. These responses, in conjunction with other systemic risk factors, collaboratively promote the development of AS (Chatzizisis et al. 2007).

(1) **Impact of High Shear Stress** As early as 1968, research proposed that the formation of AS is attributed to localized endothelial cell injury caused by elevated shear stress. This injury leads to the deposition of lipids and aggregation of platelets, eventually resulting in plaque formation. The shear stress levels estimated to damage endothelial cells were above 400 dyn/cm^2; however, within the normal cardiovascular system, arterial shear stress does not exceed 100 dyn/cm^2. Consequently, contemporary researchers widely disfavor this hypothesis.

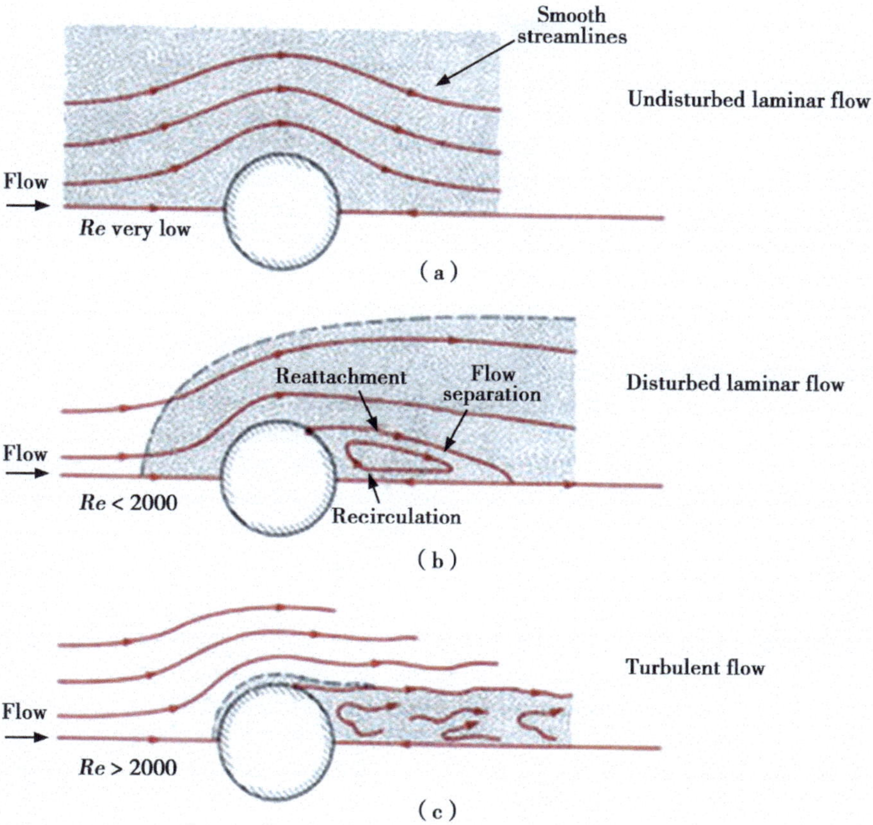

Fig. 5.6 Characteristics of Flow Patterns. Schematic figure illustrating the characteristics of flow patterns. **a** Undisturbed laminar flow is a smooth streamlined flow characterized by concentric layers of blood moving in parallel along the course of the artery; **b** disturbed laminar flow is characterized by reversed flow (i.e., flow separation, recirculation, and reattachment to forward flow); **c** in turbulent flow the blood velocity at any given point varies continuously over time, even though the overall flow is steady. Re = Reynolds number. Cited from Chatzizisis et al. (2007)

In a study conducted by Wei et al., it was observed that in regions of high shear stress, the increased shear stress results in the enlargement of intercellular gaps within endothelial cells. Which facilitates the deposition of lipids and inflammatory cells beneath the endothelium, ultimately triggering the development of intimal hyperplasia and atheromatous plaques. This process is characterized as an acute response and contributes to the instability of plaques. Conversely, the distal end of narrowed arteries experiences oscillatory low shear stress. The accumulation of lipids in this region impacts the permeability of endothelial cells, leading to lipid deposition beneath the endothelium and inducing the migration and proliferation of smooth muscle cells. Consequently, this process initiates and progresses to intimal hyperplasia in a chronic manner (Wei et al. 2007).

Recent studies have demonstrated that in proximity to arterial narrowings, shear stress increases as one approaches the narrowed segment. However, the occurrence of AS is limited within a specific range of high shear stress. Experimental evidence indicates that heightened endothelial permeability within the high shear stress region can be attributed to shear stress exceeding physiological levels, as endothelial cell functionality becomes compromised beyond 400 dyn/cm^2. Given the increased permeability in the vicinity of proximal narrowings and the rapid escalation of shear stress, the expression of two types of tight junction proteins is notably reduced. Nevertheless, this reduction remains significantly higher compared to regions with increased permeability to albumin and lipids, particularly in distal areas. This discrepancy highlights that, in comparison to low shear stress conditions, high shear stress indeed impedes the occurrence of arterial sclerosis.

Yusaku Fukumoto and colleagues propose that localized high shear stress can trigger the rupture of the atherosclerotic fibrous cap. Elevated blood flow shear stress can lower VEC proliferation and apoptosis rates while enhancing the production of vasodilatory chemical mediators, paracrine growth inhibitors, tissue plasminogen activator, and antioxidants. Simultaneously, it inhibits the generation of vasoconstrictive agents, paracrine growth stimulants, inflammatory mediators, and adhesion molecules (Fukumoto et al. 2008).

(2) **Effects of Low Shear Stress** In 1978, Caro proposed the hypothesis of low shear stress involvement in AS, suggesting that AS damage occurs in regions with relatively low wall shear stress. Currently, it is widely accepted that low shear stress constitutes a critical hemodynamic factor in arterial diseases. Arterial segments subjected to low shear stress or characterized by turbulent flow are associated with disrupted local endothelial cell function and the formation of AS plaques. Research by Masanori Yoshizumi and colleagues indicated that AS predominantly develops in regions of turbulent flow and low shear stress, while laminar flow and high shear stress reduce the expression of CD36 (receptor for ox-LDL) and monocyte adhesion, decrease ET-1 production, and enhance NO generation, thereby exerting an anti-atherogenic effect (Yoshizumi et al. 2003).

Studies by Jiang Zonglai utilizing rabbit carotid AS models and in vitro stress-cultured pig carotid arteries demonstrated that low shear stress significantly promotes the formation and progression of arterial AS. Under conditions of low shear stress, arterial reconstruction occurs prominently, with both proliferation and apoptosis of VSMCs playing pivotal roles in this remodeling process. Low shear stress inhibits NO secretion by VECs, promoting increased VSMC proliferation and reduced apoptosis. In the context of low shear stress, the expression of the anticancer gene protein p53 and the apoptosis-inhibiting gene protein Bcl-2 undergoes alteration, potentially modulating VSMC apoptosis. In the signal transduction processes related to VSMC proliferation and apoptosis governed by shear stress, the proto-oncogenes c-fos and c-myc might play regulatory roles, thus participating in vascular remodeling.

Some studies suggest that the continuous maintenance of a low shear stress state for over 48 h leads to significant endothelial cell proliferation, often accompanied by morphological changes and reduced elasticity of endothelial cells, diminished elastic fibers, heightened monocyte adhesion, and increased expression of vascular cell adhesion molecules. Additionally, decreased blood flow shear stress can prompt endothelial cell apoptosis, disrupt vascular endothelial integrity, and contribute to local thrombus formation. Research by Chatzizisis and colleagues have posited that the magnitude of low shear stress determines the complexity and heterogeneity of AS plaques. It can be regarded as an independent predictor for the occurrence and development of high-risk AS plaques characterized by substantial lipid deposition, severe inflammation, thin fibrous caps, substantial internal elastic lamina degeneration, and excessive expansive remodeling (Chatzizisis et al. 2007, 2008, 2010).

5.1.5 Impact of Shear Stress-Regulated Gene Expression on Atherosclerosis

Blood flow shear stress directly modulates the expression of multiple genes within VECs. Currently, over 40 genes in endothelial cells have been identified as subject to shear stress regulation. Based on their biological characteristics, these genes can be broadly categorized into various groups, including vascular active substances, growth factors, adhesion molecules, chemotactic factors, coagulation factors, and proto-oncogenes, among others. These shear stress-regulated genes are involved in numerous aspects of cellular processes, including proliferation, differentiation, maintenance of vascular tone, thrombosis, cell–matrix and cell–cell adhesion, as well as the regulation of inflammation or the immune system.

At the arterial level, shear stress exceeding 15 dyn/cm^2 induces the expression of anti-atherosclerotic genes in endothelial cells, while low shear stress areas, less than 4 dyn/cm^2, are more susceptible to AS and stimulate the expression of AS-related genes, thereby accelerating AS formation. Previous research has demonstrated that laminar shear stress enhances VECs' anti-inflammatory capabilities, effectively preventing monocyte adhesion, proliferation, apoptosis, and the expression of inflammatory cell genes. Therefore, laminar shear stress is a critical factor for regulating the physiological and pathological aspects of the vascular wall through the modulation of gene expression in endothelial cells.

Many early studies have indicated that various shear stress response elements (SSREs) exist in the promoters of certain endothelial genes, and they regulate their expression through interactions with various transcription factors under the influence of shear stress. For instance, an upstream 6-bp sequence, GAGACC, in the PDGF-B promoter is associated with shear stress induction. Alterations in this sequence result in reduced or abolished endothelial cell responsiveness to stress. Studies have suggested that SSREs bind specifically to endothelial cell DNA, thereby upregulating or downregulating gene products. Such genes include eNOS, ET-1, proto-oncogenes such as C-Fos and C-Jun, transforming growth factor-beta (TGF-β), and monocyte chemoattractant protein-1 (MCP-1), among others. Early research has found that NO regulates shear stress-mediated Egr-1 gene expression, with this regulation occurring through the extracellular signal-regulated kinase (ERK) pathway in endothelial cells. Steady flow and NO generated by NO donors suppress the time-dependent expression of MCP-1 and PDGF-A under the influence of shear stress by downregulating NF-κB and Egr-1, respectively. Shear stress is one of the triggering factors for the upregulation of tissue factor gene expression in endothelial cells, and its triggering mechanism is mediated by the transcription factors Egr1 and Sp1.

Shear stress can directly regulate the synthesis and secretion of bioactive substances by endothelial cells, including the induction of interleukin-8 (IL-8) production, which depends on the duration of shear stress exposure. AS is often associated not only with hemodynamic changes but also with inflammatory responses. Under fluid shear stress, endothelial cells can express multiple chemotactic cytokines. IL-8, MCP-1, and growth-regulated oncogene (GRO) are chemotactic cytokines that play significant roles in regulating the inflammatory response process.

5.1.6 Stress Regulation of Transcription Factors in Atherosclerosis and Anti-Atherosclerosis

Key elements of the rapid response to shear stress include the opening of ion channels in the luminal membrane, activation of integrins within local adhesive complexes, and activation of adhesive molecule complexes located at cell–cell junctions. Subsequently, downstream signaling responses to shear stress are activated, including various kinase phosphorylation events, GTPase activation, and reactive oxygen species production, with most converging into the MAPK pathways. Ultimately, these pathways lead to the activation of pro-AS transcription factors such as the inflammatory factor NF-κB and activator protein-1 (AP-1) or the activation of anti-AS transcription factors such as the anti-inflammatory factor Krüppel-like factor 2 (KLF2) and the antioxidant nuclear factor erythroid 2-like 2 (Nrf2) (Fig. 5.7). This allows them to bind to multiple target genes, thereby triggering or inhibiting transcriptional regulation of gene expression, which in turn modulates cellular function and the progression of diseases (Parmar et al. 2006).

The delicate balance between these pro-AS or anti-AS protective transcription factors is governed by multiple molecular releases that dictate vascular physiology/pathophysiology and the development of AS. Specifically, pro-AS shear stress weakens the bioavailability of NO by reducing eNOS mRNA and protein expression, ultimately impairing NO-dependent vasodilation and atherosclerotic protection. Additionally, pro-AS shear stress activates NF-κB-dependent proinflammatory gene expression, including adhesion molecules such as ICAM-1, VCAM-1, and E-selectin, chemokines such as MCP-1, and proinflammatory cytokines such as TNF-α, IL-1, and IFN-γ. Furthermore, pro-AS shear stress also leads to extracellular matrix degradation by increasing enzyme degradation and reducing matrix synthesis, all of which are regulated by NF-κB. Pro-AS shear stress enhances the accumulation of macrophages and vascular smooth muscle cells within plaques while upregulating the

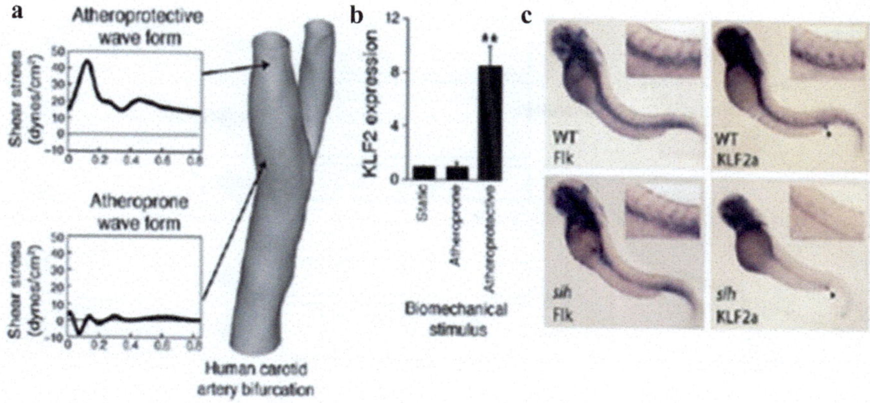

Fig. 5.7 Flow-dependent expression of KLF2 and its regulation by a MEK5/ERK5/MEF2 pathway. **a** Archetypal atheroprotective and atheroprone shear stress waveforms derived from a human carotid artery. These 2 shear stress waveforms were recreated using a dynamic flow system and applied to cultured HUVECs. **b** HUVECs were cultured under static (no flow), atheroprone, or atheroprotective flow conditions for 24 h, and KLF2 mRNA expression was measured by RT–PCR ($n = 3$; mean ± SEM). **c** Whole-mount in situ hybridization of WT or sih mutant embryos at 48 h, probed for Flk or KLF2a. Inserts show close-ups of the trunk vasculature. Anal sphincter staining is indicated by arrowheads. Cited from Parmar et al. (2006)

expression of proinflammatory cytokines (TNF-α, IL-1, and IFN-γ), stimulating them to produce numerous proteases associated with extracellular matrix degradation, such as MMP-2 and MMP-9, while inhibiting extracellular matrix synthesis, such as TGF-β and NO-mediated collagen synthesis (Liu et al. 2021). Furthermore, activation of Nrf2 upregulates antioxidant proteins such as HO-1 and NQO-1, resulting in reduced oxidative stress responses and inflammation. Simultaneously, KLF2 binds to target genes such as eNOS to promote NO production and downregulate ET-1, collectively enhancing vasodilation. KLF2 also directly or indirectly activates anti-thrombotic proteins such as thrombomodulin or pro-coagulant proteins such as tissue factor, thereby modulating coagulation (Zhou et al. 2014).

More importantly, current evidence supports the transcriptional balance between pro-AS and anti-AS transcription factors in response to blood flow, which in turn balances pro-AS and anti-AS effects. NF-κB and AP-1 are typical pro-AS transcription factors that promote intense inflammation and matrix degradation, impairing NO-dependent vascular dilation function. In contrast, KLF2 and Nrf2 are typical anti-atherosclerotic transcription factors that reduce oxidative stress, lower coagulation function, and enhance vasodilation (Figs. 5.8 and 5.9).

Atherosclerotic lesions tend to occur on the inner walls of arterial branch points, bifurcations, and curved segments, which are often associated with pro-AS blood flow patterns. At arterial bifurcations and curved regions, unstable blood flow promotes the upregulation of inflammatory genes and downregulation of anti-inflammatory genes. In contrast, straight arterial segments experience high laminar shear stress, are less

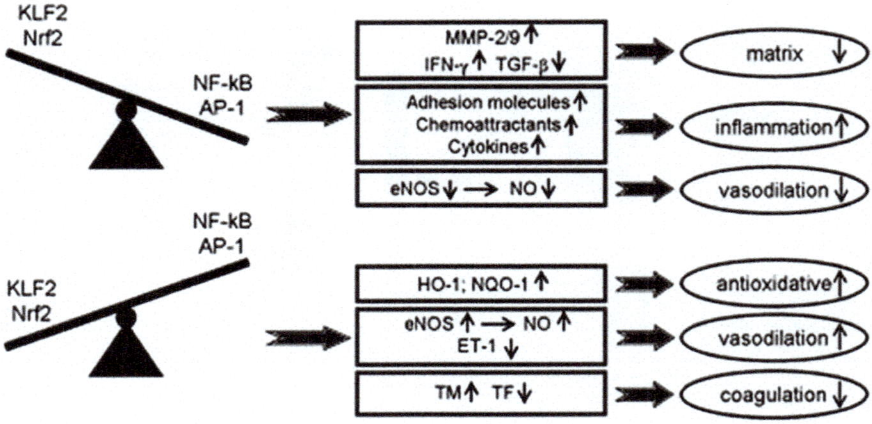

Fig. 5.8 Biomechanical control of atherogenic and atheroprotective transcription factors. Abbreviations: MMP, matrix metalloproteinase; INF, interferon; TGF, transforming growth factor; ICAM, intracellular adhesion molecule; VCAM, vascular cell adhesion molecule; TNF, tumor necrosis factor; MCP, monocyte chemoattractant protein; IL, interleukin; HO-1, heme oxygenase-1; NQO1, NAD(P)H dehydrogenase quinone 1; ET-1,endothelin-1; TM, thrombomodulin; TF, tissue factor; VSMCs, vascular smooth muscle cells. Cited from Zhou et al. (2014)

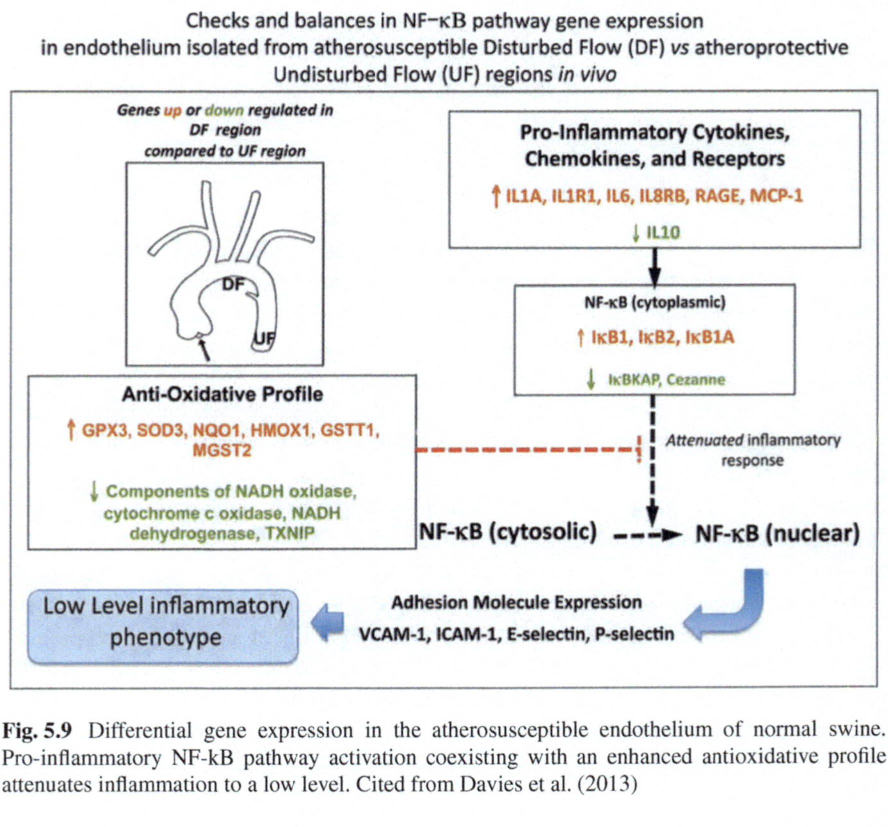

Fig. 5.9 Differential gene expression in the atherosusceptible endothelium of normal swine. Pro-inflammatory NF-kB pathway activation coexisting with an enhanced antioxidative profile attenuates inflammation to a low level. Cited from Davies et al. (2013)

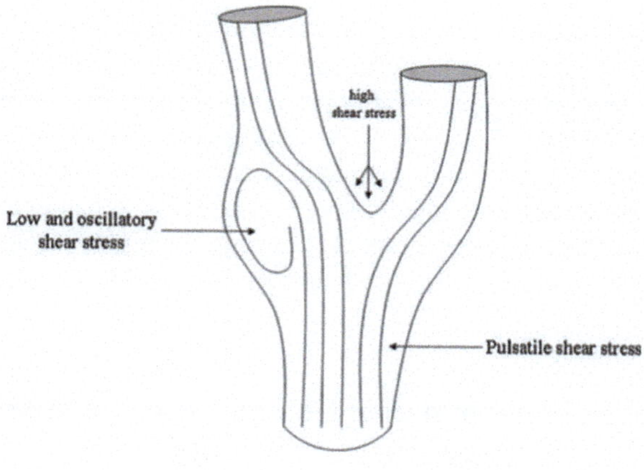

Fig. 5.10 Distribution of wall shear stress in the carotid artery bifurcation. Cited from Cecchi et al. (2011)

affected by pro-AS blood flow, and exhibit inhibited inflammatory gene expression with an upregulation of anti-inflammatory or antioxidant genes (Fig. 5.10).

5.1.7 The Clinical Implications of Shear Stress for the Treatment of Clinical Atherosclerosis

The formation and progression of AS plaques are directly related to shear stress. However, in the study of AS plaques, there is still a gap in the understanding of the mechanical and biological characteristics of blood vessels and their shear stress properties, and an effective correlation has not been established. In the process of studying the mechanical mechanisms of AS plaque formation and development, previous research has mainly focused on shear stress, with less emphasis on the mechanical properties of blood vessels themselves. Nevertheless, the composition and mechanical properties of the vascular wall are of considerable significance for the proliferation and differentiation of vascular cells, especially in the study of plaque rupture. Factors such as vascular wall thickness and elastic modulus should not be overlooked in regard to thrombus formation triggered by plaque rupture. Currently, the treatment of AS plaques primarily involves medication and interventional therapy. Both medication and stent intervention can, to some extent, achieve therapeutic or inhibitory effects on AS plaques by altering the mechanical microenvironment of blood vessels. Other clinical treatments, such as those reducing stenosis and lipid-lowering drugs, also directly or indirectly achieve therapeutic goals by modifying the mechanical microenvironment of blood vessels or altering the composition of plaques. Therefore, the future focus should be on the use of various in vitro and in vivo experimental models to simulate various forces and complex flows to study their impact on plaque stability. Enhancing fundamental clinical research on the mechanical environment of arterial plaques is a long and challenging task. Clinical treatment relies on an in-depth understanding of the mechanisms of plaque formation and rupture. In conclusion, studying the formation and development of AS plaques from the perspective of shear stress is a significant opportunity but also poses severe challenges. These challenges mainly come from two aspects: understanding and technological capability. Researchers must always remember that the cardiovascular system is dynamic, and changes in local shear stress can lead to alterations in the entire cardiovascular system's shear stress. When employing methods such as drug therapy and interventional treatment, changes not only affect local shear stress but also have an impact on shear stress throughout the entire cardiovascular system.

5.2 Hemodynamic Changes and the Stability of Atherosclerotic Plaques

Whether the appearance of clinical symptoms occurs after the formation of AS lesions depends not only on the degree of arterial lumen stenosis but, more importantly, on the nature of the plaque itself and the presence of secondary complications such as thrombosis. Clinical observations have shown that in the majority of patients with acute coronary syndromes, the narrowing of the coronary artery lumen is not severe (less than 50%), and only mild stenosis is observed during angiography. In this section, starting from the perspective of hemodynamic factors inducing the formation, development, and rupture of atherosclerotic plaques, we analyze and review existing research on the relationship between hemodynamics and plaque rupture mechanisms. This information aims to assist in the search for rational clinical treatment and prevention strategies.

5.2.1 Characteristics of Vulnerable Plaques

Pathological studies have indicated that vulnerable plaques typically exhibit the following pathological features: (1) a soft, lipid-rich lipid core (with the lipid core volume exceeding 40% of the plaque volume); (2) a thin fibrous cap; (3) infiltration of a large number of inflammatory cells, including macrophages, T lymphocytes, and mast cells; (4) minimal smooth muscle cells; (5) eccentric plaque shape; and (6) extensive neovascularization within the plaque. In late-stage AS plaques, neovascularization facilitates the accumulation of a large number of inflammatory cells within the plaque, making it more prone to intraplaque hemorrhage, which lead to plaque rupture.

Atherosclerotic plaques with characteristics conducive to rupture often do not exhibit severe stenosis but are associated with expansive vascular remodeling and acute coronary syndromes. These intrinsic factors contribute to plaque rupture, while changes in hemodynamic factors serve as important external factors in plaque rupture.

5.2.2 *Hemodynamic Characteristics at the Plaque Site and Their Impact on Plaque Development*

5.2.2.1 Hemodynamic Environment at the Plaque Site

Vascular outward expansive remodeling implies that plaque intrusion into the lumen is not severe, which lead to the formation of low shear stress (Fig. 5.11). These areas promote plaque development and rupture. Expansive vascular remodeling under low shear stress is a crucial factor leading to plaque formation (Wang et al. 2009). Studies using animal models of AS plaques have also identified low shear stress and oscillatory flow as primary mechanical factors in AS development. While an increase in shear stress at the leading edge of vascular narrowing may lead to fibrous cap rupture, plaque rupture is not solely attributed to shear stress, as wall shear stress in these areas is much smaller than the tensile stress generated due to periodic blood pulsations.

Circumferential stress, which refers to the tension exerted on the vessel wall due to repeated blood pulsations, may be a major factor leading to plaque rupture and intravascular thrombosis. An increase in plaque thickness does not significantly alter local shear stress, but it substantially increases tensile stress at the shoulders of the plaque. Clinical data show that plaque rupture occurs with a ratio of 63:37 between the shoulders and the center of the plaque. Therefore, an increase in circumferential stress is likely a major factor contributing to plaque rupture and thrombus formation (Qiu et al. 2009). Additionally, blood pressure plays a significant role in the formation and progression of plaques, especially when there is a synergistic interaction between hypertension and hemodynamics, which has a highly noticeable impact on plaque stability.

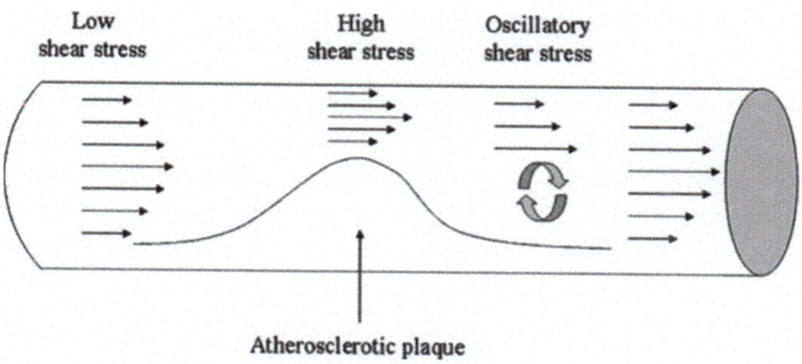

Fig. 5.11 Differential distribution of shear stress in a straight arterial segment proximal to a lumen-protruding atherosclerotic plaque. Cited from Cecchi et al. (2011)

5.2.2.2 Shear Stress and the Formation and Development of Vulnerable Plaques

1) **Low Shear Stress and Formation of Vulnerable Plaques**: Low shear stress and turbulent flow conditions contribute to plaque development, leading to changes in plaque composition. In areas of low shear stress, plaque composition alterations include decreased VSMC content, enlargement of the lipid pool, thinning of the fibrous cap with reduced collagen fiber content, and the accumulation of numerous macrophages. Conversely, plaques formed under oscillatory flow conditions tend to be smaller and more stable. The primary distinction between vulnerable and stable plaques lies in the substantial accumulation of inflammatory cells in vulnerable plaques (Qiu et al. 2011). The major factor responsible for inflammatory cell accumulation under low shear stress is the upregulation of adhesion molecules on the surface of VECs, reducing the rolling velocity of inflammatory cells within the endothelial layer. Animal studies have shown selective upregulation of VCAM-1 and ICAM-1 expression under low shear stress conditions. In vivo observations revealed increased monocyte adhesion to endothelial cells in low shear stress areas (Souilhol et al. 2020). Low shear stress activates integrins through chemical factors, allowing endothelial cells to capture inflammatory cells. Subsequent inflammatory responses stimulate ox-LDL, promoting EC capture of monocytes, which then differentiate into macrophages. The abundant presence of macrophages reduces tissue strength, facilitating plaque formation (Silva et al. 2007). Additionally, low shear stress can activate matrix metalloproteinases (MMPs) in VSMCs. This activation can lead to the degradation of the internal elastic lamina and extracellular matrix, resulting in the formation of soft, lipid-rich plaques with significant accumulation of inflammatory cells, a thin fibrous cap structure, and pronounced vulnerability characteristics. In summary, low shear stress increases the levels of inflammatory molecules during the AS process, thereby altering plaque composition and promoting the formation of vulnerable plaques.
2) **The Impact of High Shear Stress on Plaque Distribution**: Plaque rupture primarily occurs in the upstream regions of narrowed blood vessels and is mainly attributed to the accumulation of macrophages in this area and subendothelial thrombosis. The combination of computer-based three-dimensional reconstruction techniques and computational fluid dynamics has revealed substantial accumulations of ox-LDL, foam macrophages, and MMPs in the high shear stress region near the proximal end of the narrowed vessel. The degradation of the extracellular proteoglycan matrix is attributed to the secretion of serine proteases (including plasmin) by endothelial cells induced by high shear stress. These studies strongly suggest a significant correlation between high shear stress and the longitudinal distribution of vulnerable plaques. High shear stress promotes the upregulation of endothelial cell and platelet adhesion, thereby facilitating AS progression.

Research has shown that in high shear stress regions, the gradient of high shear stress damages endothelial cells, increasing their permeability. This makes it easier for lipids and inflammatory cells to enter and deposit beneath the endothelium. This ultimately leads to intimal hyperplasia and atheromatous plaque formation, which is an acute process and contributes to plaque instability. Conversely, in the distal end of the stenosis where low shear stress prevails due to oscillation, it influences lipid deposition in the area and, consequently, endothelial cell permeability. The increased endothelial permeability results in lipid deposition beneath the endothelium, inducing the migration and proliferation of smooth muscle cells, leading to the development of intimal hyperplasia, which is a chronic response. Recent research by Lingbo et al. demonstrated that under conditions of a high cholesterol diet, in a murine model of carotid artery stenosis, severe plaque obstruction was observed in the high shear stress region near the proximal end of the model group's vessels. The plaque area occupies the entire lumen, with significant immune cell and lipid infiltration, as well as pronounced hemorrhage and neovascularization trends (Fig. 5.12a, b). Conversely, in the distal low shear stress region of the model group, plaques were less prominent. there was some intimal hyperplasia, but smaller plaque areas and no significant immune cell and lipid infiltration and less pronounced hemorrhage and neovascularization trends (Fig. 5.12c, d). Therefore, the synergistic effects of high shear stress and high lipid levels promote the formation and development of vulnerable arterial plaques.

It should be noted that due to the lack of animal models of AS plaques that closely resemble mature human plaques, as well as suitable in vitro research equipment and methods, further research is needed to elucidate the relationship between hemodynamic abnormalities and AS plaque rupture. The clarification of this relationship will undoubtedly provide new drug targets for preventing and treating of cardiovascular diseases caused by AS plaque rupture.

3) **High Shear Stress Promotes Vulnerable Plaque Formation**

① High shear stress leads to intraplaque hemorrhage and plaque necrosis. Plaques with outward remodeling tend to gradually progress into high-risk plaques with thin fibrous caps. To evaluate whether high shear stress induces vascular outward remodeling, it was observed that following vascular stenosis, remodeling was more pronounced in the distal end of the stenotic vessels, while there was significant outward remodeling in the proximal end of the stenotic vessels. In other words, high shear stress induces significant vascular outward remodeling, with the cross-sectional area of vessels in the high shear stress region increasing by 0.8 times compared to the low shear stress region, where there is evidence of inward remodeling. Further research in our laboratory examined vessels in high and low shear stress regions, and the results from HE-stained sections showed that intraplaque hemorrhage and substantial plaque necrosis predominantly occurred in high shear stress regions. In particular, intraplaque hemorrhage could occur independently of significant plaque necrosis in high shear stress regions, and intraplaque hemorrhage coexisted with significant plaque necrosis in high shear stress regions. In contrast, there was no significant intraplaque hemorrhage or

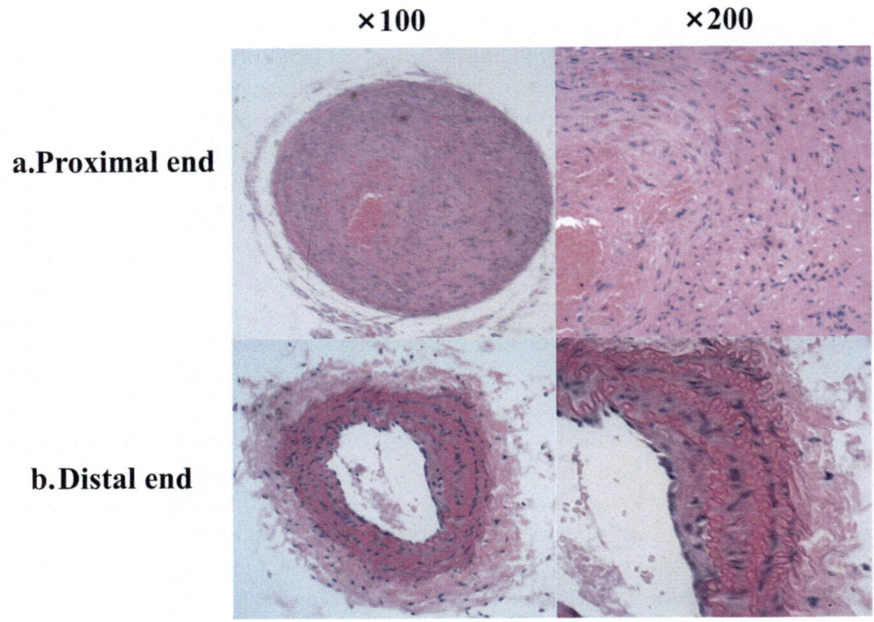

Fig. 5.12 Under high-fat diet for 8 weeks, the distribution of plaques in the proximal and distal end of the model group. **a** The plaque in proximal end (the high shear region) is very obvious, intimal hyperplasia is very serious, and the lumen is completely blocked by plaque (100×); There are large immune cells and lipid infiltrate in the plaque (200×). **b** The plaque in distal end (low shear region) is not obvious, intimal hyperplasia is slight, and the vessel is unobstructed (100×); The infiltration of immune cells and lipid in the plaque are not obvious (200×). Cited from Ling (2015)

plaque necrosis in the low shear stress region at the distal end of the stenosis, where severe intimal hyperplasia, abundant cellular components, and extracellular matrix deposition were observed, but no symptoms of vulnerable plaque, such as intraplaque hemorrhage, were present (Fig. 5.13c). Statistical analysis revealed that the likelihood of intraplaque hemorrhage and plaque necrosis occurring in high shear stress regions near the proximal end was 83%, while the likelihood of intraplaque hemorrhage occurring in the stenotic distal end region was as low as 17%, with no evidence of plaque necrosis (Fig. 5.13d). Therefore, high shear stress significantly promotes the formation of vulnerable plaques.

② High Shear Stress Induces Collagen Fiber Degradation. In the high shear stress region at the proximal end of the narrowed vessel, collagen content is lower than that in the low shear stress region (Fig. 5.14). From the perspective of collagen fibers, this indicates that high shear stress in the proximal end of narrowed vessels induces the formation of vulnerable plaques. Both type I and type III collagen are the major collagen proteins within blood vessels and are responsible for withstanding mechanical stress and maintaining vascular stability. In the high shear stress region, there is nearly no collagen fiber in the plaque intima, and the content of collagen fibers in the outer wall is extremely low. Type I and

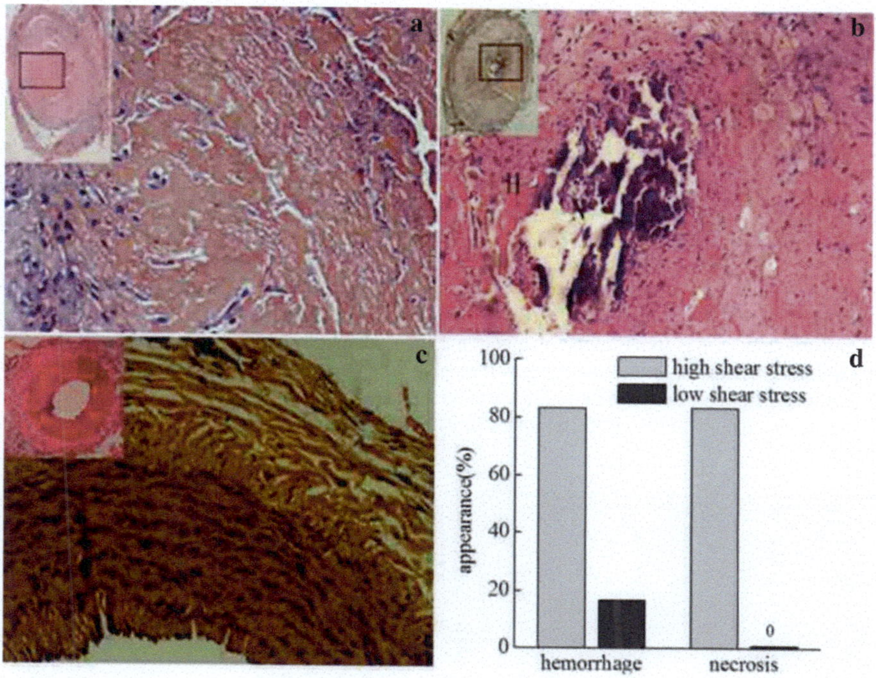

Fig. 5.13 Hemorrhaging and visible necrosis mainly localize in the high shear stress region. **a** Hemorrhages may exist without visible necrosis in the high shear region. **b** Hemorrhages can also exist with visible necrosis in the upstream direction of the stenosis with high shear stress. **c** No hemorrhaging or visible necrosis exists in low shear stress regions. Low shear stress regions were characterized by intimal hyperplasia. **d** On gross examination, frequent hemorrhaging and necrosis (83%) emerged in the upstream direction of carotid collar placement (n = 6), while occasional hemorrhaging (17%) and the absence of visible necrosis (0) were observed in the downstream direction of carotid collar placement. H represents hemorrhaging and N represents necrosis. Cited from Qiu (2011)

type III collagen are sparsely distributed. In contrast, in the low shear stress region, collagen fiber content is higher, and there are varying degrees of intimal hyperplasia. Moreover, in the outer wall of blood vessels, type I and type III collagen are interwoven, forming a stable mesh-like structure.

③ High Shear Stress Induces Elastin Fiber Fracture (Fig. 5.15). The medial layer of blood vessels is primarily composed of elastin fibers. Due to the expansive remodeling caused by high shear stress at the proximal end of narrowed vessels, elastin fibers are significantly stretched and partially degraded. This results in straightening of the elastic lamina within the elastin fibers at the proximal end of narrowed vessels, leading to partial defects and fractures in some regions.

④ High Shear Stress Region Experiences Significant VSMC Apoptosis. Previous studies have shown that VSMC apoptosis is a critical step in the transition from stable to vulnerable plaques in AS (Clarke et al. 2006). The apoptosis of smooth muscle cells was analyzed through transmission electron microscopy

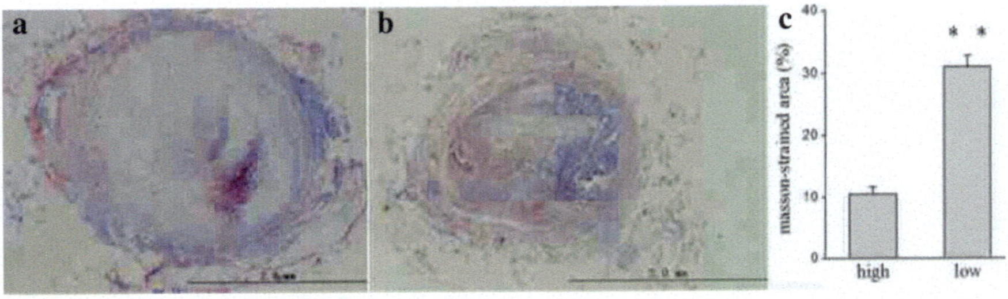

Fig. 5.14 Low collagen content is found in high shear stress regions. Conditions: high-cholesterol and high-fat diet, 8 weeks after cast placement, Masson straining for collagen (40× magnification). **a** Almost no collagen is found in the plaque in high shear stress regions. **b** Collagen on the intimal hyperplasia is localized in low shear stress regions. **c** Quantification of the percentage of areas in the intima positive for collagen and high shear stress administration decreases intimal collagen in atherosclerotic lesions (**P < 0.01, relative to the high shear stress regions). Cited from Qiu (2011)

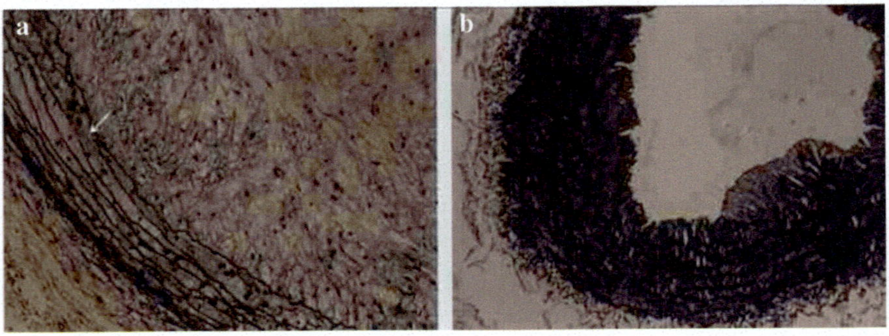

Fig. 5.15 Internal elastic lamina (IEL) defects in the high shear stress region. The results were representative of a high-cholesterol and high-fat diet, 8 weeks after cast placement. **a** VeRhoeff's van Gieson staining demonstrates that the IEL in the upstream direction was defective. **b** The IEL in the downstream direction was normal (100× magnification). A white arrow indicates the defects in the upstream aorta of hyperlipidemic mice. Cited from Qiu (2011)

and TUNEL staining. Ultrastructural analysis of vessels at 8 weeks using the cuff model revealed prominent features of apoptosis in the nuclei and cytoplasm of VSMCs in the high shear stress region at the proximal end of the narrowed vessel. Chromatin exhibited condensation, forming a crescent-shaped structure, and nuclei showed severe agglomeration of nuclear material with highly concentrated chromatin that displayed marginalization and vacuolization, among other characteristics. These results indicate early signs of apoptosis in VSMCs in the high shear stress region. In contrast, VSMCs located in the low shear stress region displayed a significant increase in mitochondrial quantity, as well as an increase in the number of Golgi apparatus and rough endoplasmic reticulum. They exhibited a prominent synthetic phenotype (Fig. 5.16).

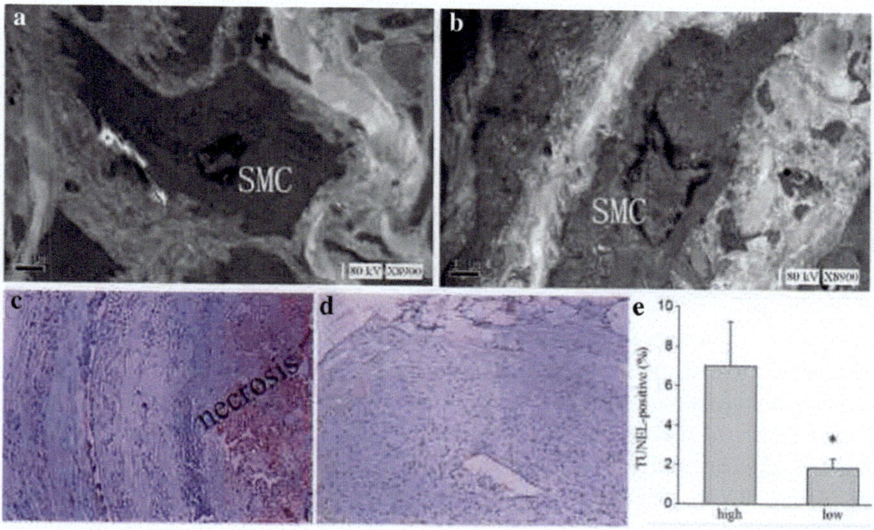

Fig. 5.16 VSMCs have a high apoptotic rate in high shear stress regions. Transmission electron micrographs and immunohistochemical staining with TUNEL demonstrated the apoptosis of VSMCs in the plaque in high and low shear stress regions. **a** Electron micrographs from an artery 8 weeks after carotid collar placement showed nuclear and cytoplasmic abnormalities characteristic of apoptotic VSMCs in high shear stress regions. Condensed chromatin forming crescents in apposition to the nuclear envelope is characteristic of the early stages of apoptosis. **b** Cells localized in the downstream direction contain an increasing number of mitochondria, rough endoplasmic reticulum, and Golgi apparatus in the cell stomata and cytoplasmic processes (8900× magnification). **c** Cell apoptosis was detected by TUNEL-staining (100× magnification). **d** Fewer apoptotic cells are found in low shear stress regions. **e** Quantification of the percentage of positive areas for TUNEL-staining, and high shear stress administration increases apoptosis of the intimal VSMCs in atherosclerotic lesions (*$P < 0.05$, relative to the high shear stress regions). Cited from Qiu (2011)

5.2.2.3 Circumferential Stress and Plaque Development

Our laboratory's research has revealed that changes in shear stress and the stiffness of blood vessels, result in corresponding alterations in the circumferential stress experienced by vascular wall cells also undergoes. This is because shear stress induces changes in the composition and structure of blood vessels, leading to modifications in the elastic modulus of the vascular wall. Sites of vascular pathology, while experiencing low shear stress, also exhibit high vascular tension, which might be a significant factor in the development of vascular AS. Clinical magnetic resonance data indicate that vascular wall thickness is a crucial factor in the progression of AS. However, low wall shear stress alone does not appear to be the primary predictive factor for AS.

5.2.3 Mechanical Mechanisms of Atherosclerotic Plaque Rupture

After elucidating the development of vulnerable plaques, numerous studies have shifted their focus towards understanding the specific locations and mechanical mechanisms of plaque rupture. Clinical research has indicated that the likelihood of plaque rupture occurring at the shoulders of the plaque is significantly higher than that at the center of the plaque. Furthermore, changes in local shear stress within vulnerable plaques are not substantial, but there is a significant increase in tensile stress at the shoulders of the plaque. Circumferential stress in vulnerable plaques is highly correlated with plaque stability, and the region of plaque rupture aligns closely with stress concentration areas. In addition to studies on the morphological characteristics of plaque cross-sections, investigations into the longitudinal distribution of plaques along arteries, based on autopsies and clinical pathology, reveal that plaque rupture primarily occurs in the upstream region of narrowed vessels. In areas upstream of high shear stress plaques, there is a decrease in blood pressure at the site of vascular injury, leading to a significant increase in uniaxial tensile stress. This sudden drop in pressure forces the plaque to deform rapidly and generates a noticeable uniaxial strain. However, in 75% of the narrowed vessel regions, the wall shear stress is considerably lower than the tensile stress generated by periodic blood pulsations. Consequently, the increased tensile stress at the front end of the narrowed vessel is likely a major factor in the rupture of the fibrous cap. Although wall shear stress alone may not directly damage the fibrous cap structure, the localized increase in shear stress is likely an initiating factor in fibrous cap rupture. Therefore, when the extracellular lipid pool exceeds 45% of the vessel wall, a stress concentration zone forms on the fibrous cap. In this region, if collagen content is reduced and macrophages are abundant, the fibrous cap becomes highly susceptible to rupture.

5.3 Research Progress and Prospects

Studies have shown that AS tends to occur more frequently in areas of disturbed flow (DF) within arteries, while it occurs less frequently in areas of laminar flow (UF). Extensive research has demonstrated that DF and UF can modulate various critical functions of vascular endothelial cells (ECs), including vascular inflammation, oxidative stress, vascular tension, cell proliferation, aging, mitochondrial function, and glucose metabolism. ECs possess various mechanosensors on their membranes to detect different flow patterns. Through these mechanosensors, ECs can perceive flow-induced biomechanical signals and transduce them into various mechanotransduction pathways, leading to changes in the expression/activity of mechanosensitive transcription factors (TFs), epigenetic modifying enzymes, noncoding RNAs, and genes, thereby initiating a series of biological responses (i.e., regulation of endothelial function). Here, we summarize the roles of endothelial mechanotransduction factors in the endothelium and discuss the potential of flow simulation compounds in treating endothelial dysfunction and associated atherosclerotic cardiovascular diseases.

5.3.1 *Atherosclerosis, Shear Stress, and Endothelial Function*

Extensive research has confirmed the nonuniform distribution of AS within arterial vessels. The focal nature of AS is attributed to different flow patterns playing distinct roles in regulating endothelial function and the development of AS. Laminar flow (UF) generates high wall shear stress (HSS) in areas such as the inner curvature of the aortic arch, where ECs exhibit an atheroprotective phenotype. Conversely, disturbed flow (DF) occurring in areas with branching and sharp bends generates low wall shear stress (LSS), triggering pro-inflammatory responses and oxidative stress in ECs, leading to endothelial dysfunction and AS formation (Chiu et al. 2011). Investigating the biomechanical basis of flow-induced endothelial dysfunction and AS development is a hot topic in research, as it holds the potential to identify new therapeutic targets for atherosclerotic cardiovascular diseases.

5.3.2 *Force Sensors and Sensitive Transcription Factors in Atherosclerosis*

5.3.2.1 Force Sensors in Atherosclerosis

Research has shown the presence of multiple force sensors in ECs. These include primary cilia, glycocalyx, caveolae, ion channels (such as Piezo1 and TRPV4), G protein-coupled receptor 68 (GPR68), and the recently discovered Plexin-D1. Here, we provide a brief overview of these mechanosensors.

Primary cilia are rod-shaped organelles that transmit mechanical and chemical stimuli from the extracellular environment into the cell. Studies in mice modified with the IFT88 gene (lacking cilia) have confirmed that primary cilia protect endothelial function and suppress the development of AS.

The endothelial glycocalyx is believed to attenuate the impact of shear stress on endothelial cells. Its disruption leads to an impaired endothelial response to shear stress, increased inflammation, and accelerated AS development.

The heterotrimeric G protein Gαq/11 binds to platelet endothelial cell adhesion molecule-1 (PECAM-1) on ECs, and this binding rapidly dissociates under shear stress. The interaction of Gαq/11 with PECAM-1 is also linked to the presence of glycocalyx. G protein-coupled receptors (GPCRs) are also mechanosensitive molecules.

TRPV4 regulates vasodilation by modulating nitric oxide production in ECs in response to shear stress. TRPV4 is activated by arachidonic acid (ARA), which is synthesized in response to shear stress by phospholipase A2 (PLA2).

Caveolae structures are primarily found in cell types exposed to mechanical stress and play a crucial role in buffering the cell membrane against mechanical stress (Potje et al. 2021) transmission electron microscopy images show an increased density of caveolae on the luminal surface of endothelial cells subjected to shear stress. Furthermore, shear stress causes the translocation of the caveolae protein Cav1 to the upstream edge of ECs, consistent with the hypothesis that caveolae buffer the cell membrane against high mechanical stress. In AS models, Cav1 mediates endothelial inflammation leading to AS. Knockout of Cav1 disrupts the formation of stress fibers in ECs, and the activation of integrin β1 in response to shear stress in ECs depends on Cav1.

Members of the Tie family of receptor tyrosine kinases are also associated with shear stress sensing. Tie1 is expressed in ECs under disturbed flow (DF) conditions. The loss of Tie1 increases the activation of endothelial nitric oxide synthase and reduces endothelial inflammatory signaling, indicating that Tie1 is a key regulator in endothelial responses to disturbed shear stress.

Piezo1 is a force-sensitive nonselective cation channel located on the endothelial cell membrane. It can be activated by shear stress, leading to the activation of endothelial nitric oxide synthase and nitric oxide release in laminar flow areas, as well as endothelial inflammation and AS through the NF-κB pathway in turbulent flow regions (Blythe et al. 2010; Douguet et al. 2019).

Plexin D1 (PLXND1) can directly sense shear stress and modulate site-specific distribution of vascular function and AS. On ECs, PLXND1 forms a complex with neuropilin-1 and VEGFR2 to respond to blood flow (Mehta et al. 2020).

5.3.2.2 Mechanosensitive Transcription Factors in Atherosclerosis

In the development of AS, the mechanical stimulation exerted by blood flow on vascular wall cells plays a pivotal role. Shear stresses, both pro-atherogenic (oscillatory or turbulent, nonunidirectional) and anti-atherogenic (stable or pulsatile, unidirectional laminar flow), stimulate the production of reactive oxygen species (ROS) and reactive nitrogen species (RNS), which participate in signal transduction for gene expression (Warabi et al. 2007). Nonunidirectional shear stress induces pro-atherosclerotic genes encoding adhesion molecules and chemokines through a mechanism reliant on superoxide and nitric oxide production. Stable or pulsatile laminar shear stress induces the expression of genes encoding enzymes for glutathione biosynthesis and detoxification, conferring cellular protection. Transcription factors sensitive to mechanical stimuli play crucial roles in the prevention, development, and treatment of AS, irrespective of the mechanical environment stimulus. We review and summarize mechanosensitive transcription factors and their inhibitors/activators in AS, shedding light on the mechanobiological mechanisms governing the regulation of AS development.

Current research, based on previous studies, acknowledges that under disturbed flow conditions, the promotion of AS (primarily pro-inflammatory) can be enhanced by mechanosensitive transcription factors. Examples include YAP/TAZ, HIF1α, NF-kB, AP1, and TLR4 (Chang et al. 2018; Mehta et al. 2021; Meli et al. 2020; Walther et al. 2021). Conversely, under laminar flow conditions, the inhibition of AS development can be achieved by activating mechanosensitive transcription factors associated with stabilizing vascular endothelial cell function and suppressing oxidative stress and inflammatory responses (Petzold et al. 2009). These include TFEB, KLF2/KLF4, and Nrf2 (Table 5.1).

With the widespread application of systems biology techniques such as RNA sequencing, single-cell RNA sequencing, proteomics, lipidomics, and metabolomics, an increasing number of mechanosensitive factors have been identified. The future directions in this field include: ① Utilizing systems biology technologies to identify novel force sensors that regulate fluid-mediated physiological and pathological responses. ② Determining the precise roles of identified mechanosensitive factors in atherosclerotic cardiovascular diseases. ③ Identifying new physiological ligands (e.g., signaling ligands) that bind to and regulate force sensors, thereby modulating endothelial function. ④ Conducting high-throughput drug screening using known mechanosensitive factors or modulators to identify new compounds (flow-simulating drugs). ⑤ Investigating the pharmacological, toxicological, and efficacy profiles of these flow-simulating compounds in AS through systematic studies in preclinical animal models and human patients. To address these and other questions, interdisciplinary research involving fields such as cell physiology, biomedical engineering, pharmacology, and functional genomics must be promoted. This interdisciplinary approach will facilitate the exploration of translational prospects in mechanobiology and the development of relevant pharmaceuticals.

5.3.3 *Synergistic Effects of Hemodynamic Factors and Systemic Risk in Atherosclerosis Formation*

In individuals with conditions such as hyperlipidemia, hypertension, and diabetes, a higher incidence of cardiovascular diseases has been attributed not only to factors closely related to individual age, gender, and lifestyle, such as physical inactivity, smoking, and obesity but also to their combined effects. It is noteworthy that the independent contributions of systemic risk factors or hemodynamic factors to AS have been relatively well established. However, our understanding of how these systemic risk factors interact synergistically to collectively promote the development of AS remains limited.

Table 5.1 Activity regulation of force-sensitive transcription factors

Force sensitive transcription factor	Laminar flow (UF)	Disturbed flow (DF)	Inhibitor/activator
Yap/Taz	Suppressed	Activated; promotes inflammation	InhiBitors: verteporfin; SWI/SNF; TIAM1; VGLL4 (Chang et al. 2018)
HIF-1α	Suppressed	Activated; increases vascular permeability; recruits immune cells; promotes endothelial cell proliferation	Inhibitors: Benzofuran-1,2,3-triazole; BIX01294; Cardenolide; Glycerol; MPT0G157; Etherfungin analogs
NF-κB	Suppressed	Activated; recruits epilepsy cells, promotes inflammation, induces enhanced HIF-1α transcription, and is mediated by TLR-4	Inhibitors: Erythromycin; BMS-06; Fluoxetine; Flavonoids; Quercetin
AP-1	Suppressed, protects endothelial cells	Activated; impaires blood vessel cell function	Inhibitors: JNK-Specific Inhibitor SP600125; Protein Kinase C (PKC); PNRI-299
TFEB	Activated, inhibits NF-κB activation	Suppressed, endothelial cell dysfunction	Activators: PKC; Mammalian Target of Rapamycin (mTOR); extracellular signal-regulated Kinase 2 (ERK2); Curcumin; Trehalose
KLF2/KLF4	Activated, inhibit pro-inflammatory transcription factors NF-κB and AP-1	Suppressed	Activators: HMG-CoA reductase inhibitor; resveratrol; histone deacetylase inhibitor (SAHA); Tannic acid (TA); Endothelial nitric oxide synthase transcription activator ERK5
NRF2	Activated, inhibits oxidative stress	Downregulated, promotes oxidative stress	Activators: resveratrol; KKPA4026; TBHQ; Keap1

5.3.3.1 Hyperlipidemia

Hyperlipidemia stands as a primary clinical risk factor representative of AS and cardiovascular diseases. In the progression of AS, LDL infiltrates the arterial wall. The permeation and accumulation of LDL within the endothelial layer of arterial vessels constitute one of the earliest and most prominent events in AS development. Systemic hyperlipidemia enhances endothelial permeability to smaller particles, such as through LDL receptor-mediated endocytic activity, changes in interendothelial cell gap junctions, or basement membrane remodeling, thereby facilitating the accumulation of LDL beneath the endothelium. Lipids retained on the arterial wall further activate monocytes and macrophages, mediating an inflammatory response. Simultaneously, oxidative stress collaborates in AS-prone regions, promoting the oxidation modification of LDL into ox-LDL. Oxidized LDL is more likely to induce endothelial dysfunction and AS-related inflammation. Given that vascular angiogenesis is a crucial factor in promoting the transformation of stable plaques into vulnerable ones and that lipoprotein accumulation plays a significant role in AS plaque development, it is essential to elucidate the role of LDL, particularly ox-LDL, in regulating vascular neogenesis-mediated plaque development.

Oxidized low-density lipoprotein (ox-LDL) can inhibit VEC proliferation and accelerate VEC apoptosis, thus explaining the apparent endothelial dysfunction observed in the formation of AS plaques. Based on these results, it is evident that ox-LDL exerts highly detrimental effects on endothelial cell survival, proliferation, and migration, all crucial processes in vascular neogenesis. As mentioned above, there exists a negative correlation between ox-LDL and angiogenesis (Yu et al. 2011). Therefore, ox-LDL is detrimental to the expansion of neovessels within plaques with high ox-LDL concentrations, affecting the relationship between lipid pools and characteristic budding angiogenesis within vulnerable plaques clinically characterized by intraplaque hemorrhage. Our group had demonstrated that low doses (less than 20 μg/mL) of ox-LDL can promote angiogenesis (Qiu et al. 2011). Differences in research results may be attributed to ox-LDL concentration. Earlier reports often used high doses (20–500 μg/mL), which greatly exceed physiological concentrations found in normal human serum. Nevertheless, from existing reports, it is equally challenging to precisely associate clinically relevant intraplaque neovascularization with lipid accumulation under pathological conditions. The challenge primarily lies in the fact that while circulating LDL concentrations can be accurately measured and defined, the exact ox-LDL concentration within atherosclerotic plaques remains poorly understood. Hence, it is necessary to determine the close relationship between lipid content within specific plaques and angiogenesis. Moreover, our research indicate that atherosclerotic plaques locally experience blood flow disturbances and hypercholesterolemia. This suggests a positive correlation between biomechanical factors and hyperlipidemia in AS. Low shear stress at AS plaque sites has been proven to promote LDL absorption and synthesis, increasing the endothelial layer's permeability to LDL. This confirms a significant interaction between mechanical factors and LDL in exacerbating AS. This is primarily due to shear stress's ability to instantly activate sterol regulatory element-binding proteins (SREBPs), critical transcription factors regulating multigene expression involved in cholesterol and fatty acid synthesis, as well as receptor-mediated LDL absorption. Conversely, low shear stress induces delayed SREBP activation. Thus, in the presence of hyperlipidemia, shear stress-mediated SREBPs promote LDL absorption and synthesis, continuously promoting LDL accumulation. This occurs primarily at the sites of specific AS lesions in regions of branching and arterial curvature due to blood flow disturbances.

Furthermore, disturbed flow significantly increases the permeability of LDL through VECs, thereby promoting the penetration and retention of LDL in regions vulnerable to AS and characterized by complex flow patterns. This shapes a mechanism dependent on SREBPs based on LDL absorption and synthesis. Shear stress in AS can accelerate the turnover of endothelial cells, specifically reducing intercellular connections between endothelial cells during processes such as mitosis and apoptosis. This accelerates the permeation of LDL beneath the endothelial layer and the development of AS (Cancel et al. 2010, 2011). Electron microscopy analysis of AS lesion sites or immunohistochemistry of single-layer endothelial cell junction proteins confirms that the increased distance between endothelial cell junctions is associated with the accelerated turnover of endothelial cells. These activities primarily occur in regions characterized by low shear stress and oscillatory shear stress. In addition to participating in the regulation of LDL permeability through cell turnover, shear stress may also be involved in other transport pathways (as shown in Fig. 5.17) that mediate the accumulation of LDL on the arterial wall. These pathways include vesicular endocytosis, facilitated by tight junction proteins such as occludin and ZO-1, or adhesive connections such as VE-cadherin and α-catenin, β-catenin, and matrix remodeling mediated by p21-activated kinase (Tarbell et al. 2010).

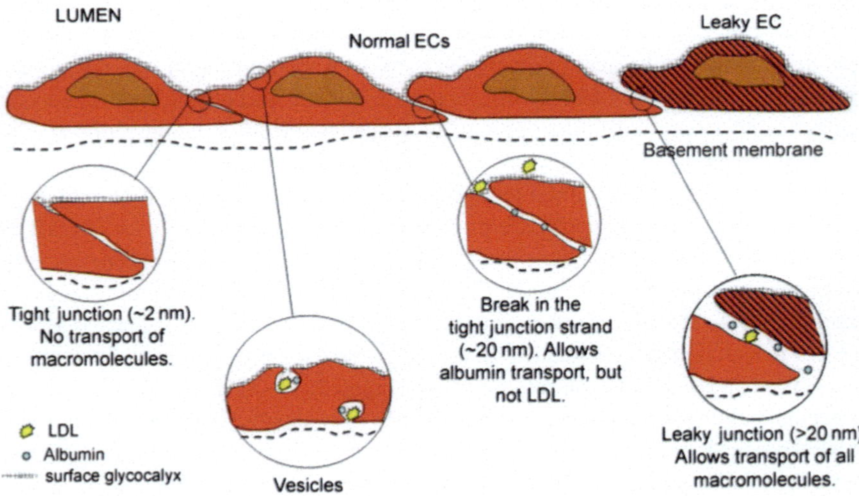

Fig. 5.17 Transport pathways across the endothelium. The major transport pathways are: the tight junctions, breaks in the tight junctions, vesicles, and leaky junctions. The surface glycocalyx covers the entrance to all but the leaky junctions. Cited from Tarbell (2010)

5.3.3.2 Hypertension

Blood pressure is a recognized integral component in the development of AS, although there are exceptions. In general, mouse models with induced hypertension accelerate AS progression. Hypertension is a multifactorial pathology involving various processes in AS, including endothelial dysfunction, vasoconstriction, vascular remodeling, thrombosis, and potentially increased vascular resistance due to inflammatory responses. In-depth studies on the specific role of elevated systolic pressure in AS have led to the conclusion that blood pressure per se is not the decisive factor in promoting AS. Instead, several processes involved in the body's response to hypertension seem to have a more significant impact on AS. These processes in AS are fundamentally similar and driven by mechanical forces. Shear stress is closely associated with endothelial function regulation, vasodilation, vascular remodeling, thrombosis, and inflammation. Shear stress, which causing AS, leads to endothelial dysfunction and long-term inhibition of vasodilator release, such as NO and prostacyclin, while promoting the production of vasoconstrictors, such as ET-1. Most of these factors not only result in increased vascular tone but also foster vascular remodeling and neointima formation. Additionally, disturbed flow promotes platelet aggregation near the vascular wall or downregulates the production of antithrombotic agents (eNOS, prostacyclin), thereby facilitating thrombosis and thus the progression of hypertension. Furthermore, flow-mediated inflammatory cell infiltration might also play a role, but its exact mechanisms remain unknown. All these observations suggest that shear stress and hypertension act synergistically in AS through by regulating endothelial function, vascular tone, thrombosis, and possible inflammation.

Essentially, AS occurs in areas of arterial hypertension, whereas there are no AS lesions in lower-pressure venous regions, and blood pressure decreases with reduced vessel diameter. This indicates a clear correlation between mechanical disturbances within the biomechanical system and hypertension in AS. Essentially, mechanical disturbances (hypertension and lack of physical exercise) combined with chemical disturbances (hyperlipidemia, diabetes, and smoking) alter vascular stiffness, thereby disrupting the vessel's response to normal or hypertensive loading (Hsu et al. 2019).

5.3.3.3 Diabetes

Diabetes is a complex disease closely associated with hyperglycemia or high glucose levels, and it is a major risk factor contributing to AS. Although the interactions between biomechanics and hyperglycemic factors in influencing AS remain poorly understood, recent hypotheses concerning endothelial function regulation, inflammation, and oxidative stress have been proposed to potentially explain their synergistic effects. Pathological shear stress reduces the utilization of NO, thereby promoting inflammation and oxidative stress reactions. In the context of high blood glucose levels, advanced glycation end-products (AGEs) are generated, which can sequentially induce inflammatory responses, oxidative stress reactions, and endothelial dysfunction through receptor-mediated pathways. In fact, the interaction between AGEs and the receptor for advanced glycation end-products (RAGE) has been one of the most extensively studied regulators in the process of AS induced by diabetes. The synthesis of advanced glycation end-products directly interacts with blood glucose levels, at least in part, through the RAGE signaling pathway, which mediates inflammation and oxidative stress reactions. Specifically, low shear stress can activate RAGE and promote RAGE-dependent inflammatory responses, while high shear stress diminishes these effects. Low shear stress protects against high glucose-induced inflammatory activity, while high shear stress exerts inhibitory effects, consistent with previous findings. Additionally, laminar shear stress counteracts the actions of high glucose, thereby inhibiting oxidative stress reactions and lipid peroxidation. In summary, these findings suggest that hyperglycemia, in synergy with shear stress leading to AS, collectively regulates inflammation, oxidative stress responses, and other processes contributing to AS, thereby accelerating AS lesions at specific sites in the disturbed flow regions of the arterial tree.

References

Blythe N M, Muraki K, Ludlow M J, Stylianidis V, Gilbert H, Evans EL. Mechanically activated Piezo1 channels of cardiac fibroblasts stimulate p38 mitogen-activated protein kinase activity and interleukin-6 secretion[J]. Journal of Biological Chemistry. 2010, 294: 17395–408.

Cancel LM, Tarbell JM. The role of apoptosis in LDL transport through cultured endothelial cell monolayers[J]. Atherosclerosis, 2010, 208(2): 335–341.

Cancel LM, Tarbell JM. The role of mitosis in LDL transport through cultured endothelial cell monolayers[J]. Am J Physiol Heart Circ Physiol, 2011, 300(3): H769–H776.

Cecchi E, Giglioli C, Valente S, Lazzeri C, Gensini GF, Abbate R, Mannini L. Role of hemodynamic shear stress in cardiovascular disease[J]. Atherosclerosis. 2011, 214(2): 249–56.

Chang L, Azzolin L, Di Biagio D, Zanconato F, Battilana G, Lucon Xiccato R, Aragona M, Giulitti S, Panciera T, Gandin A, Sigismondo G, Krijgsveld J, Fassan M, Brusatin G, Cordenonsi M, Piccolo S. The SWI/SNF complex is a mechanoregulated inhibitor of YAP and TAZ[J]. Nature. 2018, 563(7730): 265–269.

Chatzizisis YS, Coskun AU, Jonas M, Edelman ER, Feldman CL, Stone PH. Role of endothelial shear stress in the natural history of coronary atherosclerosis and vascular remodeling: molecular, cellular, and vascular behavior[J]. J Am Coll Cardiol. 2007, 49(25): 2379–93.

Chatzizisis YS, Jonas M, Coskun AU, Beigel R, Stone BV, Maynard C, Gerrity RG, Daley W, Rogers C, Edelman ER, Feldman CL, Stone PH. Prediction of the localization of high-risk coronary atherosclerotic plaques on the basis of low endothelial shear stress: an intravascular ultrasound and histopathology natural history study[J]. Circulation. 2008, 117(8): 993–1002.

Chatzizisis YS, Giannoglou GD. Shear stress and inflammation: are we getting closer to the prediction of vulnerable plaque[J]. Expert Rev Cardiovasc Ther, 2010, 8(10): 1351–1353.

Chiu JJ, Chien S. Effects of disturbed flow on vascular endothelium: pathophysiological basis and clinical perspectives[J]. Physiol Rev, 2011, 91(1): 327–387.

Clarke MC, Figg N, Maguire JJ. Apoptosis of vascular smooth muscle cells induces features of plaque vulnerability in atherosclerosis[J]. Nat Med, 2006, 12(9):1075–1080.

Cunningham KS, Gotlieb A1. The role of shear stress in the pathogenesis of atherosclerosis[J]. Lab Invest, 2005, 85(1): 9–23.

Davies PF, Civelek M, Fang Y, Fleming I. The atherosusceptible endothelium: endothelial phenotypes in complex hemodynamic shear stress regions in vivo[J]. Cardiovasc Res. 2013, 99(2): 315–327.

Davies J E, Lopresto D, Apta B, Lin Z, Ma W, Harper M T. Using Yoda-1 to mimic laminar flow in vitro: a tool to simplify drug testing[J]. Biochemical Pharmacology. 2019, 168: 473–480.

Douguet D, Patel A, Xu A, Vanhoutte P M, Honore E. Piezo ion channels in cardiovascular Mechanogy[J]. Trends in Pharmacological Sciences, 2019, 40: 956–970.

Fang C, Schmaier AH. Novel anti-thrombotic mechanisms mediated by mas receptor as result of balanced activities between the kallikrein/kinin and the renin-angiotensin systems[J]. Pharmacol Res. 2020, 160: 105096.

Fisslthaler B, Boengler K, Fleming I, Schaper W, Busse R, Deindl E. Identification of a cis-element regulating transcriptional activity in response to fluid shear stress in bovine aortic endothelial cells[J]. Endothelium. 2003, 10(4–5): 267–275.

Fukumoto Y, Hiro T, Fujii T, Hashimoto G, Fujimura T, Yamada J, Okamura T, Matsuzaki M. Localized elevation of shear stress is related to coronary plaque rupture: a 3-dimensional intravascular ultrasound study with in vivo color mapping of shear stress distribution[J]. J Am Coll Cardiol. 2008, 51(6): 645–550.

Gijsen FJ, Wentzel JJ, Thury A, Lamers B, Schuurbiers JC, Serruys PW, van der Steen AF. A new imaging technique to study 3-D plaque and shear stress distribution in human coronary artery bifurcations in vivo[J]. J Biomech. 2007, 40(11): 2349–2357.

Herrmann J, Lerman L, Lerman A. Simply say yes to no? Nitric oxide (NO) sensor-based assessment of coronary endothelial function[J]. Eur Heart J, 2010, 31(23): 2834–2836.

Hsu P L, Chen J S, Wang C Y, Wu H L, Mo F E. Shear-induced CCN1 promotes atheroprone endothelial phenotypes and atherosclerosis[J]. Circulation, 2019, 139(25): 2877–2891.

Huang J, Pu Y, Zhang H, Xie L, He L, Zhang CL, Cheng CK, Huo Y, Wan S, Chen S, Huang Y, Lau CW, Wang L, Xia Y, Huang Y, Luo JY. KLF2 mediates the suppressive effect of laminar flow on vascular calcification by Inhibiting endothelial BMP/SMAD1/5 signaling[J]. Circ Res. 2021, 129(4):e87–e100.

Koskinas KC, Chatzizisis YS, Papafaklis MI, Coskun AU, Baker AB, Jarolim P, Antoniadis A, Edelman ER, Stone PH, Feldman CL. Synergistic effect of local endothelial shear stress and systemic hypercholesterolemia on coronary atherosclerotic plaque progression and composition in pigs[J]. Int J Cardiol. 2013, 169(6): 394–401.

Ling B. The high shear stress and oxLDL by VEGF-VEGFR2 signaling pathway induced angiogenesis involved in rupture-prone plaque formation. Chongqing: PhD thesis of Chongqing University, 2015.

Ling B, et al. Shear stress and oxidized LDL regulates endothelial cell tube formation through VEGF signaling. MCB, 14 (4); 2017: 197–211.

Liu S, Jiang X, Cui X, Wang J, Liu S, Li H, Yang J, Zhang C, Zhang W. Smooth muscle-specific hur knockout induces defective autophagy and atherosclerosis[J]. Cell Death Dis. 2021, 12(4): 385.

Mehta V, Pang K L, Rozbesky D, Nather K, Keen A, Lachowski D. The guidance receptor plexin D1 is a mechanosensor in endothelial cells[J]. Nature, 2020, 578: 290–295.

Mehta V, Pang KL, Givens CS, Chen Z, Huang J, Sweet DT, Jo H, Reader JS, Tzima E. Mechanical forces regulate endothelial-to-mesenchymal transition and atherosclerosis via an Alk5-Shc mechanotransduction pathway[J]. Sci Adv. 2021, 7(28): eabg5060.

Meli VS, Atcha H, Veerasubramanian PK, Nagalla RR, Luu TU, Chen EY, Guerrero-Juarez CF, Yamaga K, Pandori W, Hsieh JY, Downing TL, Fruman DA, Lodoen MB, Plikus MV, Wang W, Liu WF. YAP-mediated mechanotransduction tunes the macrophage inflammatory response[J]. Sci Adv. 2020, 6(49): eabb8471.

Parmar KM, Larman HB, Dai G, Zhang Y, Wang ET, Moorthy SN, Kratz JR, Lin Z, Jain MK, Gimbrone MA Jr, García-Cardeña G. Integration of flow-dependent endothelial phenotypes by Kruppel-like factor 2[J]. J Clin Invest. 2006, 116(1): 49–58.

Petzold T, Orr AW, Hahn C, Jhaveri KA, Parsons JT, Schwartz MA. Focal adhesion kinase modulates activation of NF-kappaB by flow in endothelial cells[J]. Am J Physiol Cell Physiol. 2009, 297(4): C814–C822.

Potje SR, Paula TD, Paulo M, Bendhack LM. The Role of Glycocalyx and Caveolae in Vascular Homeostasis and Diseases[J]. Front Physiol. 2021, 11:620840.

Qiu JH, Wang GX, Lei DX. Hemodynamics and atherosclerotic plaque stability and its mechanism[J]. Chin J Arterioscler, 2009, 17(06): 495–497.

Qiu JH. The mechanobiological mechanisms of Id1-induced angiogenesis in rupture-prone plaque formation [D]. Chongqing University, 2011.

Qiu J, Wang G, Zheng Y, Hu J, Peng Q, Yin T. Coordination of Id1 and p53 activation by oxidized LDL regulates endothelial cell proliferation and migration[J]. Ann Biomed Eng. 2011, 39(12): 2869–2878.

Qiu J, Peng Q, Zheng Y, Hu J, Luo X, Teng Y, Jiang T, Yin T, Tang C, Wang G. OxLDL stimulates Id1 nucleocytoplasmic shuttling in endothelial cell angiogenesis via PI3K pathway[J]. Biochim Biophys Acta. 2012, 1821(10): 1361–1369.

Silva RF, Chambaz C, Stergiopulos N, Hayoz D, Silacci P. Transcriptional and posttranscriptional regulation of preproendothelin-1 by plaque-prone hemodynamics[J]. Atherosclerosis. 2007, 194(2): 383–390.

Souilhol C, Serbanovic-Canic J, Fragiadaki M, Chico TJ, Ridger V, Roddie H, Evans PC. Endothelial responses to shear stress in atherosclerosis: a novel role for developmental genes[J]. Nat Rev Cardiol. 2020, 17(1): 52–63.

Tarbell JM. Shear stress and the endothelial transport barrier[J]. Cardiovasc Res, 2010, 87(2): 320–330.

Verhamme P, Yi BA, Segers A, Salter J, Bloomfield D, Büller HR, Raskob GE, Weitz JI, ANT-005 TKA Investigators. Abelacimab for prevention of venous thromboembolism[J]. N Engl J Med. 2021, 385(7):609–617.

Walther BK, Rajeeva Pandian NK, Gold KA, Kiliç ES, Sama V, Gu J, Gaharwar AK, Guiseppi-Elie A, Cooke JP, Jain A. Mechanotransduction-on-chip: vessel-chip model of endothelial YAP Mechanogy reveals matrix stiffness impedes shear response[J]. Lab Chip. 2021, 21(9): 1738–1751.

Wang GX, Cai SX, Wang PQ, Ouyang KQ, Wang YL, Xu SR. Shear-induced changes in endothelin-1 secretion of microvascular endothelial cells[J]. Microvasc Res. 2002, 63(2): 209–217.

Wang GX. Changes in shear stress and atherosclerotic plaque formation and rupture[J]. Chin J Arterioscler, 2009 ,17(8): 625–628.

Wang W, Ha CH, Jhun BS, Wong C, Jain MK, Jin ZG. Fluid shear stress stimulates phosphorylation-dependent nuclear export of HDAC5 and mediates expression of KLF2 and eNOS[J]. Blood. 2010, 115(14): 2971–2979.

Warabi E, Takabe W, Minami T, Inoue K, Itoh K, Yamamoto M, Ishii T, Kodama T, Noguchi N. Shear stress stabilizes NF-E2-related factor 2 and induces antioxidant genes in endothelial cells: role of reactive oxygen/nitrogen species[J]. Free Radic Biol Med. 2007, 42(2): 260–269.

Wei DH, Wang GX, Wang Z. The effect of feature of wall shear stress on intimal hyperplasia and atherogenesis[J]. Chin J Arterioscler, 2007, 15(6): 410–414.

Xu S, Xu Y, Liu P, Zhang S, Liu H, Slavin S, Kumar S, Koroleva M, Luo J, Wu X, Rahman A, Pelisek J, Jo H, Si S, Miller CL, Jin ZG. The novel coronary artery disease risk gene JCAD/KIAA1462 promotes endothelial dysfunction and atherosclerosis[J]. Eur Heart J. 2019, 40(29): 2398–2408.

Yoshizumi M, Abe J, Tsuchiya K, Berk BC, Tamaki T. Stress and vascular responses: atheroprotective effect of laminar fluid shear stress in endothelial cells: possible role of mitogen-activated protein kinases[J]. J Pharmacol Sci. 2003, 91(3): 172–176.

Yu S, Wong SL, Lau CW, Huang Y, Yu CM. Oxidized LDL at low concentration promotes in vitro angiogenesis and activates nitric oxide synthase through PI3K/Akt/eNOS pathway in human coronary artery endothelial cells[J]. Biochem Biophys Res Commun. 2011, 407(1): 44–48.

Zhou T, Zheng Y, Qiu J, Wang G. Endothelial mechanotransduction mechanisms for vascular physiology and atherosclerosis[J]. J Mech Med Biol, 2014, 14(5): 1–31.

Chapter 6
Mechanobiological Mechanism by Which Id1 Regulates Lipid Absorption in Atherosclerotic Plaque Formation

Atherosclerosis (AS) refers to a disease with lipid deposits in the intima and subintima of arteries and their branches, accompanied by the migration and proliferation of smooth muscle cells in the middle layer to the subintima, resulting in intimal thickening and the formation of atherosclerotic or fibrolipid plaque lesions. AS poses a serious threat to human health and is the main cause of a variety of cardiovascular diseases (Cohn et al. 2006; Verwoert et al. 2014). Many studies and clinical results show that atherosclerotic plaques tend to occur in low oscillatory shear stress areas with blood flow disturbances such as vessel bifurcation and bending, indicating that mechanics are an important influencing factor for the formation of AS. At the same time, some studies have shown that the formation of AS is also closely related to lipoproteins (Qiu et al. 2011). Therefore, further exploration of the relationship between low oscillatory shear stress, lipid absorption and the occurrence of AS will be of great significance to further understand the occurrence and development mechanism of AS and put forward new treatment measures.

DNA binding inhibitors, also known as inhibitors of differentiation (Id), belong to the family of helix-loop-helix transcription factors. The Id protein family includes Id1, Id2, Id3 and Id4. Id1 is the most widely studied protein in this family and is involved in the whole process of embryonic cardiovascular and cerebrovascular system development (Ruzinova et al. 2003).

The Id1 protein has a variety of biological functions, plays an especially important role in angiogenesis, and has been a hot spot in cancer angiogenesis research (Ciarrocchi et al. 2007). Id1 can participate in tumor angiogenesis by inducing cell proliferation and migration and accelerating the cell cycle (Maw et al. 2009). Because tumors and atherosclerosis have many similarities, the role of Id1 in AS has received much attention in recent years.

Previous in vitro experiments in our laboratory showed that shear stress can regulate the expression of Id1 (Qiu et al. 2011). Wwith increasing shear stress, the expression of Id1 also gradually increased. Ni et al. (2010) constructed a low oscillatory shear stress (OSS) model by ligating the carotid artery of mice. Gene microarray analysis revealed that OSS inhibited the expression of the Id1 gene in vascular endothelial

G. Wang, *The Mechanobiology in Atherosclerosis*,
https://doi.org/10.1007/978-981-97-6512-6_6

cells. However, subsequent studies on the effects of OSS regulation of Id1 expression on endothelial function remain unexplored. In addition, studies have found that Id1 is highly expressed in adipose tissue, suggesting that Id1 may participate in the process of lipid metabolism, but the specific mechanism is still unclear (Farmer et al. 2006; Rosen et al. 2000; Satyanarayana et al. 2012; Seale et al. 2009). Based on these findings, it is presumed that OSS may promote atherosclerotic plaque formation by regulating lipid metabolism, and Id1 plays an important role in this process (Patil et al. 2014).

6.1 Proteomic Analysis of Low Oscillatory Shear Stress Model Blood Vessels

Proteomics is a comprehensive set of proteins from specific tissues in a specific environment as the research object. It encompasses protein analysis, identification, expression level, posttranslational modification, protein function, protein interaction, and more. This technology aims to identify the impact of diseases and drugs on life processes, as well as to explain the mechanism of gene expression regulation. In this section, through the $ApoE^{-/-}$ mouse carotid artery low oscillatory shear stress model (Fig. 6.1), proteome technology is applied to study the expression differences of vascular proteins under different shear stresses, and GO analysis and KEGG signaling pathway analysis methods are used to analyze the signaling pathways that the differentially expressed proteins may participate in and the biological effects caused, which will contribute to a deeper understanding of how hemodynamics affect the occurrence and development of atherosclerosis.

After the $ApoE^{-/-}$ mice were treated with left carotid artery ligation and fed for 48 h, the blood flow velocity of the left common carotid artery (LCA) and the right common carotid artery (RCA) were detected by small animal ultrasound (Visualsonics, Canada). As shown in Fig. 6.2, there are two blood flows in opposite directions in the left carotid artery; that is, low oscillatory flow (LCA1 and LCA2) is formed. At this time, the shear stress in the vessel is reciprocating low oscillatory shear stress, while the right carotid artery is laminar flow in a single direction, and the shear stress in the vessel is laminar flow shear stress. The blood flow velocity of the left carotid artery is lower than that of the right carotid artery. The statistical results

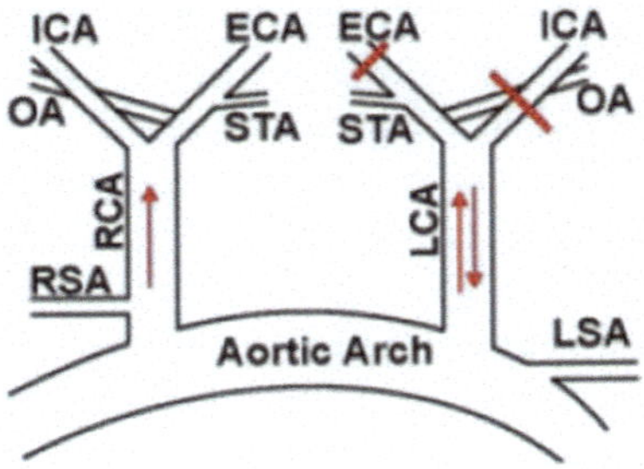

Fig. 6.1 The model of partial ligation. Cited from Zhang (2018)

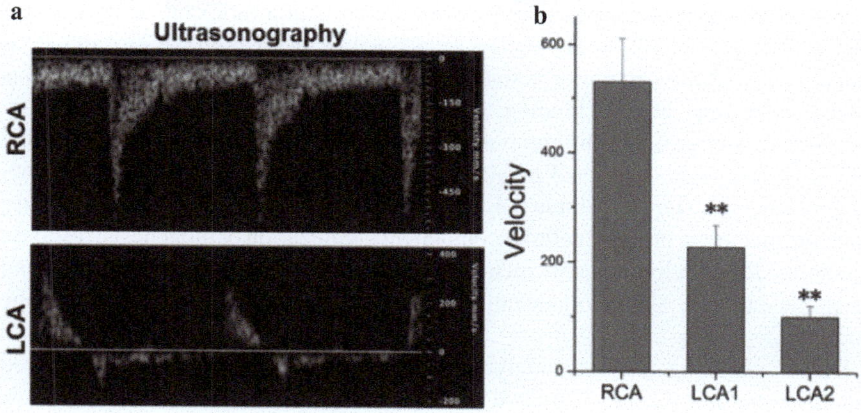

Fig. 6.2 The ligation model can be induced with OSS 48 h after partial ligation. **a** Ultrasonography was used to detect the right and left carotid artery blood flow velocities 48 h after building the model. **b** Quantitative analysis of the velocities of the RCA and LCA (n = 3). ** $p < 0.001$ versus the RCA controls. (The Mechanobiological mechanisms of Id1-regulated lipid uptake in atherosclerotic plaque formation. Chongqing: Ph.D. thesis of Chongqing university, 2018)

are shown in Fig. 6.2b. The results showed that the low oscillatory shock shear stress mouse model was successfully constructed.

Two mouse samples named LCA (OSS) and RCA (LSS) were identified by isotopic tags for relative and absolute quantification (iTRAQ). Differentially expressed proteins were defined as those with a 1.5-fold change (mean value of the comparison group) and a P value (t test of the comparison group) less than 0.05. The analysis revealed there were 168 differentially expressed proteins in LCA (OSS) compared with RCA (LSS), of which 18 proteins were downregulated and 150 proteins were upregulated. The results are shown in Table 6.1.

GO enrichment and KEGG pathway enrichment analyses were carried out on the differentially expressed proteins to identify the biological processes and related signaling pathways that may be involved. The differentially expressed proteins were annotated and enriched by GO function. Except for the proteins that were not annotated, other differentially expressed proteins were significantly enriched in 46 cellular components (corrected p value < 0.05). The six most enriched GO analysis groups are shown in Table 6.2: macromolecular complex, nonmembrane bound organelle Intracellular nonmembrane bound organelle, extracellular region, cytosol, and ribonucleoprotein complex. In the molecular function analysis, as shown in Table 6.3,

Table 6.1 Information on differentially expressed proteins

Compared group	Up-regulated	Down-regulated	All-regulated
LCA(OSS)-VS-RCA(LSS)	150	18	168

the top 10 GO terms of differential proteins are structural molecule activity, structural constitution of ribosome, nucleic acid binding, RNA binding, enzyme regulator activity, enzyme inhibitor activity, glycosaminoglycan binding, carbohydrate derivative binding, heparin binding, and sulfur compound binding. In the biological process analysis, the differential proteins were significantly enriched in 452 biological processes (corrected p-value < 0.05), and the 10 most enriched biological processes are shown in Table 6.4, including single organization process, macromolecule metabolic process, cellular component organization or biogenesis, response to stimuli, localization, protein metabolic process, transport, nitrogen compound metabolic process, single organization transport, and cellular protein metabolic process.

KEGG signaling pathway analysis revealed that the differentially expressed proteins were significantly enriched in 20 signaling pathways (corrected p-value < 0.05). The results are shown in Fig. 6.3, including complement and coagulation pathways, hematopoietic cell lineage, phagosome, fat digestion and absorption, and mineral absorption signaling pathways such as cell adhesion molecules, which are

Table 6.2 Differentially expressed proteins were most enriched in six cellular components (p value < 0.05)

Gene ontology term	Cluster frequency (%)	P value
Macromolecular complex	93 out of 156 genes, 59.6	5.097216×10^{-8}
Non-membrane-bounded organelle	77 out of 156 genes, 49.4	6.057551×10^{-7}
Intracellular non-membrane-bounded organelle	77 out of 156 genes, 49.4	6.057551×10^{-7}
Extracellular region	59 out of 156 genes, 37.8	1.421225×10^{-17}
Cytosol	54 out of 156 genes, 34.6	0.000121648
Ribonucleoprotein complex	46 out of 156 genes, 29.5	3.782956×10^{-18}

Table 6.3 Differentially expressed proteins were most enriched in ten molecular functions

Gene ontology term	Cluster frequency (%)	P value
Structural constituent of ribosome	42 out of 153 genes, 27.5	6.895091×10^{-37}
Structural molecule activity	53 out of 153 genes, 34.6	6.07364×10^{-27}
Heparin binding	16 out of 153 genes, 10.5	3.451969×10^{-10}
Enzyme inhibitor activity	22 out of 153 genes, 14.4	3.512298×10^{-10}
Glycosaminoglycan binding	17 out of 153 genes, 11.1	1.535196×10^{-9}
Carbohydrate derivative binding	17 out of 153 genes, 11.1	9.175853×10^{-9}
Sulfur compound binding	16 out of 153 genes, 10.5	6.468008×10^{-8}
Enzyme regulator activity	28 out of 153 genes, 18.3	3.004928×10^{-5}
Nucleic acid binding	41 out of 153 genes, 26.8	0.000876482
mRNA binding	8 out of 153 genes, 5.2	0.002123512

Table 6.4 Differentially expressed proteins were enriched in these biological process terms (p value < 0.05)

Gene ontology term	Cluster frequency (%)	P value
Single-organism process	122 out of 150 genes, 81.3	0.001326895
Macromolecule metabolic process	84 out of 150 genes, 56.0	6.350928×10^{-6}
Cellular component organization or biogenesis	84 out of 150 genes, 56.0	1.204832×10^{-9}
Response to stimulus	73 out of 150 genes, 48.7	0.0475322
Localization	71 out of 150 genes, 47.3	0.000344203
Protein metabolic process	68 out of 150 genes, 45.3	5.477249×10^{-6}
Transport	64 out of 150 genes, 42.7	7.91075×10^{-5}
Nitrogen compound metabolic process	61 out of 150 genes, 40.7	0.001366074
Single-organism transport	60 out of 150 genes, 40.0	6.444084×10^{-9}
Cellular protein metabolic process	57 out of 150 genes, 38.0	6.622942×10^{-5}

closely related to the occurrence and development of atherosclerosis, indicating that OSS may affect the formation of AS through these pathways.

Because the process of lipid metabolism is crucial for the formation of AS, we found that the expression of many factors related to lipid digestion and absorption changed after carotid vascular shear stress intervention in mice; for example, the expression of ApoA-I and ApoA-IV was upregulated, and the expression of CD36

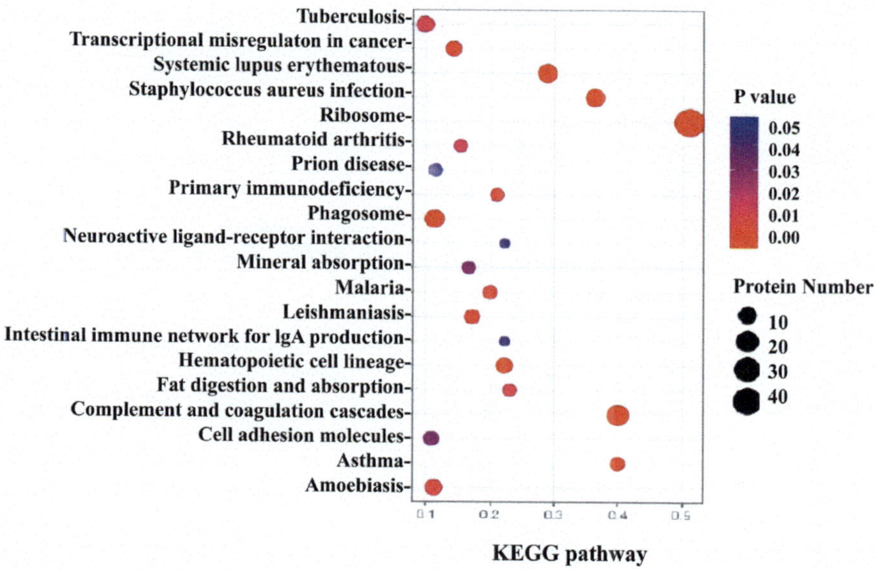

Fig. 6.3 Differentially expressed proteins were analyzed by KEGG pathway. Modified from Zhang (2018)

was downregulated. The results are shown in Fig. 6.4. Red is the upregulated protein, and green is the downregulated protein. In addition, we also found that low oscillatory shear stress can promote the expression of thrombin sensitive protein 1 (TSP1) (Fig. 6.5), which is consistent with the proteomic study of endothelial cells after mechanical loading in vitro by Tressel et al. (2007).

TSP1 is an extracellular matrix glycoprotein, and its expression can be detected in many normal and tumor cells, such as vascular endothelial cells and monocytes. Studies have shown that TSP1 is a downstream target of Id1 protein, and Id1 protein can inhibit the expression of TSP1 (Volpert et al. 2002). Our proteome identification results showed that OSS promoted the expression of TSP1 protein. KEGG signaling pathway analysis revealed that Id1 protein could inhibit the expression of TSP1 protein and participate in the RAP1 signaling pathway (Fig. 6.7). After microarray analysis of endothelial cell genes in LSS and blood vessels under OSS, Ni et al. (2010) found that OSS inhibited the expression of the Id1 gene in vascular endothelial cells. Through the combination of previous studies and our results, a possibility is drawn that OSS inhibits the expression of Id1 protein in vascular endothelial cells and then promotes its downstream protein expression.

The vascular wall is composed of endothelial cells, smooth muscle cells, fibroblasts and extracellular matrix. When shear stress acting on the vascular wall has

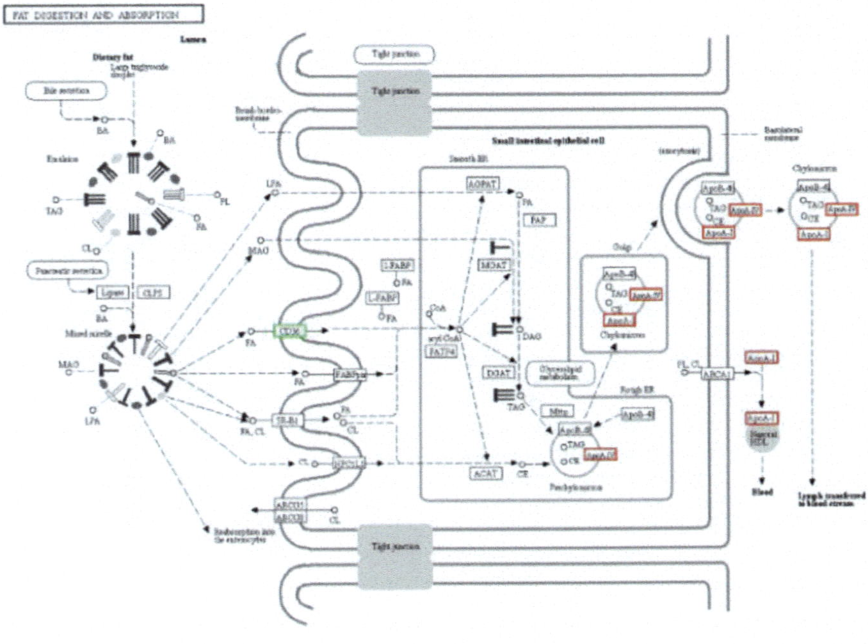

Fig. 6.4 Differentially expressed proteins involved in fat digestion and absorption. The red boxes are the upregulated proteins, and the green box is the downregulated protein. (The Mechanobiological mechanisms of Id1-regulated lipid uptake in atherosclerotic plaque formation. Chongqing: Ph.D. thesis of Chongqing university, 2018)

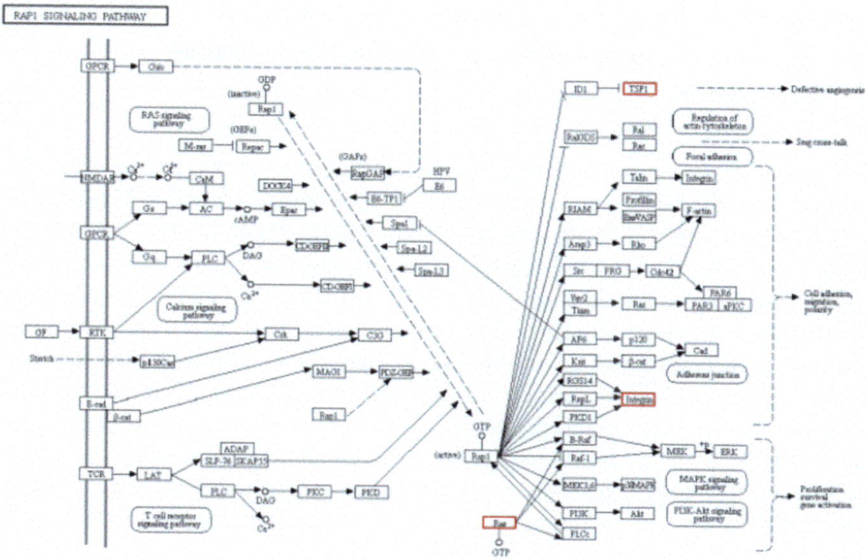

Fig. 6.5 Differentially expressed proteins involved in the RAP1 signaling pathway. The red boxes are the upregulated proteins, and the yellow box is the protein regulated by Id1. (The Mechanobiological mechanisms of Id1-regulated lipid uptake in atherosclerotic plaque formation. Chongqing: Ph.D. thesis of Chongqing university, 2018)

a dual impact. Firstly, endothelial cells can directly sense mechanical signals and transmit the signals into the cell to affect their function. Secondly, mechanical stimulation induces the release of growth factors by endothelial cells, thereby affecting smooth muscle cells and the extracellular matrix, leading to structural changes in the vascular wall (Li et al. 2005). Most proteome studies related to shear stress focus on the detection of in vitro cultured blood vessels and endothelial cells after mechanical loading. In this study, considering that blood vessels are a whole, mechanical changes will cause changes in various cells and the extracellular matrix of blood vessels; therefore, we extracted the proteins of the whole blood vessel for differential proteome analysis after changing the shear stress in the carotid artery of mice through the surgical arm, which can help us understand the biological effects caused by differentially expressed proteins in the whole blood vessel stimulated by OSS from a global perspective.

6.2 Low Oscillatory Shear Stress Causes Lipid Accumulation to Promote Atherosclerotic Plaque Formation

Intravascular lipid deposition is regulated by multiple factors. On the one hand, the atherosclerotic lipid concentration polarization hypothesis proposed by Karino et al. suggests that there is a concentration polarization phenomenon of lipids in the blood (Wada et al. 1999). Through zebrafish experiments and numerical simulation, it was confirmed that the wall lipid concentration will be higher than that in other laminar flow regions in the low turbulence flow region with a disordered blood flow field, which more easily promotes the infiltration and deposition of lipids in these regions. On the other hand, low oscillatory flow can affect the function of the endothelial layer, promote endothelial cell permeability, and aggravate lipid deposition (Wang et al. 2003; Wei et al. 2007). Moreover, after low oscillatory shear stress stimulates endothelial cells, it regulates the expression of lipid-related proteins such as cholesterol regulatory element binding protein-1 (SREBP1) in endothelial cells, which in turn accelerates the absorption of lipids by cells. However, the relevant mechanisms by which low oscillatory shear stress regulates lipid deposition and promotes plaque development are not fully elucidated at present.

Based on previous research results, the effects of OSS on lipid deposition and plaque formation were studied using a mouse low oscillatory shear stress model and an in vitro mechanical loading device.

The mice were divided into two groups: the low oscillatory shear stress group and the control group. In the low oscillatory shear stress group, the left carotid artery branch of the mice was ligated, while the contralateral right carotid artery was only treated with sham surgery. In the control group, both the left and right carotid arteries of mice were only treated with sham surgery. Two groups of mouse blood were taken for total cholesterol (TC), triglycerides (TG), LDL, and HDL content detection. The results are shown in Fig. 6.6. In the low oscillatory shear stress group and control group, the TG and HDL contents in the blood before and after surgery did not change significantly with time. The results of TC and LDL levels showed an upward trend in both the low oscillatory shear stress group and the control group after surgery, but this upward trend began to decrease again over time. There was basically no difference in TC and LDL between the 2nd, 4th, and 6th weeks before surgery, and there was no difference between the low oscillatory shear stress group and the control group. The time points with differences were mainly concentrated in the samples 24 h after surgery, and the trend of LDL changes was similar to that of TC. The slight difference is that the LDL content in the blood of the control group 24 h after surgery was significantly higher than that before surgery. We determined that this difference in TC and LDL may be due to the greater damage caused by ligation surgery compared to sham surgery.

Paraffin sections and HE staining were performed on the blood vessels in the low oscillatory shear stress area of the left carotid artery, the blood vessels in the laminar shear stress area of the contralateral right carotid artery, and the blood vessels

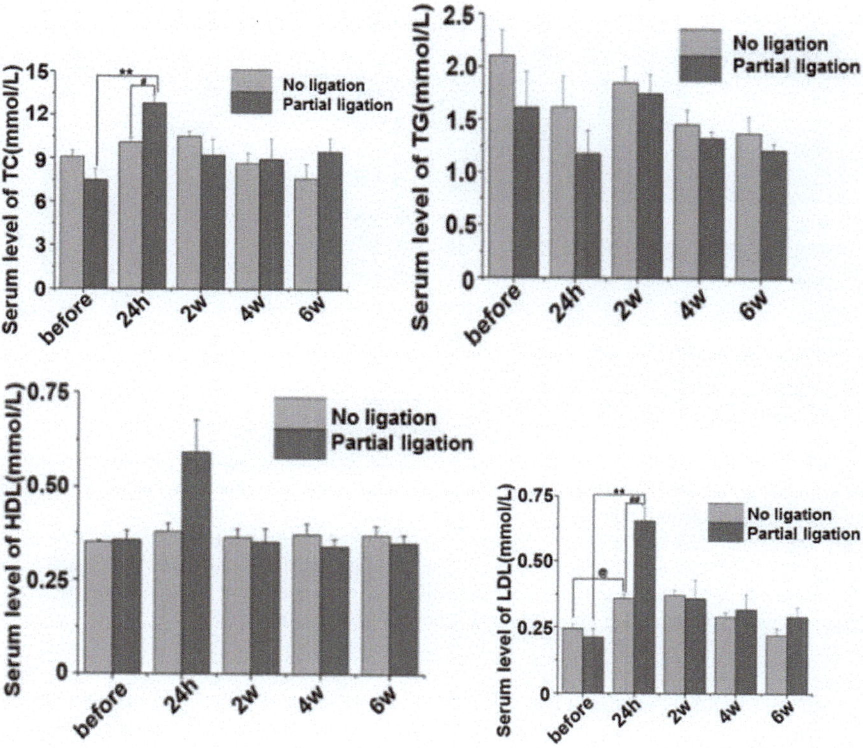

Fig. 6.6 Serum lipid profile analysis before and after surgery in the control groups and OSS groups (total cholesterol, TC; triglyceride, TG; low-density lipoprotein, LDL; high-density lipoprotein, HDL) (n = 5). ** $p < 0.001$ versus the OSS groups before surgery, $^{\#}$ $p < 0.05$ and $^{\#\#}$ $p < 0.001$ versus the control groups 24 h after surgery, $^{@}$ $p<0.05$ versus the control groups before surgery. No ligation was performed in the control groups. Partial ligation was performed in the OSS groups. (The Mechanobiological mechanisms of Id1-regulated lipid uptake in atherosclerotic plaque formation. Chongqing: Ph.D. thesis of Chongqing university, 2018)

in the left and right common carotid arteries of the control group. It was found that the intima of the left common carotid artery in the low oscillatory shear stress group began to thicken from the second week, and as the ligation time increased, the intima hyperplasia became increasingly obvious, and cholesterol crystals and lipid necrotic nuclei appeared. By the 10th week, the plaque almost blocked the entire blood vessel. However, there was no significant intimal hyperplasia or plaque formation in the vascular lumen of the contralateral right carotid artery or the control group (Fig. 6.7a, b), indicating that low oscillatory shear stress can promote intimal hyperplasia and plaque formation.

The overall oil red O staining results (Fig. 6.8) showed that lipid accumulation gradually began in the left common carotid artery of the low oscillatory shear stress group mice from the 4th week. Afterwards, with increasing time (6–10 weeks), lipid

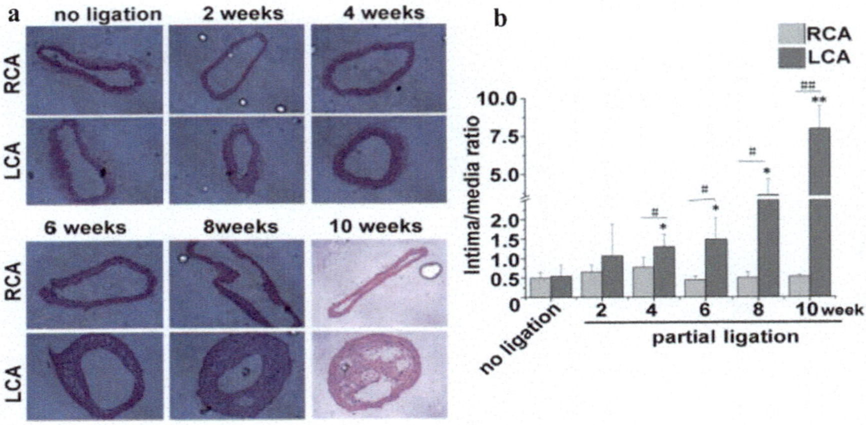

Fig. 6.7 OSS induced plaque formation. **a** HE staining was used to analyze plaque formation in the left and right carotid arteries of the OSS groups and control groups after surgery (2, 4, 6, 8, and 10 weeks). **b** Quantitative analysis of the intima/media ratio (n = 4). * $p < 0.05$ and ** $p < 0.001$ versus the LCA of the control groups. $^{\#}$ $p < 0.05$ and $^{\#\#}$ $p < 0.001$ versus the RCA of the OSS groups. No ligation was performed in the control groups. Partial ligation was performed in the OSS groups. (The Mechanobiological mechanisms of Id1-regulated lipid uptake in atherosclerotic plaque formation. Chongqing: Ph.D. thesis of Chongqing university, 2018)

deposition gradually increased, especially by the 10th week, and lipids almost spread throughout the entire lumen. However, there was no significant lipid deposition in the blood vessels of the right common carotid artery on the opposite side or in the left and right common carotid arteries of the control group mice. The results of frozen section CD31 and BODIPY fluorescence staining are shown in Fig. 6.9. The lipid absorption in the left common carotid artery intima of the low oscillatory shear stress group mice was significantly increased, while the lipid absorption in the left and right common carotid artery intima of the contralateral right common carotid artery and control group mice was very low. After statistical analysis of the lipid absorption of endothelial cells, it was found that in the low oscillatory shear stress group, the lipid absorption of left common carotid artery endothelial cells was significantly higher than that of the contralateral right carotid artery and the control group's left and right carotid arteries. Moreover, with the prolongation of ligation time, the accumulation of lipids also increased. The above results indicate that OSS can promote the absorption of more lipids by vascular endothelial cells.

HUVECs were subjected to in vitro static, laminar shear stress (12 dyn/cm^2), and low oscillatory shear stress (0.5 ± 4 dyn/cm^2) loading and treated with DiI-LDL. The lipid absorption of the cells was observed under a fluorescence microscope, as shown in Fig. 6.10. Compared with static and LSS, OSS can promote the absorption of more low-density lipoprotein by endothelial cells. The in vitro loading experiment once again proves that OSS can promote the absorption of lipids by endothelial cells.

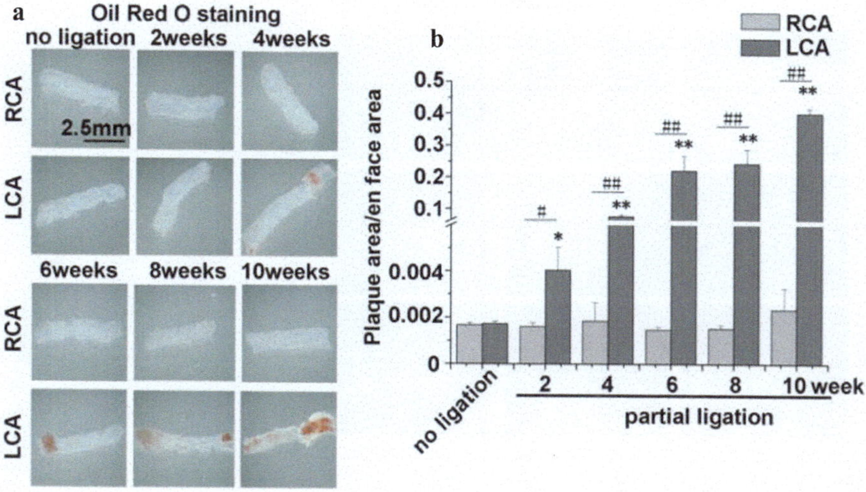

Fig. 6.8 OSS promoted lipid accumulation in vessels. **a** Oil Red O staining analysis of the time course of lipid accumulation in the left and right carotid arteries of the control groups and OSS groups (2, 4, 6, 8, and 10 weeks). **b** Quantitative analysis of the Oil red O staining areas (n = 4). * $p < 0.05$ and ** $Pp < 0.001$ versus the LCA of control groups. $^{\#}$ $p < 0.05$ and $^{\#\#}$ $p < 0.001$ versus the contralateral RCA of the OSS groups. No ligation was performed in the control groups. Partial ligation was performed in the OSS groups. (The Mechanobiological mechanisms of Id1-regulated lipid uptake in atherosclerotic plaque formation. Chongqing: Ph.D. thesis of Chongqing university, 2018)

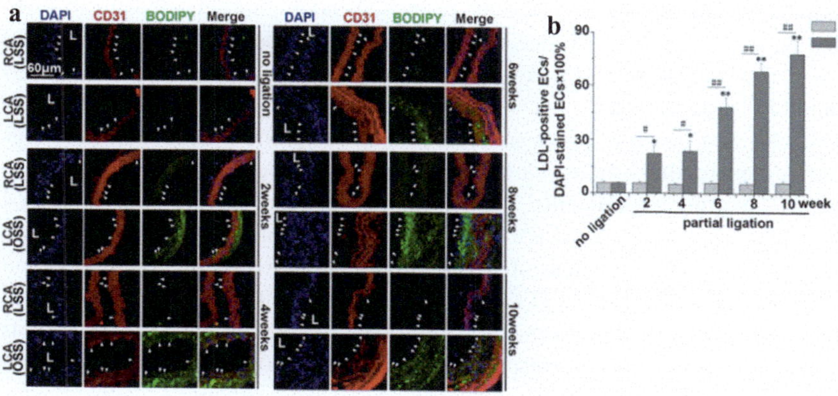

Fig. 6.9 OSS promoted endothelial cell lipid uptake in vivo. **a** Immunofluorescence staining analysis of time-course lipid accumulation in the left and right carotid arteries of the control groups and OSS groups (2, 4, 6, 8, 10 weeks); L, lumen. White arrow, endothelial cells. **b** Quantitative analysis of the percentage of BODIPY, DAPI and CD31 triple-positive cells relative to DAPI and CD31 double-positive cells in the intima (n = 4). * $p < 0.05$ and ** $p < 0.001$ versus the LCA of the control groups. $^{\#}$ $p < 0.05$ and $^{\#\#}$ $p < 0.001$ versus the RCA of the OSS groups. No ligation was performed in the control groups. Partial ligation was performed in the OSS groups. Cited from Zhang et al. (2018)

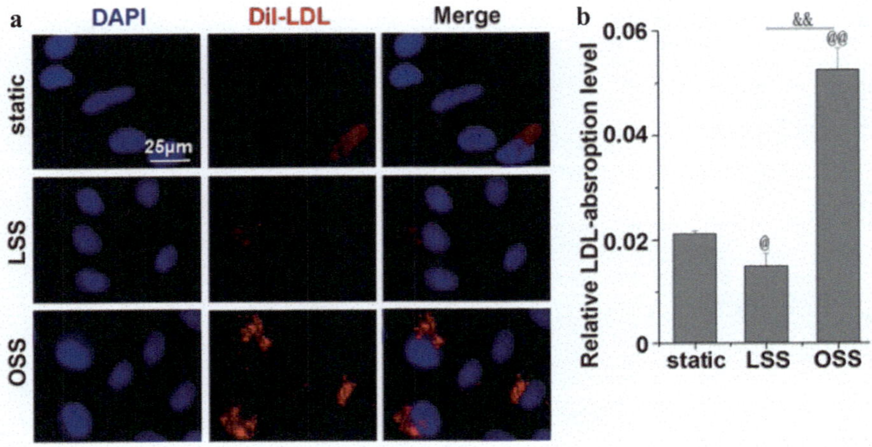

Fig. 6.10 OSS promoted endothelial cell lipid uptake in vitro. **a** Immunofluorescence analysis of LDL uptake in ECs treated with static, LSS and OSS for 24 h. **b** Quantitative analysis of the mean fluorescence intensity of LDL (n = 4). $^{@}$ $p < 0.05$ and $^{@@}$ $p < 0.001$ versus the static controls. $^{\&\&}p < 0.001$ versus the LSS controls. Cited from Zhang (2018)

Previous studies have found that low oscillatory flow can accelerate the formation of plaques by promoting the endothelial inflammatory response and EC proliferation (Cuhlmann et al. 2011; Guo et al. 2007). Endothelial cells were cocultured with BCECF-AM-labeled THP-1 cells and endothelial cells, and OSS treatment and subsequent lipid absorption significantly promoted endothelial cell adhesion to THP-1 cells. Ki67 immunofluorescence staining was used to detect the proliferation of ECs, and the results showed that OSS treatment and subsequent lipid absorption promoted EC proliferation.

In this section, the role of shear stress in atherosclerotic plaque formation was studied by constructing a mouse carotid artery ligation model and performing in vitro mechanical loading experiments. The plasma lipoproteins before and 24 h, 2, 4 and 6 weeks after operation were detected. Low oscillatory shear stress had no significant effect on the content of plasma lipoproteins. He staining and gross oil red O staining showed that the OSS area was more prone to plaque formation, and lipid deposition increased with time. Immunofluorescence staining of frozen sections found that OSS could promote vascular endothelial cells to absorb more lipids. The LSS and OSS loading of endothelial cells by an in vitro mechanical loading device confirmed that OSS could promote the absorption of lipids by endothelial cells. The adhesion test of THP-1 to HUVECs and Ki67 immunofluorescence staining showed that OSS treatment and subsequent LDL absorption could promote the proliferation of endothelial cells and adhesion to inflammatory cells.

6.3 Id1 Participates in the Lipid Absorption Process Caused by Low Oscillatory Shear Stress

In blood vessels, endothelial cells directly sense blood flow and various stresses caused by it, such as shear stress, tensile stress, and circumferential stress, through membrane receptors. When hemodynamic factors are abnormal, such as uneven distribution of shear force or local low oscillatory flow, they can cause abnormal expression of intracellular genes and proteins, thereby affecting the normal biological function of endothelial cells (Moreno et al. 2004). Once the function of endothelial cells is disordered, it will lead to the secretion of atherosclerosis-related factors, which will affect the function of smooth muscle cells and inflammatory cells, leading to the occurrence and development of atherosclerosis (Gijsen et al. 2008).

Id1 is an important transcription factor regulating the cell cycle and differentiation. It was previously reported that TSP1 is a downstream target of Id1 protein, and Id1 protein inhibits the expression of TSP1 (Volpert et al. 2002). The proteome identification results in Sect. 6.1 showed that OSS could promote the expression of TSP1 protein, and Id1 protein could inhibit the expression of TSP1 protein and participate in the process of RAP1 signaling pathway. Chih-Wen Ni et al. (2010) analyzed the RNA of vascular endothelial cells in the area of low oscillatory flow in mice by microarray chip and found that low oscillatory flow inhibited the expression of Id1 but did not perform much research on what kind of dysfunction would occur in cells after the change in Id1 expression. Based on the above results, the expression of Id1 protein in vascular endothelial cells was detected 24 h after carotid artery ligation in mice. En face staining results showed that the fluorescence intensity of Id1 protein in the ligated left carotid artery endothelial cells was significantly decreased. The Id1 protein-positive cells in the left carotid artery endothelial cells were approximately 2.5 times lower than those in the right carotid artery endothelial cells. The expression of Id1 protein in the left carotid artery endothelial cells was significantly lower than that in the right carotid artery. At the same time, the en face detection of blood vessels after 48 h of ligation revealed that the expression of Id1 protein was consistent with the 24 h trend. Because the left carotid artery after ligation has low oscillatory shear stress, while the right carotid artery of the contralateral sham operation has laminar shear stress, it is determined that low oscillatory shear stress can inhibit the expression of Id1 protein. Human umbilical vein endothelial cells were subjected to mechanical loading for 24 h, and the expression of Id1 protein was detected. Western blot results showed that laminar shear stress (12 dyn/cm^2) could promote the expression of Id1 protein, and low oscillatory shear stress (0.5 $\pm$ 4 dyn/cm^2) could instantly promote the expression of Id1 protein (Fig. 6.11).

Id1-overexpressing cells were constructed and loaded with LSS (12 dyn/cm^2) and OSS (0.5 $\pm$ 4 dyn/cm^2) to verify the role of Id1 in the OSS-mediated regulation of lipid absorption. After 24 h of mechanical loading, the cells were treated with DIL LDL, and then the absorption of DIL LDL in the cells was observed and analyzed by fluorescence. It was found that in both flag-id1 cells and flag con cells, the lipid absorbed by the cells after OSS treatment was significantly higher than that of the

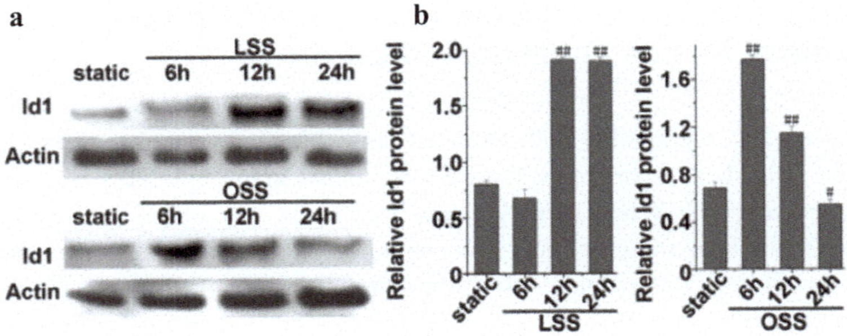

Fig. 6.11 Id1 protein expression was regulated by shear stress in HUVECs. **a** Western blot analysis of the level of Id1 in cells treated with LSS and OSS over a time course of 24 h. **b** Quantitative analysis of Id1 expression; the expression was normalized to that of Actin (n = 3). ## $p < 0.001$ versus the static controls. Cited from Zhang (2018)

LSS treatment group (Fig. 6.12a, c), indicating that OSS can promote the absorption of intracellular lipids, while the lipid absorbed by flag-id1 cells under OSS treatment was significantly lower than that of flag con cells, indicating that Id1 is involved in the regulation of endothelial cell lipid absorption by OSS, and Id1 overexpression can inhibit the effect of OSS (Fig. 6.12a, c).

To determine which proteins Id1 will interact with in atherosclerotic disease to participate in the changes in vascular function caused by OSS, we used IPA software. The results of relevant upstream regulatory factors are shown in Fig. 6.13. These predicted results show that Id1 plays an important role in atherosclerotic disease, which can be regulated by many factors, and can also regulate the expression of many factors to affect the formation of atherosclerotic plaques. The factors related to lipid metabolism that may have an effect on Id1 were also analyzed by IPA. As shown in Fig. 6.14, Id1 and LDLR, SREBP1, peroxisome proliferator activated receptor γ (PPARγ) and other lipid-related factors were identified.

En face staining showed that OSS could inhibit the expression of Id1 protein in vascular endothelial cells. An in vitro mechanical loading device was used to load vascular endothelial cells with LSS and OSS. The results showed that compared with the static state, LSS (12 dyn/cm^2) promoted the expression of Id1 protein in endothelial cells, and the protein expression increased with time, reaching a peak at 12 h. Compared with static conditions, OSS (0.5 ± 4 dyn/cm^2) promoted the expression of Id1 at the 6th hour, but with the extension of loading time, the expression of Id1 protein was inhibited, and the expression of Id1 was lower than the static level at the 24th hour. This shows that compared with the static state, both LSS and OSS can regulate the expression of Id1, but LSS and OSS have different effects on the expression of Id1. LSS promotes the expression of Id1 protein, while OSS inhibits its expression. To investigate whether the Id1 protein is involved in the regulation of lipid absorption by OSS, Id1 overexpression and control cell lines were constructed. After loading the two kinds of cells with laminar shear stress and low oscillatory shear stress, the absorption of DiI-LDL was detected. The results showed

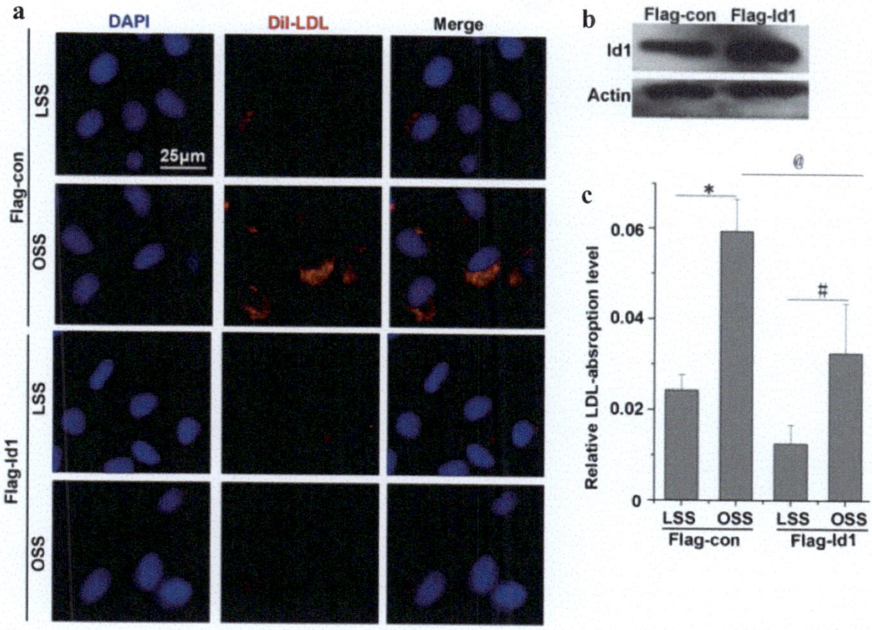

Fig. 6.12 Id1 overexpression can abolish OSS-mediated lipid uptake in ECs. **a** Immunofluorescence analysis of LDL uptake in Id1-overexpressing (Flag-Id1) and control (Flag-con) ECs treated with shear stress. **b** Western blotting was used to analyze the protein level of Id1 in Flag-Id1 and Flag-con ECs. **b** Quantitative analysis of the mean fluorescence intensity of LDL in Flag-Id1 and Flag-con ECs (n = 4). * $p < 0.05$ versus the Flag-con LSS. $^{\#}$ $p < 0.05$ versus Flag-Id1 LSS, $^{@}$ $p < 0.05$ versus Flag-con OSS. Cited from Zhang (2018)

that OSS promoted the process of lipid absorption by endothelial cells, but after Id1 overexpression, the promoting effect of OSS on endothelial lipid absorption was partially inhibited; IPA software predicts the factors that interact with Id1 in atherosclerotic diseases and the lipid-related proteins that Id1 may regulate. The results show that Id1 regulates the expression of many AS-related proteins, and there may be a regulatory relationship between Id1 and LDLR.

6.4 Id1 Participates in the Process of Endothelial Lipid Absorption Induced by Low Oscillatory Shear Stress by Regulating LDLR Expression

Atherosclerotic plaques begin with the deposition of LDL in the vascular wall, and the absorption of intracellular LDL is mainly achieved through the endocytosis of low-density lipoprotein receptor (LDLR), which is also known as the LDLR pathway (Brown et al. 1979). For the regulation of LDLR expression, many studies have

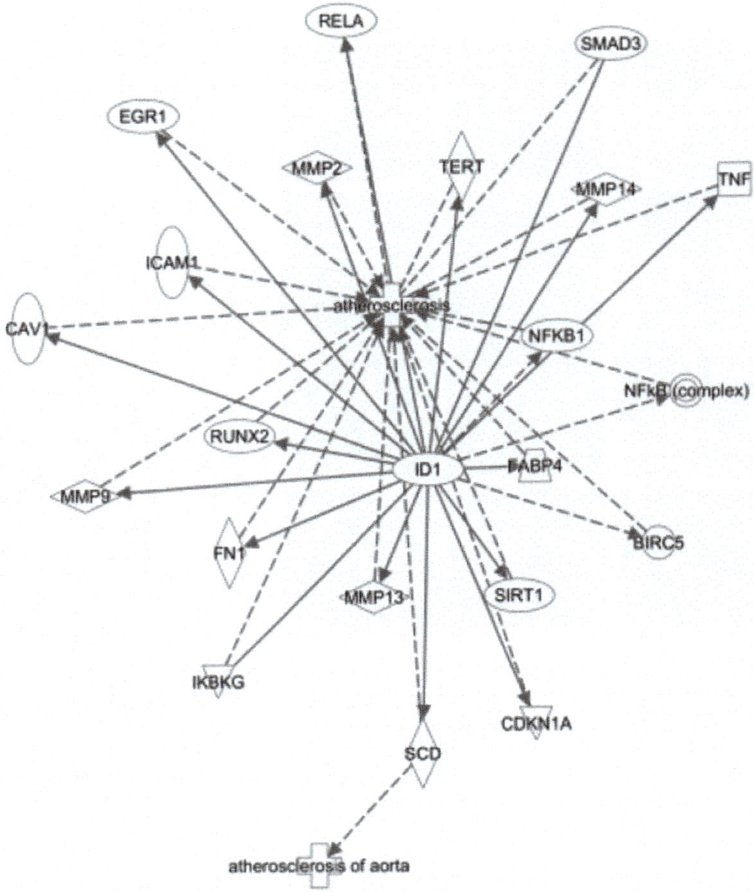

Fig. 6.13 IPA was used to analyze the molecules that are regulated by Id1 in atherosclerosis. In atherosclerotic diseases, Id1 can regulate the expression of many factors closely related to AS diseases, such nuclear factor-kB, CAV1, MMP13, etc. Cited from Zhang (2018)

been carried out at the molecular level. SREBP1 is an important transcription factor regulating the expression of LDLR and belongs to the bHLH family (Hua et al. 1993; Yokoyama et al. 1993). It can specifically bind to cholesterol regulatory element 1 (SRE-1) to regulate gene expression. The IPA analysis in this study also found that Id1 may have a regulatory relationship with LDLR, so this section studied the role of LDLR in the regulation of lipid absorption in vascular endothelial cells by OSS and the relationship between Id1 and LDLR through in vivo and in vitro methods.

The experiment was divided into the Id1 overexpression control group (Flag-con), Id1 overexpression group (Flag-Id1), Id1 interference control group (Si con), and Id1 interference group (si-Id1) to explore the expression changes of Id1 and LDLR in each group. The results showed that the expression of Id1 protein in the

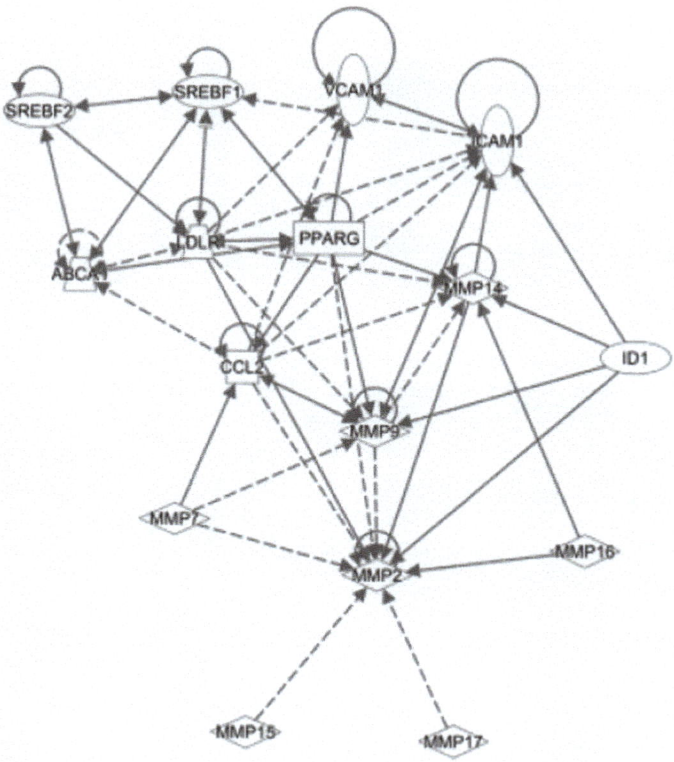

Fig. 6.14 The relationship between Id1 and lipid metabolism-related molecules. IPA prediction showed that Id1 was associated with LDLR (low density lipoprotein receptor), SREBP1 (cholesterol regulatory element binding protein-1) and other lipid-related factors. (The Mechanobiological mechanisms of Id1-regulated lipid uptake in atherosclerotic plaque formation. Chongqing: Ph.D. thesis of Chongqing university, 2018)

overexpression group was nearly twice as high as that in the overexpression control group, which was significantly higher than that in the control group (Fig. 6.15a, b), while the expression of Id1 protein in the cells in the interference group was also nearly twice as high as that in the interference control group, which showed that both overexpression and interference cells reached the overexpression and interference effects, respectively (Fig. 6.15c, d). On this basis, we detected the expression of LDLR protein. The results are shown in Fig. 6.15. When Id1 was overexpressed, the expression of LDLR protein was inhibited, while when Id1 was knocked down, the expression of LDLR was significantly increased. The expression of LDLR in Id1-t, con-t, si-Id1 and si-con cells was analyzed by fluorescence quantitative PCR. The results are shown in Fig. 6.15e, f. Similar to the protein results, the expression of LDLR decreased in Id1-overexpressing cells but increased in Id1-interfering cells. This indicates that Id1 negatively regulates the expression of LDLR.

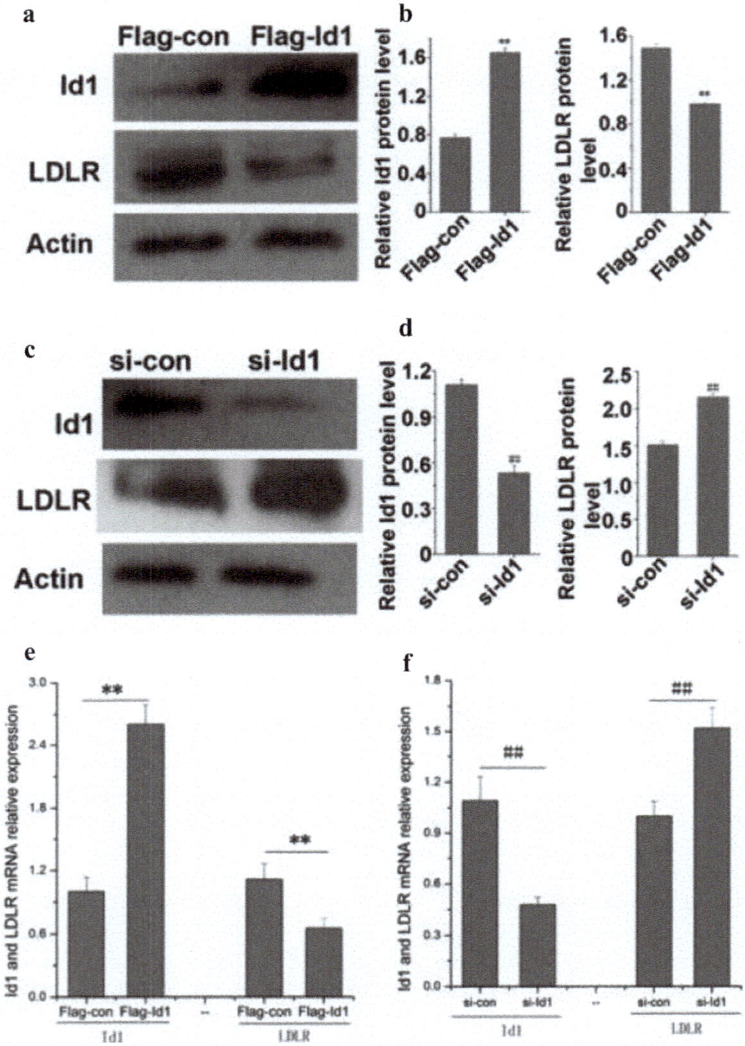

Fig. 6.15 Id1 regulates the expression of LDLR. **a** Western blotting was used to analyze the protein expression of Id1 and LDLR in Flag-Id1 and Flag-con ECs. **b** Quantitative analysis of Id1 and LDLR expression; the expression was normalized to that of Actin (n = 3). **c** Western blotting was used to analyze the expression of Id1 and LDLR protein in Id1 knockdown (si-Id1) and control (si-con) ECs. **d** Quantitative analysis of Id1 and LDLR expression; the expression was normalized to that of Actin (n = 3). **e** Real-time PCR was used to detect Id1 and LDLR expression levels in Flag-Id1 and Flag-con ECs (n = 3). **f** Real-time PCR was used to detect Id1 and LDLR expression levels in si-Id1 and si-con ECs (n = 3). ** $p < 0.001$ versus the Flag-con controls. $^{\#\#}$ $p < 0.001$ versus the si-con controls. (The Mechanobiological mechanisms of Id1-regulated lipid uptake in atherosclerotic plaque formation. Chongqing: Ph.D. thesis of Chongqing university, 2018)

To study how Id1 affects LDL uptake by cells and whether LDLR is involved in this process, Id1 and LDLR cotransfected cells (si-Id1 + si-LDLR) were constructed, and the siRNA was successfully transferred into cells according to the green fluorescence in cells of the control group (Fig. 6.16). Western blotting was used to detect the expression of Id1 and LDLR in co-transfected cells. Compared with that in the si-con group, the expression of Id1 in the si-Id1 and si-Id1 + si-LDLR groups decreased significantly, indicating that the interfering cells interfered with the expression of Id1 protein. The expression results of LDLR protein showed that compared with the si-con group, the expression of LDLR in the si-Id1 group increased, indicating that LDLR was activated after Id1 interference, while the expression of LDLR in the si-id1 + si-LDLR group was lower than that in the si-Id1 group, indicating that si-LDLR achieved the interference effect (Fig. 6.17). When Flag-con, Flag-Id1, si-con, si-Id1 and si-Id1 + si-LDLR cells were treated with DIL LDL, the results showed that the LDL absorbed by cells with Id1 overexpression was significantly lower than that of the control group (Fig. 6.18a, b). When Id1 and LDLR were simultaneously knocked down, the LDL absorption caused by Id1 interference was inhibited (Fig. 6.18c, d), indicating that Id1 regulates the LDLR-mediated lipid absorption pathway.

Then, it was investigated whether OSS could affect lipid absorption by regulating the expression of LDLR through Id1. First, whether LDLR protein in vascular endothelial cells was affected by low oscillatory shear stress was examined in a mouse low oscillatory shear stress model. The left and right carotid arteries of mice were ligated and sham operated, respectively. After 24 h, the carotid arteries were taken for en face staining detection. The results showed that compared with the LSS of the right carotid artery, the fluorescence intensity of LDLR in endothelial cells in the OSS area of the ligated left carotid artery was significantly increased (Fig. 6.19a). According to the statistics of the number of LDLR-positive cells and DAPI-positive cells, the LDLR-positive endothelial cells in the OSS region of the left carotid artery were approximately 4 times that in the LSS region of the right

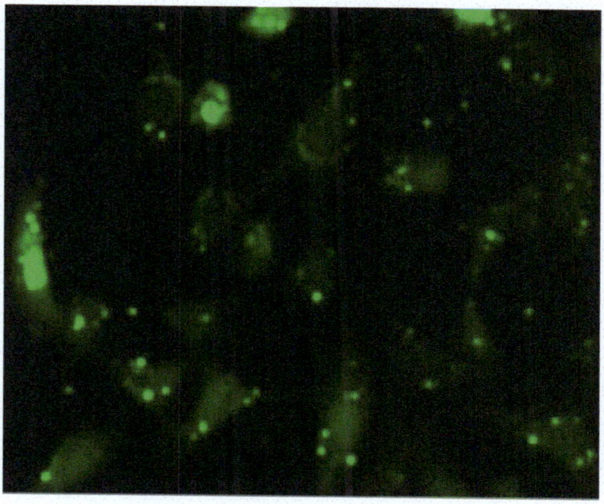

Fig. 6.16 After transfection, si-con cells had very strong green fluorescence. (The Mechanobiological mechanisms of Id1-regulated lipid uptake in atherosclerotic plaque formation. Chongqing: Ph.D. thesis of Chongqing university, 2018)

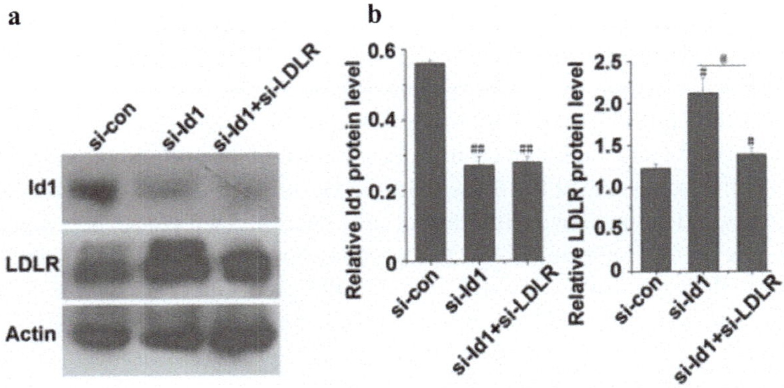

Fig. 6.17 The expression of Id1 and LDLR in si-con, si-Id1 and si-Id1 + si-LDLR ECs. **a** Western blot analysis of the protein levels of Id1 and LDLR in si-con, si-Id1 and si-Id1 + si-LDLR ECs. (F) Quantitative analysis of Id1 and LDLR expression; the expression was normalized to that of Actin. (n = 3). # $p < 0.05$ and ## $p < 0.001$ versus the si-con controls. @ $p < 0.05$ versus the si-Id1 groups. Cited from Zhang et al. (2018)

carotid artery (Fig. 6.19b). Analysis of the fluorescence gray of LDLR showed that the expression of LDLR in the OSS region was approximately 3 times that in the LSS region (Fig. 6.19b).

Human umbilical vein endothelial cells were subjected to mechanical loading, and the expression pattern of LDLR protein was detected after mechanical loading in vitro. The results are shown in Fig. 6.20. After LSS loading, compared with the static state, the expression of LDLR protein showed an upward trend with the extension of loading time, reaching a peak at 12 h, but with the increase in loading time to 24 h, the expression of LDLR began to decline and was lower than that in the static state, indicating that laminar flow instantly promoted the expression of LDLR, but after a certain time, it inhibited its expression (Fig. 6.20a). After OSS loading, compared with the resting state, the expression of LDLR protein continued to increase with increasing loading time, with the highest expression at 24 h (Fig. 6.20a). According to the statistical analysis of the Western blot results, compared with the static state, LSS transiently promoted the expression of LDLR. With increasing time, the expression of LDLR began to decline, while OSS continued to activate the expression of LDLR (Fig. 6.20b). The results in vivo and in vitro showed that OSS can promote the expression of LDLR, and in Chap. 3, we found that OSS can promote lipid absorption in endothelial cells. Therefore, we treated endothelial cells with LDLR interference (si-LDLR) to test whether OSS can affect lipid absorption by affecting the expression of LDLR. After LDLR was successfully knocked down (Fig. 6.21a), OSS was loaded on the control group cells (si-con) and si-LDLR cells, and then intracellular lipid absorption was detected. The results showed that the amount of lipid absorption in si-LDLR cells was significantly lower than that in si-con group cells (Fig. 6.21b, c). These results indicate that LDLR mediates the regulation of cellular lipid absorption by OSS.

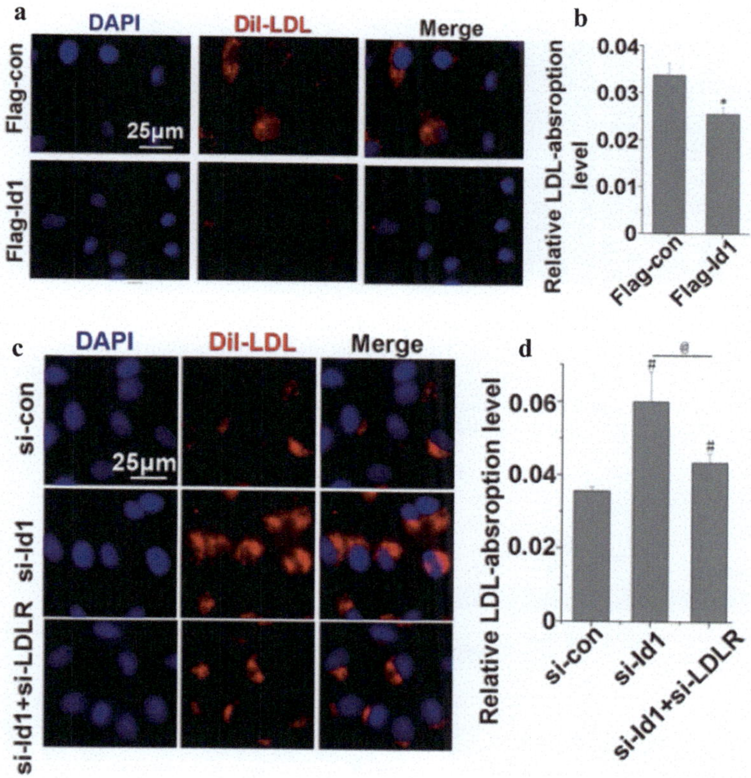

Fig. 6.18 Id1 influenced LDL uptake by regulating LDLR expression in ECs. **a** Immunofluorescence analysis of LDL uptake in Flag-Id1 and Flag-con ECs. **b** Quantitative analysis of the mean fluorescence intensity of LDL (n = 4). **c** Immunofluorescence analysis of LDL accumulation in si-con, si-Id1 and si-Id1 + si-LDLR ECs. **d** Quantitative analysis of the mean fluorescence intensity of LDL (n = 4). * $p < 0.05$ versus the Flag-con controls; $^{\#}$ $p < 0.05$ versus the si-con controls; $^{@}$ $p < 0.05$ versus the si-Id1 groups. Cited from Zhang et al. (2018)

The endothelial cells were subjected to mechanical loading for 24 h, and then the expression of Id1 and LDLR was detected. After mechanical loading, the loaded slides were removed and subjected to immunofluorescence staining. The results showed that compared with static and LSS, the fluorescence intensity of Id1 after OSS treatment was significantly reduced. Statistical analysis of the fluorescence intensity showed that OSS could inhibit the expression of Id1 (Fig. 6.22a, b). However, the fluorescence intensity of LDLR increased significantly after OSS treatment. Statistical analysis results showed that OSS could promote the expression of LDLR (Fig. 6.22c, d). The total protein of LSS- and OSS-treated cells was extracted and detected by Western blot. The results are shown in Fig. 6.23. Compared with static and LSS, the expression of Id1 protein was significantly decreased after OSS treatment, while the expression of LDLR protein was significantly increased. The expression patterns of the two proteins were opposite. The Western blot results were consistent with the

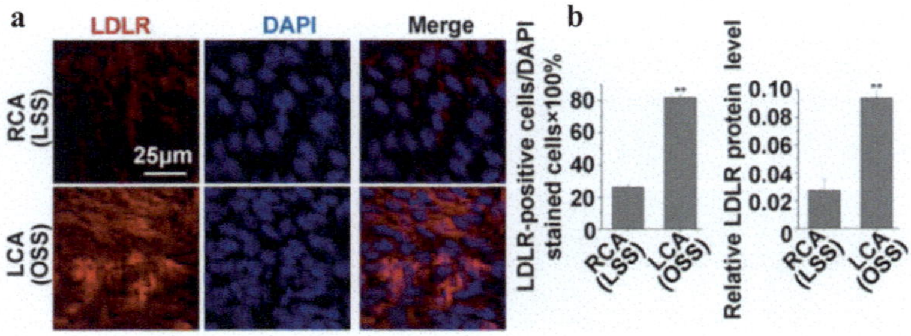

Fig. 6.19 LDLR expression in the left and right carotid arteries was detected 24 h after ligation. **a** En face staining analysis of the level of LDLR. The fluorescence intensity in LCA was significantly higher than that in RCA. **b** Quantitative analysis of the percentage of LDLR-positive cells relative to DAPI-stained cells and the mean fluorescence intensity of LDLR (n = 3). ** $p < 0.001$ versus the RCA (LSS) controls. Cited from Zhang et al. (2018)

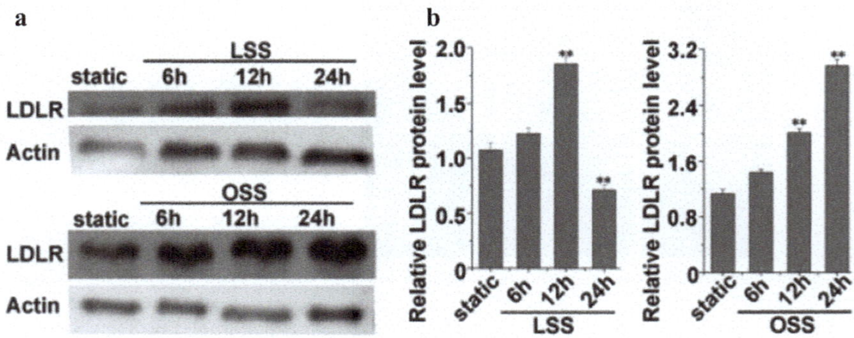

Fig. 6.20 LDLR protein expression was regulated by shear stress in HUVECs. **a** Western blotting was used to analyze the protein level of LDLR treated with LSS and OSS over a time course of 24 h. **b** Quantitative analysis of LDLR expression; the expression was normalized to that of Actin (n = 3). ** $p < 0.001$ versus the static controls. The Mechanobiological mechanisms of Id1-regulated lipid uptake in atherosclerotic plaque formation. Chongqing: Ph.D. thesis of Chongqing university, 2018)

immunofluorescence results and confirmed our previous speculation that Id1 and LDLR were both regulated by OSS and showed a negative correlation trend.

OSS inhibits the expression of Id1 and promotes the expression of LDLR, and Id1 can regulate the expression of LDLR. Is Id1 involved in the regulation of LDLR by OSS? Id1-overexpressing and control cells were subjected to mechanical loading for 24 h, the protein was extracted, and the expression of LDLR was detected by Western blot. The results are shown in Fig. 6.24. The expression of LDLR in Flag-con cells treated with OSS was significantly higher than that in Flag-con cells treated with LSS; Flag-Id1 was loaded with OSS for 24 h, and then the expression of LDLR was detected. The expression of LDLR in Flag-Id1 cells was significantly lower than that in Flag-con cells under OSS treatment (Fig. 6.24), indicating that Id1 is involved in

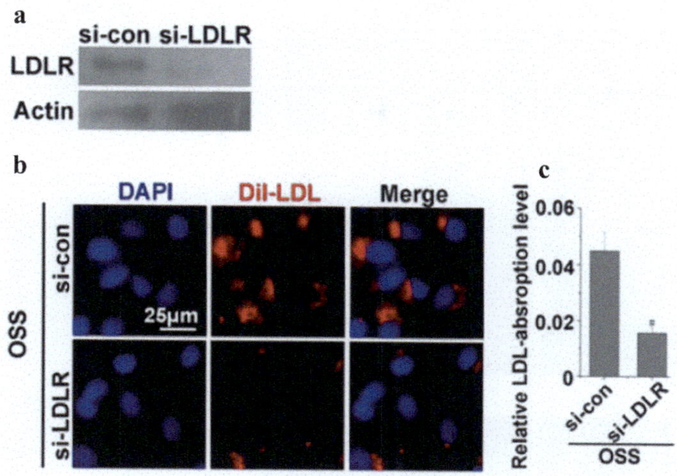

Fig. 6.21 OSS influenced LDL uptake through LDLR in ECs. **a** Western blot analysis of the protein level of LDLR in si-con and si-LDLR ECs. **b** Immunofluorescence analysis of LDL uptake in si-con and si-LDLR ECs treated with OSS for 24 h. **c** Quantitative analysis of the mean fluorescence intensity of LDL (n = 4). * $p < 0.05$ versus si-con controls. Cited from Zhang (2018)

the regulation of LDLR by OSS. OSS promotes the expression of LDLR by inhibiting Id1.

Previous studies have shown that Id1 may inhibit the function of SREBP1 by binding to SREBP1, thereby inhibiting its expression. To test this idea, first, the protein of Flag-Id1 cells was extracted, and coimmunoprecipitation experiments were carried out with Flag and SREBP1 antibodies to verify whether Flag-Id1 interacts with SREBP1. The results are shown in Fig. 6.25a. When the protein bound to SREBP1 was pulled with SREBP1 antibody and verified by Western blot with Flag antibody, it was found that exogenous Id1 could bind to SREBP1. In contrast, the protein bound to Flag-Id1 was pulled down with a Flag antibody, and Western blot verification with an SREBP1 antibody revealed that SREBP1 could bind to exogenous Id1. The results of co-immunoprecipitation showed that the foreign Id1 protein interacted with the SREBP1 protein. The colocalization of Id1 and SREBP1 was detected by the fluorescence colocalization method. The results are shown in Fig. 6.25b. Green fluorescence represents the Id1 protein, red fluorescence represents the SREBP1 protein, and blue represents the nucleus. After synthesizing green and red fluorescence, it was found that green Id1 protein and red SREBP1 protein colocalized (indicated by white arrow). The results of coimmunoprecipitation combined with fluorescence colocalization showed that Id1 physically bound to SREBP1 in cells. In addition, we extracted Flag-con, Flag-Id1, si-con, and si-Id1 cellular proteins and analyzed the expression of SREBP1 by Western blot. The statistical results showed that the abnormal expression of Id1 did not affect the expression of SREBP1 protein (Fig. 6.25c, d). These results indicate that Id1 can form a complex with SREBP1.

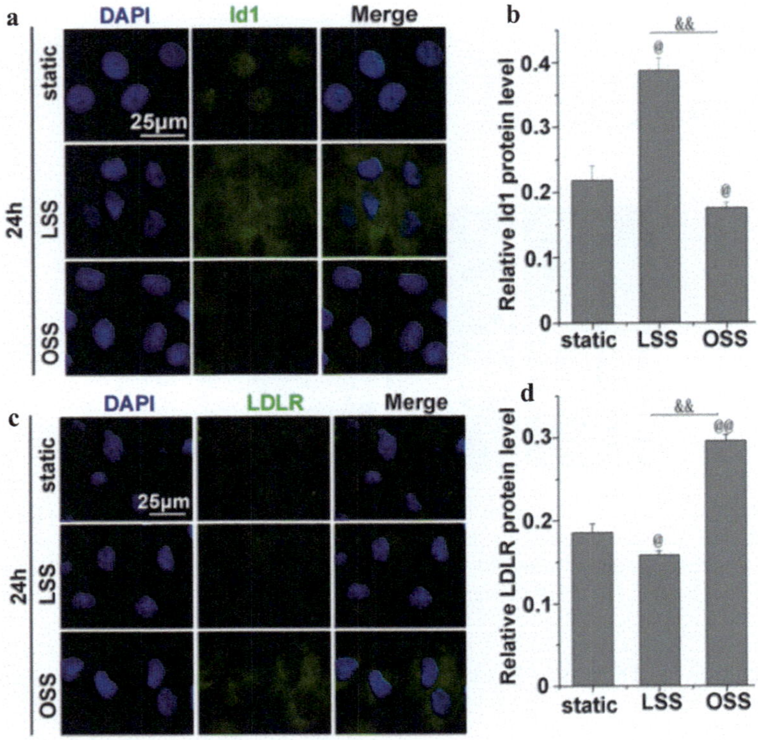

Fig. 6.22 Expression of Id1 and LDLR in ECs treated with static, LSS and OSS. **a** Immunofluorescence analysis of Id1 expression in ECs treated with static, LSS and OSS for 24 h. **b** Quantitative analysis of the mean fluorescence intensity of Id1 (n = 3). **c** Immunofluorescence analysis of LDLR expression in ECs treated with static, LSS and OSS for 24 h. **d** Quantitative analysis of the mean fluorescence intensity of LDLR (n = 3). $^{@}$ $p < 0.05$ and $^{@@}$ $p < 0.001$ versus the static controls; $^{\&\&}p < 0.001$ versus the LSS controls. (The Mechanobiological mechanisms of Id1-regulated lipid uptake in atherosclerotic plaque formation. Chongqing: Ph.D. thesis of Chongqing university, 2018)

In this chapter, the molecular mechanism of the Id1 protein involved in the regulation of lipid absorption by OSS was studied through cell molecular biology, biochemistry and other methods: the expression and lipid absorption of LDLR were detected after Id1 overexpression and interference cells were constructed. The results showed that Id1 overexpression inhibited the expression of LDLR protein and the absorption of LDL by endothelial cells. In contrast, Id1 upregulated the expression of LDLR protein and promoted the absorption of LDL by endothelial cells. At the same time, the interference of Id1 and LDLR inhibited the lipid absorption process caused by the interference of Id1 alone. Id1 affects cellular lipid absorption by regulating the expression of LDLR. In vivo en face staining and in vitro Western blotting showed that OSS could inhibit the expression of Id1 protein and continuously promote the expression of LDLR. After interfering with LDLR, the lipid absorption process caused by OSS was inhibited, indicating that OSS affects the intracellular lipid absorption

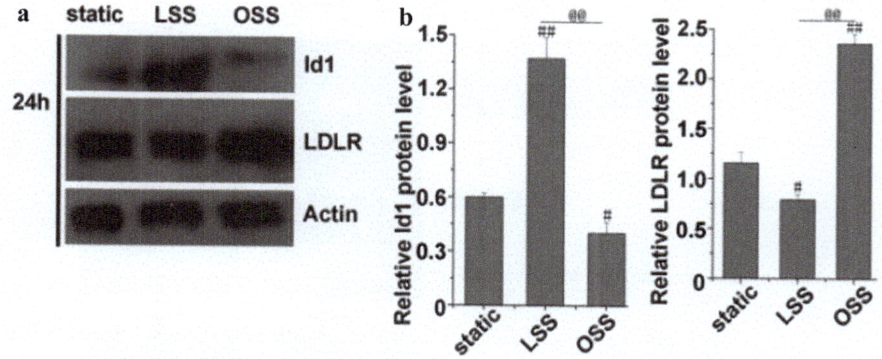

Fig. 6.23 Expression of Id1 and LDLR in ECs treated with static, LSS and OSS. **a** Western blot analysis of the protein levels of Id1 and LDLR after treatment with static, LSS and OSS for 24 h. **b** Quantitative analysis of Id1 and LDLR expression; the expression was normalized to that of Actin (n = 3). $^{\#}$ $p < 0.05$ and $^{\#\#}$ $p < 0.001$ versus the static controls; $^{@@}$ $p < 0.001$ versus the LSS controls. Cited from Zhang et al. (2018)

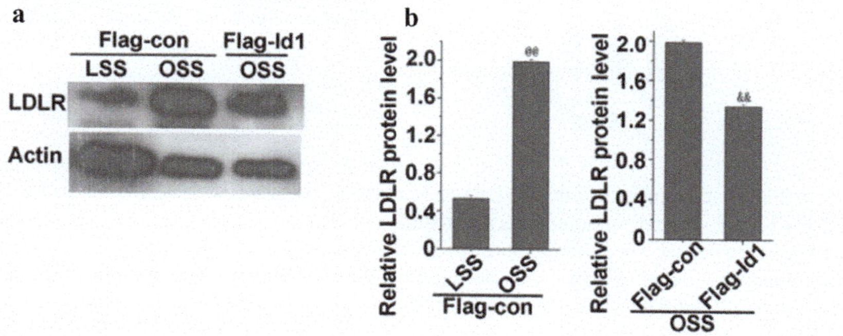

Fig. 6.24 Id1 was involved in OSS-regulated LDLR expression. **a** Western blot analysis of LDLR expression in Flag-Id1 and Flag-con ECs treated with LSS and OSS. **b** Quantitative analysis of LDLR expression; the expression was normalized to that of Actin (n = 3). $^{@@}$ $p < 0.001$ versus the Flag-con LSS controls. $^{\&\&}p < 0.001$ versus the Flag-con OSS controls. Cited from Zhang et al. (2018)

process by regulating the expression of LDLR. The cells in the Id1 overexpression and control groups were subjected to mechanical loading, and the expression changes of LDLR in cells were detected by Western blot. The results showed that Id1 overexpression inhibited the LDLR-promoting effect of OSS, which indicated that OSS regulated LDLR expression through Id1. The relationship between Id1 and SREBP1 was detected by co-immunoprecipitation, fluorescence colocalization and Western blotting. Id1 could combine with SREBP1 to form a complex, which may play an important role in the regulation of LDLR expression by Id1.

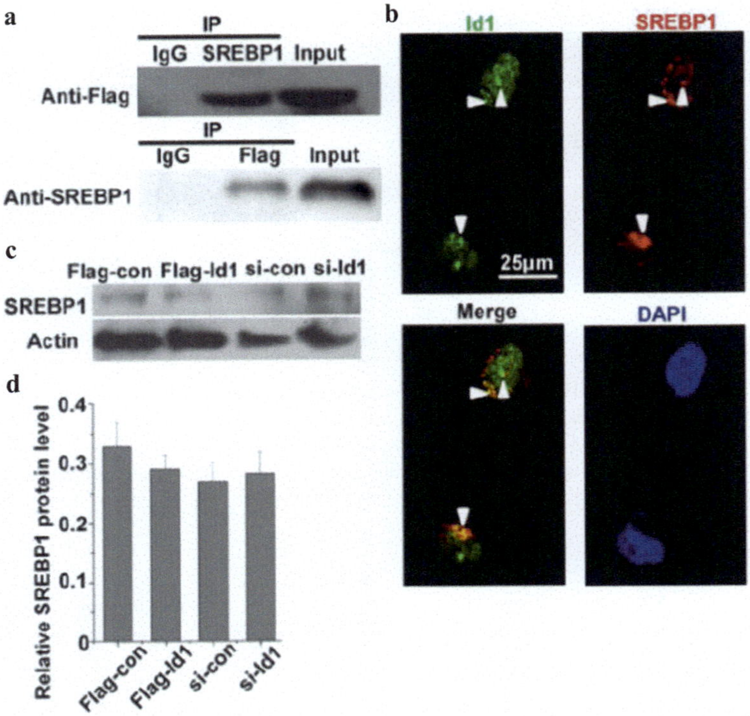

Fig. 6.25 Id1 can interact with SREBP1. **a** Immunoprecipitation was carried out using an anti-SREBP1 antibody, and coimmunoprecipitated Id1 was detected by Western blot using an anti-Flag antibody (top panel). Then, immunoprecipitation was carried out using an anti-Flag antibody, and coimmunoprecipitated SREBP1 was detected by Western blot using an anti-SREBP1 antibody (bottom panel). As a negative control, mouse control IgG was used for immunoprecipitation (n = 3). **b** Colocalization of SREBP1 and Id1 in ECs. ECs were stained with a rabbit anti-Id1 polyclonal antibody (green) or a mouse anti-SREBP1 monoclonal antibody (red). The white arrow indicates the colocalization of red fluorescence and green fluorescence (n = 5). **c** Western blotting was used to analyze the protein level of SREBP1 in Flag-con, Flag-Id1, si-con and si-Id1 cells. **d** Quantitative analysis of SREBP1 expression in the Flag-con, Flag-Id1, si-con and si-Id1 groups (n = 3). Cited from Zhang et al. (2018)

6.5 Research Progress and Prospects

AS has the characteristics of high incidence and many complications, and AS plaque is produced due to the deposition of lipids in the late stage of the disease. AS plaque rupture and thrombosis will lead to acute coronary syndrome, myocardial infarction, heart failure, ischemic stroke and other serious cardiovascular and cerebrovascular diseases. Shear stress runs through the whole process of the occurrence and development of AS. Studying the molecular signaling pathways that regulate AS in response to shear stress can provide a new idea for the treatment of AS.

Because the inhibitor of differentiation (Id) protein family lacks the basic domain for binding to DNA, Id proteins play a role by binding to basic helix-loop-helix (bHLH) transcription factors, thereby inhibiting bHLH binding to DNA and affecting cell functions by regulating the expression of bHLH transcription factor downstream genes. At present, most studies on Id1 are focused on tumors. Id1 plays pivotal roles in tumor angiogenesis and tumor cell migration (Castañón et al. 2017; Hu et al. 2017; Jin et al. 2018). Id1 in tumor cells can affect the expression of many factors, such as integrin, vascular cell adhesion molecule (VCAM), and intercellular cell adhesion molecule (ICAM) (Ni et al. 2010). A force-sensitive factor-related omics study found that the expression of Id1 in endothelial cells was affected by changes in shear stress. Coincidentally, some researchers have also detected the expression of Id1 protein after stimulating endothelial cells with different sizes of shear force, confirming that Id1 can respond to mechanical signals (Qiu et al. 2011).

6.5.1 Id1 Affects Endothelial Cell Uptake of LDL

Id1 is involved in the formation and metabolism of lipids and is highly expressed in the adipose tissue of mice and humans. The absorption of lipids by vascular endothelial cells is likely regulated by Id1. Id1 interacts with alkaline HLH transcription factors and inhibits their transcriptional activity (Seale et al. 2009). The transcription factor SREBP1 in the HLH family is the main regulator of lipid synthesis and is involved in the synthesis of fats, cholesterol, and LDL receptor (LDLR)-mediated absorption. LSS in the blood instantly activates SREBP1 in the endothelial cells of the vascular wall, while OSS causes a sustained activation state of SREBP1. This hemodynamic change not only increases the synthesis of fatty acids and cholesterol in cells but also upregulates the expression of LDLR, thereby increasing LDL uptake (Alessia et al. 2007). Therefore, Id1 interacts with SREBP1 and regulates LDLR expression, affecting endothelial cell uptake of LDL when receiving signals of hemodynamic changes.

6.5.2 Id1 Participates in the Formation of Atherosclerosis

The Id1 family has multiple biological functions and is particularly important for angiogenesis. For example, $Id1^{\pm}$ and $Id3^{-/-}$ double knockout mice may die due to vascular system abnormalities, while $Id1^{\pm}$ $Id3^{-/-}$ mice can survive. Id1 plays an important role in angiogenesis and inflammatory reactions in various gastric cancers, breast cancers, and intestinal cancers. Many studies on Id1 have focused on its ability to participate in tumor angiogenesis by inducing cell proliferation and migration and accelerating the cell cycle (Maw et al. 2009). This function is highly coincident with the occurrence of atherosclerosis. Mechanics mediating the Id1-p53 signaling pathway can promote the formation of vulnerable atherosclerotic plaques by regulating angiogenesis (Qiu et al. 2011). Many stimulaing factors within arterial plaques, such as oxidative stress, inflammatory cell infiltration, hyperlipidemia, vascular hypoxia inducible factors, and vascular endothelial growth factor (VEGF), can upregulate the expression of Id1 to regulate the development of plaques. In human umbilical vein endothelial cells, hypoxia inducible factor 1 can upregulate the expression of VEGF and Id1, and upregulation of Id1 can also promote VEGF expression.

Id1 is also involved in the infiltration and phagocytosis process of inflammatory cells. NADPH oxidase promotes the upregulation of Id1 expression, and tumor necrosis factor (TNF-α) participates in the expression of Id1 physiological function, while inhibiting Id1 expression reduces the aggregation of white blood cells and macrophages. Id1 can be expressed in endothelial cells in normal mouse kidneys and is necessary for a normal response to injury (Sharma et al. 2020). In Id1 knockout mice, glomerular membrane and myofibroblast proliferation, matrix deposition, and proteinuria are increased. The electron microscopy results showed that capillary EC injury and lumen narrowing were observed in Id1 knockout mice. The aging pathway in cells is activated. Therefore, upregulation of Id1 in endothelial cells can prevent microvascular damage, aging, and other processes. Polyacrylamide hydrogels with different hardnesses were used to analyze whether the hardness changes would affect the ability of endothelial cells to adhere to THP-1 cells. It was found that the Id1 molecule was involved in the regulation process. After the overexpression of Id1 in HUVECs, the cell adhesion factors ICAM-1 and VCAM-1 and the chemotactic factor MCP-1 also showed high expression (Zhang et al. 2016). Therefore, changes in basal hardness may affect the expression of Id1, which may be related to the development of atherosclerosis.

The risk factors inducing atherosclerosis are complex and diverse, such as hyperlipidemia, diabetes, hypertension, smoking, and family genetics, but regardless of which of the above factors is the main factor of atherosclerosis, the final lesion location of atherosclerosis has a certain preference. Low oscillatory flow can promote the occurrence and development of atherosclerosis, while high shear stress or laminar flow will not aggravate the occurrence and development of atherosclerosis. Id1 can affect the adhesion process of endothelial cells to monocytes and the uptake of lipids in response to mechanics, thus playing a role in the occurrence and development of atherosclerosis. Although many mechanosensitive factors have been found at present, most of them only remain in the experimental stage, and there are few drugs developed for mechanosensitive factors and even a lack of mechanical methods to treat atherosclerosis. Therefore, it is of great significance to develop corresponding drugs or methods to prevent and treat atherosclerosis from mechanosensitive targets.

References

Alessia C, Vladimir J, Yuval S, Daniel JN, Vivek M, Robert S K, Stephen DN, Robert B. Id1 restrains p21 expression to control endothelial progenitor cell formation. PloS One, 2007, 2(12): e1338.

Brown MS, Goldstein JL. Receptor-mediated endocytosis: insights from the lipoprotein receptor system[J]. Proc Natl Acad Sci U S A. 1979, 76(7): 3330–3337.

Castañón E, Soltermann A, López I, Román M, Ecay M, Collantes M, Redrado M, Baraibar I, López-Picazo JM, Rolfo C, Vidal-Vanaclocha F, Raez L, Weder W, Calvo A, Gil-Bazo I. The inhibitor of differentiation-1 (Id1) enables lung cancer liver colonization through activation of an EMT program in tumor cells and establishment of the pre-metastatic niche[J]. Cancer Lett. 2017, 402: 43–51.

Ciarrocchi A, Jankovic V, Shaked Y, Nolan DJ, Mittal V, Kerbel RS, Nimer SD, Benezra R. Id1 restrains p21 expression to control endothelial progenitor cell formation[J]. PLoS One. 2007, 2(12): e1338.

Cohn JN. Arterial stiffness, vascular disease, and risk of cardiovascular events[J]. Circulation. 2006, 113(5): 601–603.

Cuhlmann S, Van der Heiden K, Saliba D, Tremoleda JL, Khalil M, Zakkar M, Chaudhury H, Luong le A, Mason JC, Udalova I, Gsell W, Jones H, Haskard DO, Krams R, Evans PC. Disturbed blood flow induces rela expression via c-jun n-terminal kinase 1: a novel mode of NF-kappaB regulation that promotes arterial inflammation[J]. Circ Res. 2011, 108(8): 950–959.

Farmer SR. Transcriptional control of adipocyte formation[J]. Cell Metabolism. 2006, 4(4):263.

Gijsen FJ, Wentzel JJ, Thury A, Mastik F, Schaar JA, Schuurbiers JC, Slager CJ, van der Giessen WJ, de Feyter PJ, van der Steen AF, Serruys PW. Strain distribution over plaques in human coronary arteries relates to shear stress[J]. Am J Physiol Heart Circ Physiol. 2008, 295(4): H1608–1614.

Guo D, Chien S, Shyy JY. Regulation of endothelial cell cycle by laminar versus oscillatory flow: distinct modes of interactions of AMP-activated protein kinase and akt pathways[J]. Circ Res. 2007, 100(4): 564–571.

Hu XM, Lin T, Huang XY, Gan RH, Zhao Y, Feng Y, Ding LC, Su BH, Zheng DL, Lu YG. ID1 contributes to cell growth invasion and migration in salivary adenoid cystic carcinoma[J]. Mol Med Rep. 2017, 16(6): 8907-8915.

Hua X, Yokoyama C, Wu J, Briggs MR, Brown MS, Goldstein JL, Wang X. SREBP-2, a second basic-helix-loop-helix-leucine zipper protein that stimulates transcription by binding to a sterol regulatory element[J]. Proc Natl Acad Sci U S A. 1993, 90(24): 11603–11607.

Jin X, Jin X, Kim LJY, Dixit D, Jeon HY, Kim EJ, Kim JK, Lee SY, Yin J, Rich JN, Kim H. Inhibition of ID1-BMPR2 intrinsic signaling sensitizes glioma stem cells to differentiation therapy[J]. Clin Cancer Res. 2018, 24(2): 383–394.

Li YS, Haga JH, Chien S. Molecular basis of the effects of shear stress on vascular endothelial cells[J]. J Biomech. 2005, 38(10): 1949–1971.

Maw MK, Fujimoto J, Tamaya T. Overexpression of inhibitor of DNA-binding (ID)-1 protein related to angiogenesis in tumor advancement of ovarian cancers[J]. BMC Cancer. 2009, 9: 430.

Moreno PR, Purushothaman KR, Fuster V, Echeverri D, Truszczynska H, Sharma SK, Badimon JJ, O'Connor WN. Plaque neovascularization is increased in ruptured atherosclerotic lesions of human aorta: implications for plaque vulnerability[J]. Circulation. 2004, 110(14): 2032–2038.

Ni CW, Qiu H, Rezvan A, Kwon K, Nam D, Son DJ, Visvader JE, Jo H. Discovery of novel mechanosensitive genes in vivo using mouse carotid artery endothelium exposed to disturbed flow[J]. Blood. 2010, 116(15): e66–e73.

Patil M, Sharma BK, Satyanarayana A. Id transcriptional regulators in adipogenesis and adipose tissue metabolism[J]. Front Biosci (Landmark Ed). 2014, 19: 1386–1397.

Qiu JH. The mechanobiological mechanisms of Id1-induced angiogenesis in rupture-prone plaque formation [D]. Chongqing University, 2011.

Rosen ED, Walkey CJ, Puigserver P, Spiegelman BM. Transcriptional regulation of adipogenesis[J]. Genes & Development. 2000, 14(11): 1293–1307.

Ruzinova MB, Schoer RA, Gerald W, Egan JE, Pandolfi PP, Rafii S, Manova K, Mittal V, Benezra R. Effect of angiogenesis inhibition by Id loss and the contribution of bone-marrow-derived endothelial cells in spontaneous murine tumors[J]. Cancer Cell. 2003, 4(4): 277–289.

Satyanarayana A, Klarmann KD, Gavrilova O, Keller JR. Ablation of the transcriptional regulator Id1 enhances energy expenditure, increases insulin sensitivity, and protects against age and diet induced insulin resistance, and hepatosteatosis[J]. Faseb j. 2012, 26(1): 309–323.

Seale P, Kajimura S, Spiegelman BM. Transcriptional control of brown adipocyte development and physiological function--of mice and men[J]. Genes Dev. 2009, 23(7): 788–797.

Sharma S, Plotkin M. Id1 expression in kidney endothelial cells protects against diabetes-induced microvascular injury[J]. FEBS Open Bio 2020, 10(8): 1447–1462.

Tressel SL, Huang RP, Tomsen N, Jo H. Laminar shear inhibits tubule formation and migration of endothelial cells by an angiopoietin-2–dependent mechanism[J]. Arteriosclerosis Thrombosis & vascular Biology. 2007, 27(10): 2150–2156.

Verwoert GC, Franco OH, Hoeks AP, Reneman RS, Hofman A, Cm VD, Sijbrands EJ, Witteman JC, Mattace-Raso FU. Arterial stiffness and hypertension in a large population of untreated individuals: the rotterdam Study[J]. Journal of Hypertension. 2014, 32(8): 1606–1612.

Volpert OV, Pili R, Sikder HA, Nelius T, Zaichuk T, Morris C, Shiflett CB, Devlin MK, Conant K, Alani RM. Id1 regulates angiogenesis through transcriptional repression of thrombospondin-1[J]. Cancer Cell. 2002, 2(6): 473–483.

Wada S, Karino T. Theoretical study on flow-dependent concentration polarization of low density lipoproteins at the luminal surface of a straight artery[J]. Biorheology. 1999, 36(3): 207–223.

Wang G, Deng X, Guidoin R. Concentration polarization of macromolecules in canine carotid arteries and its implication for the localization of atherogenesis[J]. J Biomech. 2003, 36(1): 45–51.

Wei DH, Wang GX, Tang CJ. Concentration polarization of low density lipoprotein at the distal end of carotid stenosis promotes atherogenesis[J]. Acta Physiologica Sinica, 2007(06): 831–839.

Yokoyama C, Wang X, Briggs MR, Admon A, Wu J, Hua X, Goldstein JL, Brown MS. SREBP-1, a basic-helix-loop-helix-leucine zipper protein that controls transcription of the low density lipoprotein receptor gene[J]. Cell. 1993, 75(1): 187–197.

Zhang K, Chen YD, Zhang T, LU Huang L, Wang Y, Yin TY, Qiu JH, Hans G, Wang GX. A novel role of Id1 in regulating oscillatory shear stress-mediated lipid uptake in endothelial cells[J]. Ann Biomed Eng. 2018 Jun, 46(6):849–863.

Zhang T. The effects of substrate stiffness on HUVEC adhesion with THP-1 cells and molecules associated with adhesion[D]. Chongqing University, 2016.
Zhang K. The mechanobiological mechanisms of Id1-regulated lipid uptake in atherosclerosis plaque formation[D]. Chongqing University, 2018.

Chapter 7
Mechanical Mechanisms of In-Stent Restenosis and Atherosclerosis Following the Implantation of Intravascular Stents

In recent years, the number of individuals worldwide suffering from cardiovascular diseases (CVDs) caused by atherosclerosis has been on the rise. The annual global mortality rate due to coronary heart disease exceeds 14 million, surpassing that of other diseases. It is projected that by 2030, the annual global deaths from cardiovascular diseases will reach 23 million (World health statistics 2020 2020). Recently, percutaneous coronary intervention (PCI) has emerged as a highly mature and effective treatment modality for cardiovascular diseases. PCI offers advantages such as simplicity, safety, efficacy, and minimal invasiveness, revolutionizing traditional treatment approaches for coronary heart disease. Stent implantation procedures are primarily performed using PCI techniques (Garg & Serruys 2010; Gori 2021; Stefanini et al. 2014; Yelamanchili et al. 2020). However, in-stent restenosis (ISR) and late-stage vessel thrombosis have remained significant challenges accompanying stent technology advancement (Yang et al. 2021; Wang et al. 2018). From bare metal stents (BMS) to drug-eluting stents (DES) and currently biodegradable stents, researchers have extensively investigated this issue from various perspectives, including materials, mechanics, and molecular mechanisms (AbuRahma 2018; Borhani et al. 2018; Dangas et al. 2010; Ge et al. 2019; Nicolais et al. 2018; Rykowska et al. 2020; Yang et al. 2017; Zhao et al. 2018). With the continuous progress of science and technology, research methods have become increasingly rich and reliable, leading to a deeper understanding of the formation mechanisms and treatment strategies for in-stent restenosis and atherosclerosis (Ge et al. 2019; Jang et al. 2019; Li et al. 2017; Nicolais et al. 2018; Yang et al. 2021; Zhao et al. 2018).

The occurrence of ISR involves the following aspects:

① Lower wall shear stress (WSS) facilitates the deposition of lipid molecules and inflammatory factors on the vascular endothelium (Samady et al. 2011; Thondapu et al. 2018).

② Accelerated proliferation of smooth muscle cells in the vascular media (Curcio et al. 2011).

G. Wang, *The Mechanobiology in Atherosclerosis*,
https://doi.org/10.1007/978-981-97-6512-6_7

③ Synthesis of a large amount of collagen and elastin, leading to increased elastic fibers in the intimal layer, resulting in intimal thickening and stiffening (Chaabane et al. 2013; Suna et al. 2018).

Numerous studies have indicated that changes in the mechanical microenvironment within the blood vessel after stent implantation are significant factors contributing to ISR (Cheng & Zhang 2019; Fereidoonnezhad et al. 2017; He et al. 2020; Hung et al. 2016; Koskinas et al. 2012). In addition to pathophysiological reasons such as inflammation, drug resistance, thrombosis, rejection, and extracellular matrix effects, as well as patient-specific factors, local hemodynamic changes following stent implantation are also important and cannot be overlooked. ISR, especially in regions with low WSS values, is caused by excessive disruption of the vessel wall and abnormal hemodynamics within the stented vessel (Shishido et al. 2016).

7.1 Construction of the Rabbit Carotid Artery Stenosis Model and Stent Implantation Experiment

Animal experiments play a pivotal role in the development of PCI, as animal models are practically useful in almost every technical aspect of stent technology and early biological development. There are several animal models to choose from, offering various attributes suitable for specific evaluations (Shishido et al. 2016). Due to their unique characteristics in lipid metabolism and sensitivity to cholesterol diet, rabbits have been widely used in the study of human atherosclerosis (Fan et al. 2015). Currently, there are several methods for constructing rabbit models of atherosclerotic stenosis, including liquid nitrogen freezing injury, balloon injury, intraluminal air drying injury, microsuturing, and chemical cauterization (Liu et al. 2016). Among these methods, the first three are more commonly used due to their simplicity, modularity, and controllability of plaque sites. The balloon injury method involves repeatedly dragging an inflated balloon within the blood vessel to disrupt the endothelial structure, causing endothelial cell injury and dysfunction. Monocytes that are free in the bloodstream invade the vascular wall and transform into macrophages, promoting the accumulation of lipids and inflammatory cells at the site of injury. When combined with a high-fat diet, an atherosclerosis model can be obtained.

7.1.1 *Experimental Animal Model*

Healthy male New Zealand white rabbits weighing approximately 2 kg were used as experimental animals. The experimental animal model needed two surgeries. The first surgery involved balloon injury to induce carotid artery stenosis, and the second surgery involved stent placement in the narrowed segment of the carotid artery. For three days before the surgery and seven days after the surgery, the rabbits were fed medication through their feed to prevent infection or severe thrombotic death. Clopidogrel was administered at a dosage of 7.5 mg/kg/day, and aspirin was administered at a dosage of 10 mg/kg/day.

7.1.1.1 Construction of the Rabbit Carotid Artery Stenosis Model

The experimental balloon was thoroughly cleaned using ultrasound and then exposed to ultraviolet light overnight. It was wrapped in sterilized newspaper and kept ready for use. The PLLA intravascular stent delivery system (2.0 stent delivery system) was stored in a refrigerator at 4 °C and brought to room temperature 30 min before the surgery. Surgical instruments and medical cotton were sterilized in an autoclave and dried in an oven the day before the surgery.

The operating room was disinfected and subjected to overnight ultraviolet irradiation to ensure a sterile experimental environment. The animals were fasted and weighed before the experiment. They were injected intramuscularly with 0.25 mL/kg xylazine. Once the rabbit limbs were completely relaxed, a dose of 2.5 mL/kg 10% chloral hydrate was administered intraperitoneally. After complete anesthesia, the fur on the neck of the rabbit was shaved using clippers. The rabbit was placed supine and fixed on the operating table. The skin of the neck was sterilized with iodine solution for 3 min, followed by deiodination with 75% alcohol. Heparin sodium solution was injected into the marginal vein along the rabbit's ear. After covering with sterile surgical drapes, the skin and fascia in the middle of the neck were incised. The muscles were bluntly separated using tissue forceps to expose the carotid artery. Sterile cotton threads were used to occlude blood flow proximally and distally, with a ligated segment of approximately 20 mm. A small incision was made approximately 3 mm from the proximal end of the ligated section, and a dry cotton ball was used to absorb the outflowing blood. The balloon catheter guidewire was inserted through the incision and advanced into the ligated vessel segment. The balloon was inflated at a rate of 2 atmospheres every 5 s, up to a maximum of 10 atmospheres until it fully expanded. The balloon was then slowly deflated, causing negative pressure while being dragged back and forth within the vessel 3–5 times to disrupt the endothelial structure. Subsequently, the balloon was withdrawn. Carotid artery pulsation was observed to confirm normal blood flow, and the incision of the blood vessel was sutured, followed by removal of the ligatures. The muscle fascia and skin were sutured, and the animal was removed from the operating table, kept warm until fully awake, and observed for its condition while continuing medication for 7 days. The dosage and administration method of the medication remained the same as before the surgery.

7.1.1.2 Stent Implantation Experiment

Five weeks after balloon injury, the rabbits underwent stent implantation surgery. The operating room was disinfected and subjected to overnight ultraviolet irradiation the day before the surgery to ensure a sterile experimental environment. On the day of the experiment, the animals were fasted and weighed. They were injected intramuscularly with 0.25 mL/kg xylazine. Once the rabbit limbs were completely relaxed, a dose of 2.5 mL/kg 10% chloral hydrate was administered intraperitoneally. After anesthesia and sterilization of the rabbit, sterile surgical drapes were used. Using the suture site on the blood vessels retained after balloon injury as a reference, sterile cotton threads were used to occlude blood flow proximally and distally, with a ligated segment of approximately 25 mm. A small incision was made approximately 5 mm from the proximal end of the ligated section, avoiding the previous surgical incision site. A dry cotton ball was used to absorb the outflowing blood, and the stent delivery system was inserted through the incision into the ligated vessel segment. The position of the stent was visually confirmed to be at the proximal, middle, and distal positions relative to the suture line. After confirming the position of the stent, the balloon was detached, and pulsation of the carotid artery was observed. After suturing, the animal was removed from the operating table, kept warm until fully awake, and observed for its condition while continuing medication for 7 days. The dosage and administration method of the medication remained the same as before the surgery.

7.1.1.3 Experimental Sample Processing

Five weeks after balloon injury, two rabbits from the blank control group were randomly selected. After 9 weeks, the remaining four rabbits from the blank control group were euthanized for sampling and tissue section staining. The samples were placed in a 30% saturated sucrose solution and stored in a refrigerator at 4 °C for 12 h. Then, the samples were removed and placed in a 50% saturated sucrose solution for dehydration at 4 °C for 12 h. Subsequently, the samples were processed for sectioning. The OCT-embedded blocks with adhesive were clamped on the holder of a microtome, and the section thickness was adjusted to 20 μm. The handle was rotated to perform trimming. When the vascular tissue could be seen on the OCT-embedded block, the section thickness was adjusted to 8 μm, and the handle was rotated to start sectioning. The cut vascular tissue was flattened and placed on an adhesive-coated glass slide using a soft brush. The prepared sections were placed in a slide box and stored at −20 °C. The sections were processed using a KD-2950 semiautomatic cryostat microtome from Kedi Instruments.

The uninjured vascular segments were fixed with 4% paraformaldehyde for frozen sectioning. Subsequently, H&E staining and Masson's trichrome staining were performed to observe vascular reconstruction. The stented vascular segments were fixed with 4% paraformaldehyde, and the adjacent vascular segments of the stent were used for frozen sectioning followed by H&E staining and Masson's trichrome staining to observe intimal thickening of the blood vessel wall.

7.1.2 *Animal Model Processing Results*

The image data were analyzed using ImageJ software. Experimental results are represented as the mean ± standard deviation, and GraphPad Prism v7.04 software was used to analyze the experimental data and detect significant differences between groups.

7.1.2.1 Stained Pathological Morphological Observation

(1) H&E Staining: Frozen sections were removed and fixed in 10% formaldehyde for 5 min. After rinsing with running water for 2 min, the sections were immersed in distilled water for 3 min. Hematoxylin stain was dropped onto the target section and left for 2 min, followed by rapid rinsing with tap water. Subsequently, the sections were decolorized with 0.5% hydrochloric acid ethanol for 2 s and then rapidly rinsed with distilled water. Add 0.25% ammonia water until the tissues turn blue, and rinse with tap water for 1 min. Nuclear staining was observed under a light microscope. After staining with 1% eosin for 1 min, the sections were rapidly rinsed with distilled water and then washed with 80, 90, and 95% ethanol for 15 s each. The color of the nucleus and cytoplasm was observed under a light microscope. The sections were decolorized twice with 100% ethanol for 2 min each time. Xylene was added to enhance the refraction of the cover glass, added twice for 2 min each time. The images after H&E staining are shown in Fig. 7.1a, where significant differences in vascular morphology between the injured segment and the normal group can be observed. One month after balloon injury to the left common carotid artery in rabbits, there was a significant increase in smooth muscle cell proliferation and intimal thickening in the injured segment (Chen et al. 2015). In contrast, the right common carotid artery without injury treatment exhibited normal cellular morphology, intact elastic lamina structure in the media, and no intimal thickening. Based on the H&E staining images, the statistical results of the internal intimal thickness of the common carotid arteries in the normal group and the experimental group were measured (Fig. 7.1c). The intimal thickness in the injured segment was significantly increased compared to that in the normal group ($p < 0.01$).

(2) Masson's Staining: An appropriate amount of Weigert's iron hematoxylin A solution and Weigert's iron hematoxylin B solution were mixed equally to prepare the Weigert's iron hematoxylin staining solution. The staining solution was dropped onto the specimen on the slide, and after 2 min of staining, the slide was rinsed briefly with running water. The sample was differentiated with acid ethanol for 5 s and rinsed with running water for 2 min. The bluing solution was dropped onto the specimen for 5 s to restore the blue color, and the specimen was rinsed with running water for 2 min, stained with Ponceau fuchsin staining solution for 2 min, and then rinsed with running water for 1 min. Acetic acid working solution was prepared by mixing distilled water and acetic acid solution in a certain proportion and rinsing the sections with freshly

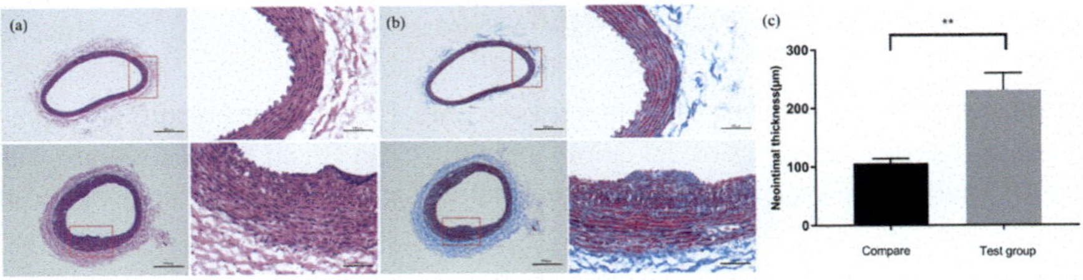

Fig. 7.1 Results of carotid artery staining in rabbits. **a** HE staining. **b** Masson staining. **c** Neointimal thickness (n = 6), **p < 0.01. Adapted from He et al. (2020)

prepared acetic acid working solution. The phosphomolybdic acid solution was dropped, and the specimen was immersed and then tilted and drained. Restain with aniline blue staining solution, tilt and drain the staining solution. Then, the sections were treated with acetic acid working solution and observed under a light microscope until no blue color was released from the sample. Dehydrate with ethanol: Rapid dehydration with 95% ethanol, followed by three rounds of dehydration with absolute ethanol. The images after Masson's staining are shown in Fig. 7.1b, where significant differences in vascular morphology between the injured segment and the normal group can also be observed. One month after balloon injury to the left common carotid artery in rabbits, an abundance of collagen fibers was found to distribute between the elastic lamina in the injured segment. It is believed that balloon injury stimulates smooth muscle cells in the vascular media to secrete a large amount of collagen protein, and the slow degradation rate of collagen protein further promotes intimal thickening. In contrast, the right common carotid artery without injury treatment exhibited normal cellular morphology, intact elastic lamina structure in the media, and no intimal thickening.

7.1.2.2 Animal Model Imaging Data

(1) Rabbit Carotid CTA Measurement: After feeding rabbits a high-fat diet for 5 weeks following balloon injury, computed tomography angiography (CTA) was performed to verify the carotid artery stenosis model. Conventional computed tomography (CT) provides good imaging of high-density tissues within the human body, but it has limited imaging capability for low-density tissues and organs such as blood vessels. By combining vascular contrast enhancement, CTA can clearly display the detailed characteristics of blood vessels. Prior to the experiment, the animals were fasted for 1 day. On the day of the experiment, rabbits were weighed first and then intramuscularly injected with 0.25 mL/kg luminol. After complete relaxation of the rabbit's limbs, an intraperitoneal injection of 10% chloral hydrate at a dose of 2.5 mL/kg was administered. With the rabbit fully anesthetized, an indwelling venous catheter was placed through the marginal ear vein and connected to the contrast agent injector of the CT machine. The contrast agent was injected at a dose of 4 mL/

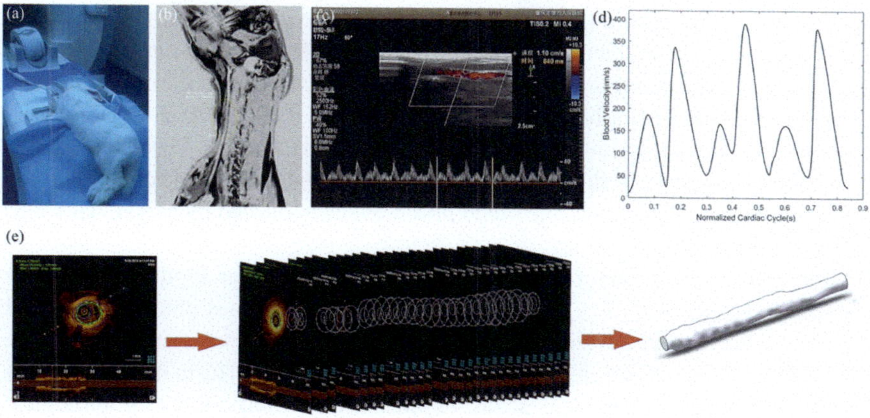

Fig. 7.2 Representative images of rabbit. **a** Rabbits were prepared for CT scanning after anesthesia. **b** Raw CTA data of the rabbit's left carotid artery. **c** Doppler ultrasound measurement. **d** Flow velocity waveform in the carotid artery derived from ultrasound. **e** Three-dimensional model reconstruction on the basis of the OCT data. Adapted from He et al. (2020)

kg and a rate of 0.7 mL/s. Subsequently, the CT scan was performed. The CT machine used in this experiment was a 64-row 128-slice helical CT machine produced by a company called LIAN YING with a slice thickness of 0.625 mm. The rabbits were positioned on the CT scanning bed until image acquisition. The image data of the scanned rabbits are shown in Fig. 7.2a. Figure 7.2b shows the left carotid artery after balloon injury, with the blue box indicating the region of vascular injury. There is significant tissue damage in that area, and the blood vessel exhibits a certain degree of stenosis. Combined with the results from histological staining, it can be concluded that the rabbit carotid artery stenosis model was successfully established.

(2) Rabbit Carotid Doppler Ultrasound Measurement: Doppler ultrasound, referred to as D-type ultrasound diagnosis, has the advantages of being safe, noninvasive, easy to operate, and cost-effective. It can provide continuous observation of organ movement and function. It is an important diagnostic tool in clinical practice for assessing anatomical structures and depth blood flow information using ultrasound. After collecting the CTA image data of the rabbits, Doppler ultrasound scanning of the carotid artery was performed to obtain blood flow velocity data in the injured vessel region. A Vvo2100 color Doppler ultrasound instrument was used for the ultrasound measurement, and the measurement results are shown in Fig. 7.2c. The ultrasound image reflects the blood flow velocity waveform during multiple cardiac cycles in the rabbit carotid artery. The red area represents the blood rendered by the ultrasound device. From the figure, it can be observed that the experimental animal has a high heart rate, with a single cardiac cycle lasting approximately 0.24 s and a heart rate of approximately 215 beats per minute. The peak systolic velocity at the end of systole is approximately 400 mm/s. The ultrasound-acquired image data are imported into SolidWorks software, and using spline curve fitting, the blood

flow waveform within three cardiac cycles (approximately 0.84 s) is obtained. The fitted curve is then imported into Getdata software in image format. The coordinate values are set based on the selected point values during ultrasound measurement, with a horizontal increment set at 0.002 s. After region-based selection, the original curve is discretized into 420 data points. The obtained inlet velocity data serve as the initial blood flow velocity value for the computational model. Figure 7.2d shows the blood flow waveform plotted by MATLAB based on the inlet velocity values.

(3) Rabbit Carotid Artery OCT Measurement: Intravascular Optical Coherence Tomography (OCT) is a novel imaging technique that has emerged in recent years following the development of CT and magnetic resonance imaging (MRI). This technique utilizes high-resolution visible light to perform highly spatially resolved three-dimensional scans of the surface morphology of biological tissues. Compared to CT, OCT has no radiation effects on the human body and overcomes the limitation of MRI in scanning metallic samples. However, it has the drawback of limited penetration depth. After 4 weeks of stent implantation in the experimental animals, OCT measurements were conducted on the stented vessel segment. The animals were euthanized in advance, and the stented vessel segments were extracted for ex vivo scanning. The OCT measurement was performed using the C7-XR system from St. Jude Medical. The measurement results are shown in Fig. 7.2e, clearly displaying the luminal contours of the vessel wall after stent implantation. The bright regions represent the vessel wall, while the shadowed areas indicate the blockage caused by the opaque stent wires. The fiber optic sensor of the OCT device had a moving speed of 15 m/s, and the system sampling frequency was set at 150 Hz. For a single vessel sample with a length of 30 mm, approximately 300 cross-sectional images, with a layer thickness of approximately 0.1 mm, were obtained. These images effectively depicted the local geometric characteristics of the vessel after stent implantation. For each sample image, approximately 40 cross-sections with remarkable geometric features were selected. The lumen contours of each image were extracted, aligned along the central axis using SolidWorks, and spatially arrayed. Subsequently, the geometric model was reconstructed through lofting operations. The reconstruction process and geometric model are illustrated in Fig. 7.2e.

7.2 Analysis of Mechanical Behavior During the Stent Expansion Process

Stent implantation is an effective method for treating arterial stenosis and occlusion, including atherosclerosis. It is estimated that over 1.5 million people worldwide undergo stent implantation procedures each year (Gori 2021; World health statistics 2020 2020; Yelamanchili et al. 2020). The general procedure involves precise localization of the diseased narrowed vessel using digital subtraction angiography (DSA), followed by femoral/radial artery puncture and deployment of the stent using a guidewire to the narrowed area. The vessel is then dilated by balloon angioplasty, and the stent delivery system is removed from the body. During this process, the stent undergoes physical expansion, resulting in plastic deformation and applying radial support forces on the vessel wall, thus restoring patency to the previously obstructed vessel. Clearly, direct contact occurs between the stent and the vessel wall during and after expansion, leading to significant deformation of the vessel wall. Therefore, investigating the mechanical environment of the vessel wall is of great significance for studying plaque rupture, neointimal mechanics, and in-stent restenosis.

With the rapid development of finite element methods and computer technology, finite element analysis (FEA) has been widely used to simulate the implantation process of stents and study various characteristics of stent design, such as smaller stent dimensions, balloon crimping characteristics, compliance during stent implantation, radial recoil after stent expansion, and higher reliability (Li et al. 2017; Khosravi et al. 2017; Wiesent et al. 2019). FEA methods can not only analyze the mechanical behavior of the stent but also investigate the interaction between the stent and the vessel, the vascular injury caused by the stent, and the fatigue life of the stent (He et al. 2020; Wei et al. 2019; Wu et al. 2007). Researchers are attempting to reveal the mechanisms of atherosclerosis in large arteries and peripheral vessels from a biomechanical perspective, improve the vascular mechanical environment, and prevent the occurrence of in-stent restenosis after percutaneous coronary intervention (PCI) (Fan & Deng 2017). Therefore, understanding the mechanical characteristics of the stent and vessel after implantation, as well as the local hemodynamic environment, is crucial for reducing future medical interventions and potential occurrences of restenosis or occlusion.

7.2.1 *Finite Element Model of Stent Expansion*

7.2.1.1 Solid Model

(1) Geometric model construction: Taking the 316 L vascular stent provided by Beijing Meizhongshuanghe Medical Devices Co., Ltd. as the research object, the parametric modeling software SolidWorks was used to construct the integrated model of the stent, balloon, vessel, and plaque. Figure 7.3a shows the simplified model of the S-ring stent. To facilitate convergence, a straight rod was used instead of the S-ring as the connecting rod in the calculation. The stent is composed of a closed-loop seven-element structure formed by a sinusoidal ring and straight rods. The geometric parameters of the stent are shown in the figure. The total length of the stent is 12 mm, the strut width is 0.1 mm, the thickness is 0.09 mm, and the initial outer diameter of the stent is 2 mm. Figure 7.3b shows the expanded structure model of the stent. A tubular surface is created inside the stent as the balloon model. The length of the tubular surface is 18 mm, and the

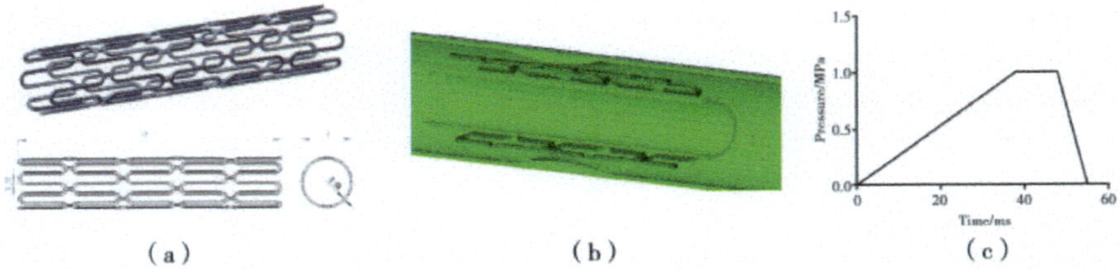

Fig. 7.3 Structural model of stent expansion. **a** Structure and geometry of the stent (mm). **b** Finite element model of stent expansion. **c** Pressure curve in the balloon. Adapted from He (2019)

initial outer diameter is 1.5 mm. The vessel and plaque models are established using B-spline curves. The maximum thickness of the plaque is 0.45 mm, the length is 6.5 mm, the outer diameter of the vessel is 3.2 mm, the length is 25 mm, and the wall thickness is 0.2 mm.

(2) Material properties: The material properties in the model refer to literature (Chiastra et al. 2013; Gastaldi et al. 2010). The stent is made of 316 L stainless steel and is simulated as a linear elastic material. The elastic modulus and Poisson's ratio are 200 GPa and 0.3, respectively. The model uses an 8-node hexahedral mesh, generating a total of 195,862 mesh elements. The vessel and plaque can be approximately considered as linear, isotropic, incompressible materials. The elastic modulus and Poisson's ratio of the vessel are 1.75 MPa and 0.499, respectively. The model uses a 10-node tetrahedral mesh, generating a total of 68,654 mesh elements. The elastic modulus and Poisson's ratio of the plaque are 2.19 MPa and 0.499, respectively. The balloon is simulated using a hyperelastic rubber material, and its material properties are described by the Mooney-Rivlin constitutive equation. The relevant parameters are shown in Table 7.1. The model uses an 8-node hexahedral mesh, generating a total of 7,328 mesh elements.

(3) Boundary conditions: To accurately simulate the expansion process of the stent, axial displacements and circumferential rotations of characteristic nodes at both ends of the stent need to be constrained. This ensures that the stent can expand properly while preventing any axial or circumferential movements. All degrees

Table 7.1 Material properties

Structure	Blood vessel	Plaque	Stent	Balloon
Material	Carotid artery	Calcified plaque	316 L	Rubber
Element type	10-node tetrahedron	10-node tetrahedron	8-node hexahedron	8-node hexahedron
Material model	Linear, isotropic	Linear, isotropic	Linear, isotropic	Hyperelastic
Density (kg mm^{-3})	–	–	7.98×10^{-6}	1.07×10^{-6}
Elastic modulus/gpa	0.00175	0.00219	200	$C_{10} = 0.0010$ $C_{01} = 0.0007$
Poisson's ratio	0.499	0.499	0.3	0.495

of freedom on the two end faces of the balloon are constrained to prevent any spatial changes during radial expansion. The inner surface of the balloon is uniformly subjected to a time-varying pressure profile, as shown in Fig. 7.3c. The pressure gradually increases from 0 to 1 MPa (approximately 10 atmospheres) over a period of 0 to 38 s and then remains constant for 10 ms. From 48 to 55 ms, the pressure inside the balloon is linearly decreased to remove the pressure.

(4) Computational procedure: There are three contact relationships during the stent expansion process, namely, stent inner surface-balloon, balloon-vessel inner wall, and vessel inner wall-stent outer surface contacts. In the ABAQUS environment, three contact surfaces are first established. The contact between the stent surface and the vessel wall is calculated using a penalty function algorithm to accurately control the motion of the contact node pairs between the two models, while the other two contact groups are set as frictionless contacts (Wang & Shao 1997). The commercial finite element software ABAQUS/Explicit v.6.13 is used to simulate the stent expansion process. Due to the presence of large deformations and complex contacts in the model during this process, explicit dynamic analysis is employed, and the convergence of the calculation is improved by reducing the time increment step size. The computational simulation process is performed on a DELL PRECISION T7910 graphical workstation with an E5-2640v4 processor.

7.2.1.2 Fluid Model

(1) Fluid properties: Blood is considered an incompressible Newtonian fluid. It is assumed to be a homogeneous continuous fluid, and the turbulent model is set to laminar flow. The motion of blood flow can be described by the Navier–Stokes (N–S) equations, which include the momentum and mass control equations as follows:

$$\rho\left[\frac{\partial u}{\partial t} + (u \cdot \nabla)u\right] + \nabla p - \mu\nabla^2 u = 0 \tag{7.1}$$

$$\nabla \cdot u = 0 \tag{7.2}$$

where u is the velocity of the blood flow, ρ is the density (1060 kg/m^{-3}), μ is the dynamic viscosity (3.5×10^{-3} m$^{-1}\cdot$s^{-1}), and p is the relative pressure with respect to atmospheric pressure.

(2) Model construction: Due to the significant deformation of the stent during expansion, the individual struts of the stent cannot tightly adhere to the vessel wall. It is challenging to obtain proper contact surfaces between the stent, plaque, and vessel. Assuming that the vibrations and deformations generated by the blood impact on the stent and vessel can be neglected, a one-way fluid-structure coupling approach is adopted. The contact surfaces between the stent, vessel, and blood are treated as rigid walls, and only the hemodynamic parameters of

the blood within this fluid domain are studied. Because the demanding requirements for contact surfaces are not necessary, a parametric modeling approach is used. A simplified hemodynamic model of vessel-stent coupling, as shown in Fig. 7.4a, was constructed using the commercial CAD software SolidWorks. In this model, the vessel and inner wall of the plaque are represented as uniform and smooth cylindrical surfaces. The stent adhered tightly to the inner wall of the vessel, and the portion of the stent inside the vessel was then removed through Boolean operations. The remaining filled blood inside the vessel serves as the computational fluid domain.

(3) Mesh generation: The hemodynamic model adopts an unstructured mesh generated using ICEM CFD v.15. After importing the geometric model of the expanded vessel with the stent in STL format, the global maximum mesh size is set. The smallest structural size in the model is approximately 0.1 mm, which corresponds to the thickness of the stent strut. Thus, the maximum mesh size is set to 0.1 mm, as shown in Fig. 7.4b. The generated mesh has high quality and can effectively represent the local details of the stent. Since our research focuses on the hemodynamic characteristics near the vessel wall, which require higher computational accuracy, after generating the tetrahedral volume mesh for the fluid domain, a boundary layer mesh needs to be generated near the wall. The boundary layer mesh is generated by extruding prismatic layers along the vessel wall based on the volume mesh. The total height of the boundary layer mesh is set to 0.25 m, with a linear growth rate of 1.2. A total of 5 layers are generated, as shown in Fig. 7.4c. After mesh generation, the mesh quality is checked. The ICEM provides various standards for mesh quality checks. The 'Quality' criterion is selected with a minimum standard set to 0.3, and the number of

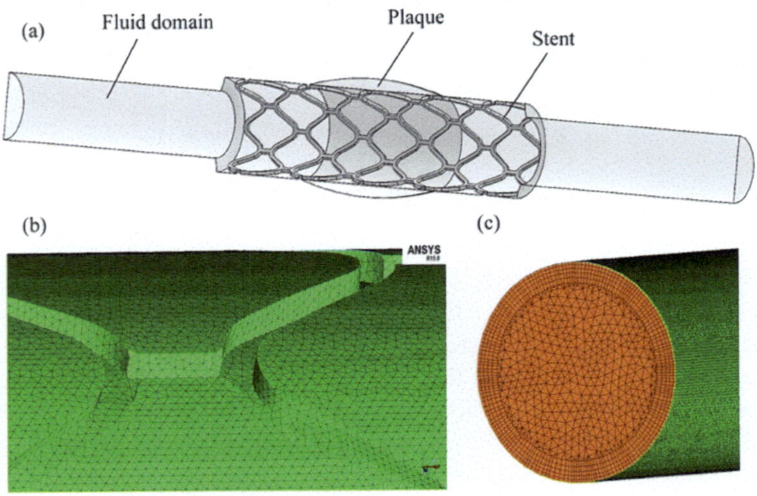

Fig. 7.4 Fluid model of stent expansion. **a** Simplified hemodynamic parameterized model. **b** Grid model near the stent. **c** Boundary layer mesh model. Adapted from He (2019)

smoothing iterations is set to 5. Ultimately, a high-quality unstructured mesh model with 3,868,515 elements is generated, and the mesh data are output using the CFX solver.

(4) Boundary and Initial Conditions: For the boundary conditions of the vessel wall, we assume it to be a rigid no-slip wall without considering wall velocity. The measured blood flow waveform is used as the inflow velocity (Fig. 7.2d). We collected velocity samples at 420 time points within three cycles. Therefore, transient analysis is performed for the calculation. The total time, time step, and initial time are set corresponding to the inflow velocity data and are determined as 0.84 s, 0.002 s, and 0 s, respectively. The inlet boundary condition is set as an open inlet, allowing normal inflow and outflow of blood at the inlet plane. The velocity is set as the velocity component in the newly established coordinate system, and the z-component is consistent with the velocity value at time 0. The outlet is set as a zero-pressure open outlet, assuming that the outlet flow has reached a steady state. The initial condition of the flow, namely, the initial velocity of the inflowing blood, is set by specifying the velocity components in the Cartesian coordinate system at the inlet. The positive direction of the z-axis represents the direction of blood inflow, and the z-component of velocity is set to the value at time 0, while the x- and y-components are set to 0.

(5) Model Computation: CFX transient analysis involves iterative calculations of the N–S equations. When the residuals of the flow rate and momentum fall below the preset values, the flow is considered to have reached a steady-state convergence. Prior to computation, residual values and maximum iteration counts need to be set. The residual type is chosen as RMS, and the residual value is set at 1×10^{-6}. Based on the convergence iteration count during steady-state computation of this model, the maximum iteration count is ultimately set to 50. To ensure better convergence of the computation, three monitoring points are selected near the vessel wall in the vicinity of the stent. The velocity and absolute pressure at each point are monitored. Convergence is considered achieved when the residual curve falls below the RMS preset value and both the velocity and pressure curves at the monitoring points reach a steady state. Finally, the computation process is performed using double-precision parallel computing on the DELL workstation PRECISION T7910E5-2640v4.

(6) Hemodynamic Parameters: Based on the parameter settings in the preprocessing, the computation converges within the third cardiac cycle. We chose data from the third cycle for postprocessing analysis. Through postprocessing operations, hemodynamic parameters of specified regions can be obtained, such as simple parameters such as wall shear stress (WSS), streamlines, shear strain rate (SSR), and complex parameters such as time-averaged wall shear stress (TAWSS), oscillatory shear index (OSI), and relative residence time (RRT).

WSS refers to the tangential tension exerted by blood flow on endothelial cells, which represents the frictional force of blood flow on the cell surface. Changes in wall shear stress, abnormal blood flow conditions such as nonlaminar flow, turbulence, and boundary layer separation may be related to the development of atherosclerosis.

In this calculation, the formula for calculating wall shear stress is:

$$\tau_w = n \cdot \underset{\text{ij}}{\overrightarrow{\overrightarrow{\tau}}} \tag{7.3}$$

Here, n is the normal vector of the flow field's wall, and $\underset{\text{ij}}{\overrightarrow{\overrightarrow{\tau}}}$ is the fluid viscous stress tensor. τ_w is a vector whose magnitude equals the viscous stress on the surface, and the direction corresponds to the direction of the viscous stress acting on the surface.

SSR refers to the rate of change of fluid flow velocity with respect to the distance from the wall. Its calculation formula is:

$$\text{SSR} = \frac{\mathrm{d}v}{\mathrm{d}r} \tag{7.4}$$

Here, v represents the axial velocity profile of the blood flow, and r represents the distance from the sampling point to the wall of the fluid domain. For Newtonian fluids, the ratio of shear stress to shear rate is the fluid viscosity. Numerous studies have shown that low shear rates near the vessel wall can lead to endothelial dysfunction, promoting lipid deposition and plaque formation.

TAWSS is the time-averaged spatial shear stress integrated over a period of time. In this context, it calculates the time-averaged wall shear stress by integrating the shear stress at each time point within a single cardiac cycle. The calculation formula is:

$$\text{TAWSS} = \frac{1}{T}\int_0^T |\tau_w|\,\mathrm{d}t \tag{7.5}$$

Here, T represents the duration of the third cardiac cycle in the ultrasound results. TAWSS allows analysis of the distribution characteristics of the average wall shear stress on the vessel wall within a single cardiac cycle.

OSI is a temporal parameter that reflects the oscillation degree of WSS within a single cardiac cycle, representing the change in shear stress direction. Clinical research has shown that a high oscillatory shear index can promote the formation of atherosclerotic plaques. The calculation formula is:

$$\text{OSI} = \frac{1}{2}\left(1 - \frac{\left|\int_0^T \tau_w\,\mathrm{d}t\right|}{\int_0^T \tau_w\,\mathrm{d}t}\right) \tag{7.6}$$

From Eqs. (7.6), it can be seen that the magnitude of the oscillatory shear index ranges from 0 to 0.5. When the oscillation is significant, the OSI approaches 0.5, and when the oscillation is low, the OSI approaches 0.

RRT is an important parameter associated with the occurrence of atherosclerosis. Its calculation formula is:

$$\mathrm{RRT} = \frac{1}{(1 - 2\mathrm{OSI}) \cdot \mathrm{TAWSS}} \tag{7.7}$$

Physically, RRT describes the relative residence time of particles on the blood vessel wall within a single cardiac cycle. It plays a crucial role in studying the distribution of low-density lipoprotein and its relation to the occurrence of atherosclerosis. It can be observed from the equation that the size of RRT is related to both OSI and TAWSS, and its distribution generally resembles the calculation results of OSI.

7.2.2 *Numerical Calculation Results Analysis*

7.2.2.1 Solid Model Results Analysis

After applying pressure inside the balloon, as shown in Fig. 7.3c, the changes in the geometry of the stent and blood vessel are illustrated in Fig. 7.5a–f. From 0 to 38 ms, the stent undergoes elastic deformation, gradually conforms to the shape of the blood vessel, and causes it to expand. From 38 to 48 ms, the pressure inside the balloon remains constant. During this stage, the stent primarily undergoes plastic deformation, exhibiting slow morphological changes, while the blood vessel extends to its maximal shape. Subsequently, the balloon pressure is relieved, and due to plastic deformation, the stent retains its shape, maintaining the expanded state of the blood vessel. It can be observed from the figures that the overall stress level on the surface of the stent is much higher than that on the blood vessel wall. As our focus is the change in the mechanical microenvironment inside the blood vessel after stent expansion, particularly the mechanical characteristics in the region of the plaque, excessive analysis of stent stress–strain is not conducted. Figure 7.6 show the distribution and variation of equivalent stresses on the blood vessel wall during the stent expansion process. When the stent initially contacts the blood vessel, due to the narrowed lumen caused by the plaque, the stress on both sides of the plaque is higher, while the stress on the plaque surface is lower. With further expansion of the stent, the stress at the plaque gradually increases, and the lumen of the blood vessel also enlarges. After complete expansion of the stent, the stress within the plaque is at a relatively low level, and high stress occurs on both sides of the plaque on the blood vessel wall, especially at the contact location between the stent and the proximal end of the plaque, as shown by the red circle in Fig. 7.6. The highest stress reaches 2.4 MPa. Additionally, since the stent model used in our study is a closed-loop mesh-like structure, it can be observed from the figure that the stress level at the direct contact position between the blood vessel wall and the stent strut is significantly higher than the stress values on the internal mesh of the blood vessel wall.

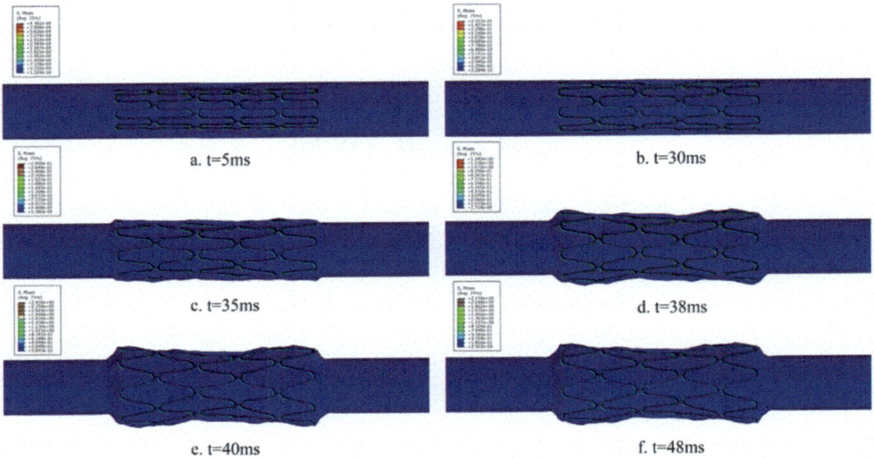

Fig. 7.5 Stent expansion process. Adapted from He (2019)

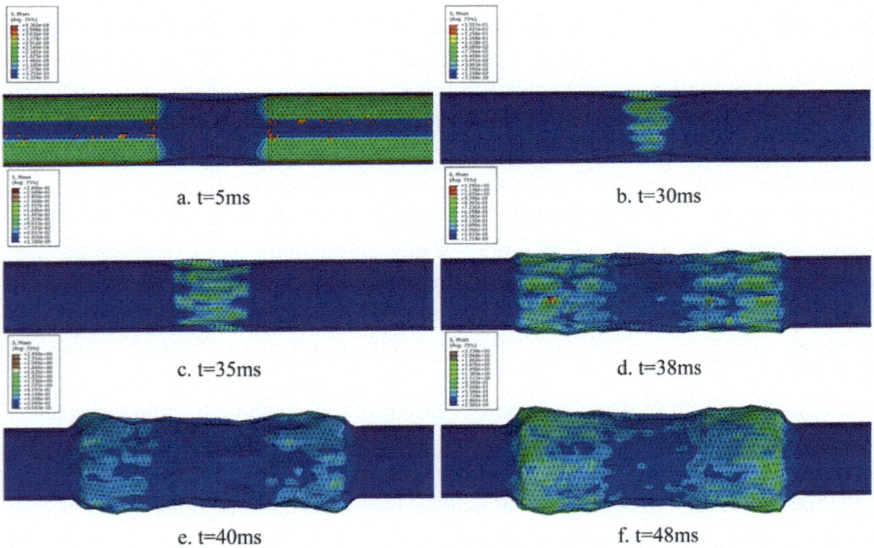

Fig. 7.6 Equivalent stress distribution of the vessel wall (MPa). Adapted from He (2019)

7.2.2.2 Hemodynamic Analysis

After parameterization modeling and transient calculations based on the finite volume method, we analyzed the hemodynamic characteristics of blood flow inside the blood vessel after stent implantation. Figure 7.7a–d display the distribution of wall shear stress (WSS), streamlines, shear strain rate (SSR), and vortices during the diastolic peak blood flow velocity. Figure 7.7e, f provide the distribution maps of the time-averaged wall shear stress (TAWSS), oscillatory shear index (OSI), and relative residence time (RRT) on the vessel wall during a single cardiac cycle. From Fig. 7.7a, it can be observed that the hemodynamic characteristics on the surface of the stented segment are distinct. Overall, the SSR is lower at the entrance of the stent, gradually increases in the direction of blood flow, and reaches its peak at the third element from the outlet. Subsequently, the WSS decreases at the outlet, indicating relatively high levels of WSS in the middle of the stented segment. The WSS near the stent struts is lower, especially near the connector strut. The WSS at the top of the stent struts is higher than that at the sides, as the struts overall influence the blood flow. The blood flow near the stent struts is slower, but the velocity increases at the top of the struts, resulting in a higher WSS in that region. Figure 7.7b shows the fitted streamline pattern, revealing that the blood flow velocity in the central axis region of the vessel adheres to the characteristics of laminar flow in a cylindrical tube. The velocity is particularly low near the vessel wall, especially near the stent struts, and significant low-speed disturbed flow occurs at the entrance of the stent.

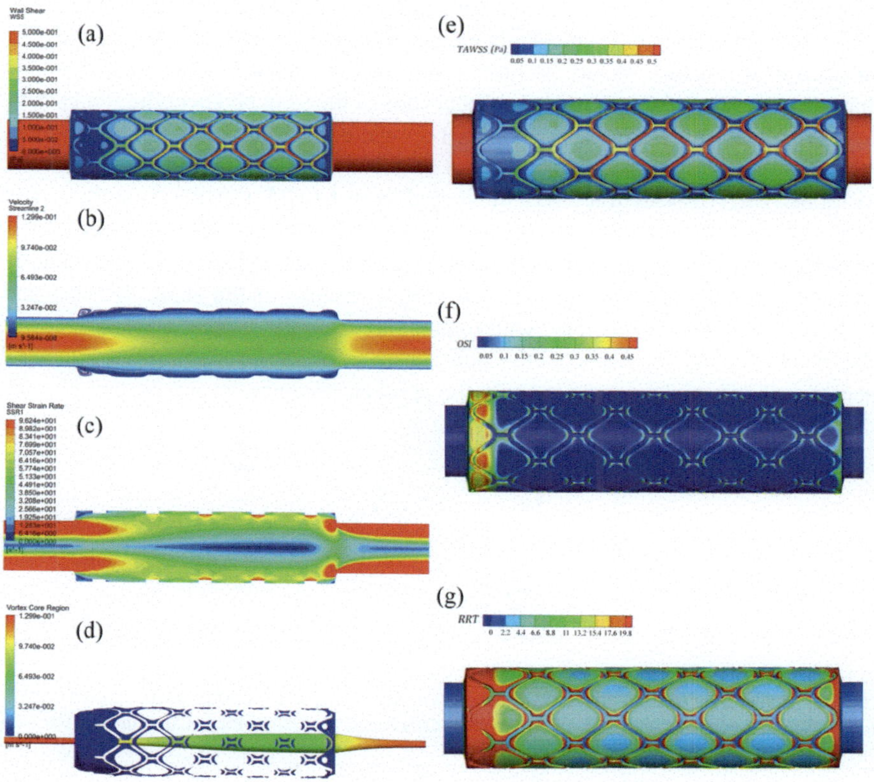

Fig. 7.7 Distribution map of hemodynamic parameters. **a** WSS. **b** Streamline. **c** SSR. **d** Vortex core. **e** TAWSS. **f** OSI. **g** RRT. Adapted from He (2019)

To further reveal the distribution of low-speed disturbed flow regions on the vessel wall, Fig. 7.7d shows that the low-speed vortex regions mainly occur at the entrance of the stent. Additionally, a small amount of low-speed vortices is observed in the outlet region, especially near the stent struts and connector struts. Figure 7.7c indicates that the low SSR near-wall region mainly appears near the front end of the stent, while the high SSR primarily occurs at the top of the stent struts and on the vessel wall at the inlet and outlet. The TAWSS distribution in Fig. 7.7e is consistent with the WSS distribution in Fig. 7.7a. The OSI distribution in Fig. 7.7f shows significant oscillations near the stent struts, and the OSI at the entrance of the stent is notably higher than in other regions, indicating significant oscillation when blood enters the stented segment, consistent with the vortex distribution shown in Fig. 7.7d. As mentioned earlier, the RRT distribution is generally consistent with the OSI. From Fig. 7.7g, it can be observed that areas with higher OSI correspondingly exhibit higher levels of RRT. In the entire stented segment, high RRT primarily occurs near the inlet of the stented vessel and near the stent struts, and there are also regions of higher RRT at the outlet of the stent.

To better analyze the hemodynamic characteristics of different regions within the stented vessel segment, we extracted several cross-sections within the stent region, as shown in Fig. 7.8a, and statistically analyzed the magnitudes of TAWSS and SSR for each cross-section. Since the previous results indicated distinct hemodynamic characteristics near the proximal and distal ends of the stent and near the stent struts, more cross-sections were placed in those regions. In the middle section of the stent, adjacent cross-sections were positioned at the stent support strut and connector strut positions. The cross-sections were numbered from 1 to 25, starting from the vessel inlet. Figure 7.8b, c display the mean values of TAWSS and SSR for each cross-section. Near the entrance of the stent, both TAWSS and SSR exhibited significantly lower values than other regions. In the middle section of the stent, TAWSS and RRT show uniform fluctuations, with higher values on the support struts and lower values at the connector struts. At the stent outlet, TAWSS drops sharply, while RRT increases significantly, consistent with the previous analysis.

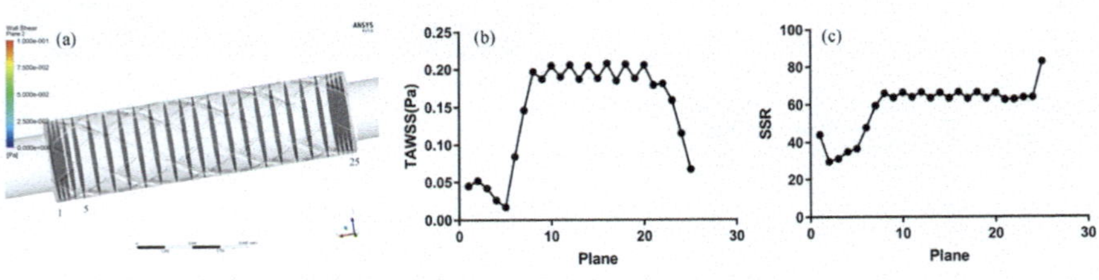

Fig. 7.8 Acquisition of hemodynamic parameters at different cross sections. **a** Cross section position. **b** The numerical values of TAWSS on each cross section. **c** The numerical values of RRT on each cross section. Adapted from He (2019)

7.3 Hemodynamic Analysis Based on OCT Imaging

With the development of computers, medical imaging technology has been widely used in the diagnosis of cardiovascular diseases. Medical imaging techniques analyze the imaging manifestations of the etiology for the diagnosis of cardiovascular diseases, providing an important basis for the objective evaluation of vulnerable plaque morphology, structure, composition, and biomarkers, as well as early prevention and treatment. Optical coherence tomography (OCT), as the latest imaging technology applied intravascularly, stands out for its high resolution, high image clarity, and easy recognition of images, meeting the requirements of clinicians to observe various tissue characteristics inside blood vessels (Olinic et al. 2019; Tearney et al. 2012). Due to its higher sensitivity in characterizing lipid-rich plaques compared to intravascular ultrasound, OCT is commonly used for detecting vulnerable plaques (Buccheri et al. 2017). OCT can also be an ideal method for determining the indications for and endpoints of rotational atherectomy of calcified lesions before stent implantation by evaluating calcium thickness and curvature (Nakajima et al. 2021). OCT has achieved optimization of stent implantation through immediate and semi-automatic quantification of stent apposition and expansion, leading to potentially improved clinical outcomes (Kurogi et al. 2021).

7.3.1 3D Model Reconstruction

Three-dimensional reverse reconstruction is a process that utilizes two-dimensional projections based on imaging data to reconstruct three-dimensional information of the physical object. It mainly includes steps such as data acquisition, point cloud processing, feature recognition and analysis, and geometric restoration. For clinical imaging data, the software Mimics (Materialise's interactive medical image control system) developed by Materialise in the United States is used for reconstruction. This software is a versatile grayscale image segmentation and visualization tool that can automatically identify coronal and sagittal planes in CT and MRI images. It enables fast segmentation, 3D rendering, and finite element analysis through operations such as threshold segmentation and region growing. Currently, this software is widely used in clinical anatomical practices, including cardiovascular system modeling and simulation, as well as bone tissue reconstruction.

7.3.1.1 Model Construction

If 316 L metal stents are used in the experiment, severe metal artifacts will occur under the routine clinical high-resolution CT scanner, which will greatly interfere with geometric reconstruction. The size of the stent structural unit is small, while the typical slice thickness of conventional CT is usually 0.625 mm. Using CT to acquire images will also result in a significant loss of geometric information in the model. Moreover, after a period of implantation, the geometric morphology of the stent has undergone obvious changes, especially the coupling interface between the stent and the blood, which is no longer a continuous and complete surface. It is difficult for CT to obtain this coupling interface. Taking into account these factors, OCT is used to obtain the inner contour of the target blood vessel, and then parametric processing of the contour is performed using SolidWorks software to finally obtain a more realistic geometric model of the stented vessel segment. OCT scans were performed on eight implanted stented vessel segments, and the processed images of each sample were used. More than 40 cross-sections with significant geometric features were selected. First, the lumen contours of each image were extracted, and then the contours were aligned along the centroid using SolidWorks and spatially arrayed. Finally, the geometric model was reconstructed through a sweeping operation, as shown in Fig. 7.3e in terms of the reconstruction process and the geometric model.

7.3.1.2 Model Setup

The meshing, boundary conditions, and computational settings of the model are described in Sect. 7.2. Tetrahedral unstructured meshes were used for the interior of the blood vessels, and prism meshes were used for the near-wall boundary layers. Transient computation type was selected, and the inlet boundary condition was set to the blood flow velocity data obtained from ultrasound measurements in the carotid arteries of the eight rabbits. Due to the relatively simple computational model, the steady-state condition calculation converged well, so the iteration number was set to 30 to reduce the computation time. The positions of stent implantation (proximal end, middle section, distal end) were considered ordinal independent variables, and the blood flow hemodynamic parameters in the pre, mid-, and poststented vessel segments were analyzed for their correlations to investigate the influence of different stent implantation positions on the distribution of blood flow hemodynamic parameters within the vessel. Figure 7.9 show the locations of stent implantation and the regions of the front, middle, and back segments of each vessel segment.

7.3.2 *Hemodynamic Analysis*

7.3.2.1 Hemodynamic Parameters

A comparative analysis of the hemodynamic parameters of the three models was conducted, and the differences in the numerical values of parameters in different regions of the blood vessels were statistically analyzed (Table 7.2), especially the hemodynamic characteristics of the primary and vulnerable high-risk regions of atherosclerotic plaques mentioned earlier.

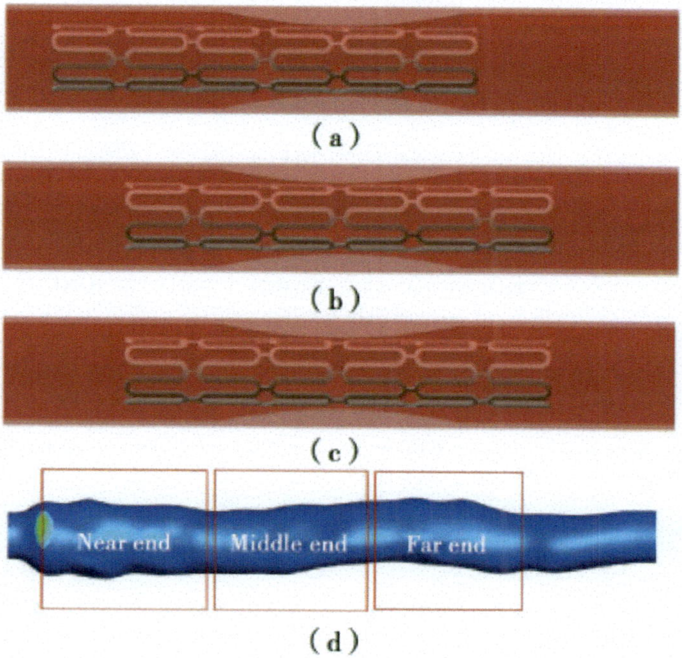

Fig. 7.9 The different positions of the stent implantation and blood vessel segment. **a** Proximal model. **b** Middle model. **c** Distal model. **d** The near-end, middle-end and far-end regions for each blood vessel segment. Adapted from He et al. (2020)

(1) TAWSS: Fig. 7.10 show the distribution heatmap of wall shear stress (TAWSS) on the vessel walls for the three models. For the proximal model, the overall level of wall shear stress is highest, with a small region of low shear force mainly distributed at the entrance of the vessel. The shear force in the middle and posterior segments of the vessel is relatively evenly distributed. For the central model, the distribution of wall shear stress is more unique, with a relatively high shear stress in the central area of the plaque in the middle segment of the vessel and obvious low shear stress regions at the inlet and outlet of the vessel. The region corresponding to the inlet of the vessel is the low shear disturbed flow region near the stent, which poses a higher risk of neointimal hyperplasia and restenosis. For the distal model, there are obvious low shear stress regions at the bifurcation, corresponding to the central area of the plaque. The lower shear force in this region may have adverse effects on plaque rupture and in-stent restenosis. The distribution of wall shear stress in the middle and posterior segments of the vessel is relatively uniform, with a clear low shear stress region at the exit of vessel number 4.
(2) OSI: Fig. 7.11 show the distribution heatmap of the oscillatory shear index (OSI) on the vessel walls for the three models. From the figure, it can be observed that the distribution of the OSI has distinct regional characteristics, with high

Table 7.2 The hemodynamic descriptors for the three models in different regions

Parameter	Model	Anterior segment	Middle segment	Posterior segment
TAWSS	Proximal end model	0.63 ± 0.11	0.79 ± 0.24	0.93 ± 0.37
	Central end model	0.32 ± 0.09	0.51 ± 0.18	0.40 ± 0.07
	Distal end model	0.12 ± 0.27	0.33 ± 0.069	0.43 ± 0.03
	p	0.0026	0.062	0.081
OSI	Proximal end model	0.00058 ± 0.00075	0.00016 ± 0.00006	0.00037 ± 0.00045
	Central end model	0.021 ± 0.022	0.00013 ± 0.00004	0.00025 ± 0.00004
	Distal end model	0.09 ± 0.014	0.00021 ± 0.00001	0.00093 ± 0.0012
	p	0.004	0.298	0.6
RRT	Proximal end model	3.432 ± 0.651	1.912 ± 0.491	1.961 ± 0.595
	Central end model	5.429 ± 0.275	2.196 ± 0.385	4.500 ± 0.701
	Distal end model	6.771 ± 1.317	3.384 ± 0.164	3.200 ± 0.283
	p	0.026	0.037	0.020

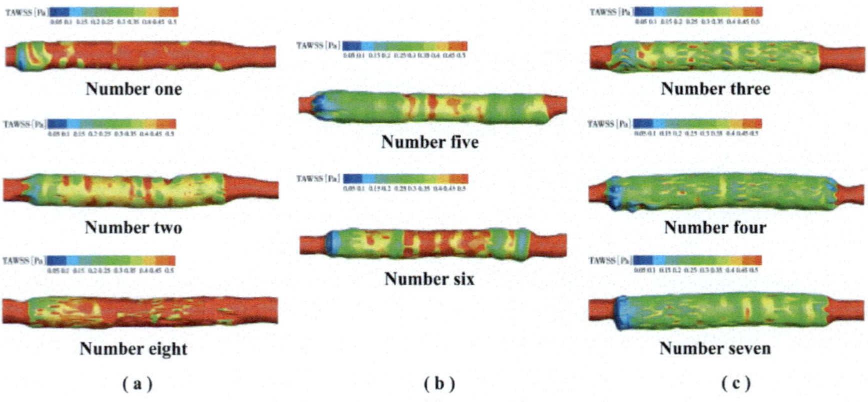

Fig. 7.10 Distribution of TAWSS in the vascular model. **a** Three proximal models. **b** Two middle models. **c** Three Distal models. Adapted from He et al. (2020)

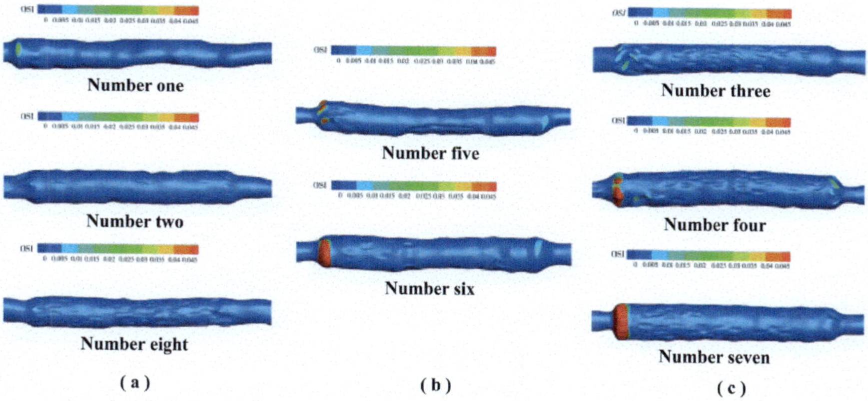

Fig. 7.11 Distribution of the OSI in the vascular model. **a** Three proximal models. **b** Two middle models. **c** Three Distal models. Adapted from He (2019)

OSI values occurring in the central and distal models. For the proximal model, the OSI remains at a relatively low level, with only a small area of high OSI at the inlet of vessel number 1 and almost no high OSI regions in vessels number 2 and 3. For the central model, there are significant high OSI regions at both the inlet and outlet of the vessel, and at the inlet, there is a significant high OSI region corresponding to the low shear stress region. For the distal model, vessel number 3 exhibits a relatively normal OSI, with only a slightly high OSI at the inlet, while vessel number 4 shows a higher OSI at the inlet and outlet, consistent with the low shear stress region, and vessel number 7 has a very high OSI at the inlet, with other regions maintaining a low level of OSI.

(3) RRT: Fig. 7.12 shows the distribution heatmap of wall-relative residence time (RRT) on the vessel walls for the three models. From the figure, it can be observed that the distribution of RRT on the vessel walls is similar to the distribution of TAWSS. Higher RRT values are typically present in regions of low shear stress and high OSI. Overall, the distal model has the highest RRT level, while the proximal model has the lowest RRT level, with a smaller area of high RRT and smaller extreme values. Higher RRT values occur in the region near the inlet of the vessel. In the central model, high RRT values are mainly present at the vessel inlet, where high OSI and low shear stress are also observed. In the distal model, high RRT values are present at the vessel inlet and near the struts of the stent. Vessel number 4 exhibits higher RRT values in the high OSI region at the outlet.

7.3.2.2 Correlation Between Hemodynamic Parameters and Stent Position

To better analyze the influence of different stent implantation positions on the size of hemodynamic parameters in the proximal segment of the vessel, multiple sample

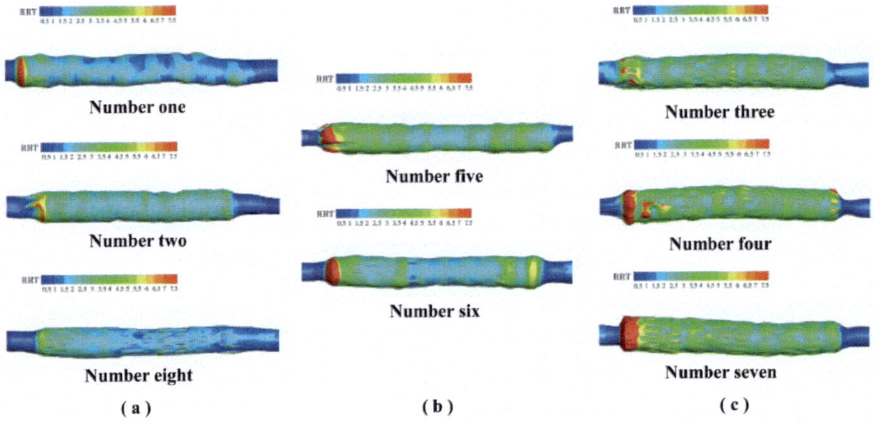

Fig. 7.12 Distribution of RRTI in the vascular model. **a** Three proximal models. **b** Two middle models. **c** Three Distal models. Adapted from He (2019)

points of TAWSS, OSI, and RRT were randomly collected in the proximal segment. Twenty sample points were selected for the proximal model, and thirty sample points were selected for the distal model, resulting in a sample size of 180. The stent implantation positions were categorized into three levels from proximal to distal as 1, 2, and 3, and the Spearman correlation analysis results are shown in Table 7.3. For the proximal model, the correlation coefficients between hemodynamic parameters (TAWSS, OSI, and RRT) and stent implantation distance are relatively high, reaching –0.718, 0.898, and 0.818, respectively, indicating good correlation with statistical significance within the confidence interval. As the stent implantation position moves farther away, the TAWSS in the proximal segment of the vessel gradually decreases, while the OSI and RRT increase. For the central model, the correlation coefficients between the three hemodynamic parameters and stent implantation position were relatively small, indicating a lower correlation. For the distal model, the correlation coefficients between the three hemodynamic parameters and stent implantation position were also relatively small, especially the correlation coefficient between OS and distance, which was only 0.055, indicating no significant correlation between the three parameters and stent implantation position.

7.3.3 Analysis of Vascular Pathological Morphology in the Stent Segment

Samples were taken from the proximal segment of the stented vessels and stained, as shown in Fig. 7.13. The HE staining results showed that one month after stent implantation, all three models exhibited intimal thickening in the proximal segment of the stents. However, compared to the middle and distal models, the proximal

Table 7.3 Spearman rank correlation analysis between the hemodynamic descriptors in different regions of the stenting segment and stent implantation positions

Model	Hemodynamic parameters	Spearman's rank correlation coefficient
Proximal end	TAWSS	−0.718**
	OSI	0.898**
	RRT	0.818**
Central end	TAWSS	−0.206*
	OSI	0.190*
	RRT	0.231*
Distal end	TAWSS	−0.213**
	OSI	0.055
	RRT	0.265**

Note Arranged from proximal to distal positions, numbered from 1 to 3, with a sample size of 180 for each calculation. * $p < 0.05$; ** $p < 0.01$

model (A) showed a lower degree of intimal thickening. The Masson staining results revealed that in the middle and distal models (B, C), a large amount of collagen fibers formed within the vessel media, disrupting the elastic lamina structure and causing severe intimal thickening. In contrast, the proximal model exhibited a more orderly arrangement of the elastic lamina. This further indicates that changes in the stent implantation position have an impact on the vascular structure in the proximal segment. Based on the HE staining images, the statistical results of intimal thickness in the proximal segment of the stented vessels for the three models were measured (Fig. 7.13c). The intimal thickening in the middle and distal models was significantly increased compared to that in the proximal model ($p < 0.01$). There was also a significant increase in intimal thickening when comparing the middle and proximal models ($p < 0.01$) and the distal and middle models ($p < 0.05$). Therefore, it can be concluded that as the stent implantation position moves farther away, the degree of intimal thickening in the proximal segment of the vessels significantly increases.

7.4 Research Progress and Prospects

Percutaneous coronary intervention (PCI) has been widely used to treat atherosclerosis. However, restenosis caused by stent implantation is one of the main drawbacks of this interventional treatment, which can lead to vascular injury, neointimal formation, and subsequent stent failure, requiring repeat interventions. According to clinical statistics, 20–30% of patients develop in-stent restenosis (ISR) within 12 months after bare metal stent implantation (AbuRahma 2018; Gori 2021). Although the use of new-generation drug-eluting stents helps suppress ISR compared to bare metal stents, the incidence of postoperative ISR remains as high as 10% (Aoki & Tanabe 2021; Nicolais et al. 2018; Rykowska et al. 2020). Therefore, it is crucial to explore

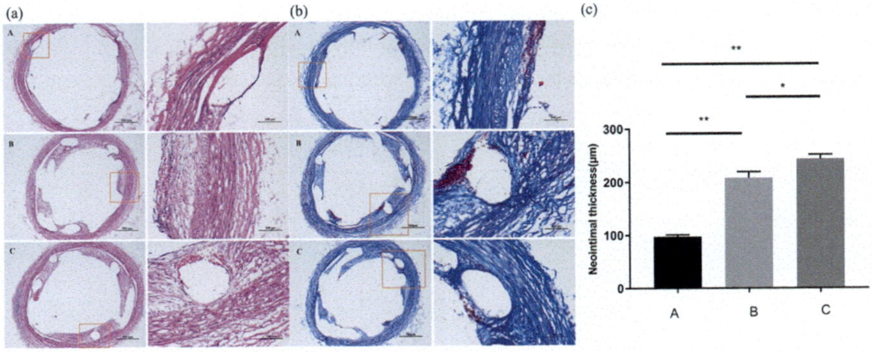

Fig. 7.13 Pathological analysis of stent segment blood vessels. **a** HE staining. **b** Masson staining. **c** Neointimal thickness. [**A** Proximal cardiac model; **B** middle cardiac model; **C** distal cardiac model. $^{**}p < 0.01$;$^{*}p < 0.05$]. Adapted from He (2019)

the causes of ISR and find better treatment options associated with achieving better therapeutic outcomes (Borhani et al. 2018; Nicolais et al. 2018). Vascular imaging during interventional procedures, coronary artery imaging, the overall condition of the patient, and intracoronary imaging of drug and ISR sites are all considered potential mechanisms for identifying ISR. Current medical approaches are not able to fully modify the biological and genetic factors of ISR. There is a need to focus on tailored treatment methods that target the mechanical mechanisms of atherosclerosis and surgical technical factors after stent implantation to reduce the occurrence of ISR.

7.4.1 Animal Models for Stent Implantation

Animal models for stent implantation provide practical testing platforms to assess the feasibility, safety, and efficacy of new devices. These animal models include small species (mice, rats), medium-sized species (rabbits), and large species (pigs, sheep, dogs, primates). Currently, the safety, toxicity, technical feasibility, and efficacy of stents have been addressed. However, the effectiveness is still limited by the lack of realistic models for atherosclerosis. This can be overcome by using animal models with metabolic syndrome or genetically modified animals (Iqbal et al. 2016). Regulatory agencies provide explicit information on the role of animal testing, and guidelines published by experts are available. The choice of animal depends on the experimental objectives, but when selecting an animal model, some general principles need to be considered:

① Animals should be readily available and affordable for maintenance, easy to handle, and appropriate in size to allow all expected experimental procedures.
② Animals should be bred in a laboratory environment with a clear genetic background.
③ The selected animal model should exhibit a high degree of similarity to human lipid metabolism and pathophysiological characteristics (Fan et al. 2018).

Rabbits are among the most commonly used experimental animals in biomedical research. Their short lifespan, short gestation period, high offspring numbers, low cost, and availability of genomics and proteomics make them an excellent species to investigate various aspects of human diseases, such as atherosclerosis (Fan et al. 2018). The pig coronary artery stenting model is considered the gold standard for safety assessment. Pigs have a long history in the development of percutaneous coronary intervention, particularly in understanding the biological properties of vascular intervention, repair, and new interventional devices (Li et al. 2020). However, due to species differences between animals and humans, these models still exhibit some variations in blood lipid levels, lesion morphology, and the development and morphological characteristics of human atherosclerosis (Sun et al. 2011). Additionally, considering the economic costs and technical difficulties, stent implantation procedures have not been performed in digital subtraction angiography. In the future, we will explore more reliable and representative animal models with greater similarity and reproducibility to human atherosclerosis.

7.4.2 *Medical Imaging Evaluation of Stent Implantation*

Advanced medical imaging techniques such as coronary angiography (CA) can determine the presence of stenotic lesions in blood vessels, provide a definitive diagnosis regarding the location, extent, severity, and condition of the vessel wall, and help decide on treatment options (interventional, surgical, or medical) as well as evaluate treatment efficacy. CA is considered a relatively safe and reliable invasive diagnostic technique and has been widely used in clinical practice as the "gold standard" for diagnosing coronary artery disease (Tonino et al. 2020). However, it is limited to two-dimensional projection of coronary artery anatomy. Coronary CT angiography (CCTA) has unique advantages in observing soft and hard plaques in coronary arteries, assessing vessel patency, evaluating coronary stents, and postoperative evaluation after coronary artery bypass grafting. It has become an important diagnostic and screening tool for evaluating coronary artery lesions. Percutaneous coronary intervention (PCI) guided by intraoperative CCTA, in comparison with PCI guided solely by angiography, can achieve comparable immediate angiographic and safety outcomes, along with similar stent sizing selection and more frequent poststent dilation (Opolski et al. 2020). Intravascular ultrasound (IVUS) provides real-time cross-sectional images of blood vessels, allowing clear visualization of the vessel wall structure, thickness, lumen size, and shape, accurate measurement of vessel diameter and cross-sectional area, and identification of lesions such as calcification, fibrosis, and lipid pools that cannot be detected by coronary angiography. It can detect early vascular lesions that are not visible in coronary angiography and has significant implications for guiding strategies, stent selection, and evaluating treatment outcomes in coronary PCI. Some randomized controlled trials and observational studies examining intracoronary imaging during and after stent implantation indicate that IVUS-guided stent implantation is associated with reduced major adverse cardiac events and improved target vessel revascularization in patients with complex coronary artery disease (including long lesions, severe calcification, bifurcation, chronic total occlusion, and left main disease) (Buccheri et al. 2017; Darmoch et al. 2020; Gao et al. 2019). Optical coherence tomography (OCT), as an emerging intravascular imaging modality based on near-infrared light reflection, offers an axial resolution approximately ten times higher than that of IVUS, enabling precise description of plaque morphology and luminal geometry (Olinic et al. 2019; Tearney et al. 2012). Recent prospective randomized multicenter trials have demonstrated the noninferiority of OCT-guided PCI and IVUS-guided PCI regarding acute and long-term clinical outcomes (Ali et al. 2016; Jones et al. 2018).

7.4.3 Finite Element Modeling of Stent Implantation

Hemodynamic abnormalities are important factors in the restenosis of different regions in the cardiovascular system. Stent implantation can restore normal blood flow in narrowed arteries but inevitably leads to unfavorable hemodynamic changes closely related to shear stress (SR) within the lumen. In recent years, numerical simulation has become a widely used method to study the interaction between blood flow and the arterial wall. Although good results have been achieved in studying the hemodynamics of stent implantation, most of the results are based on calculations using idealized models, such as Chen et al.'s simulation of the mechanical response of the vessel wall during stent expansion and errors in size and position. Assumptions are made in numerical simulations, including the common use of Newtonian blood viscosity in previous studies (Chiastra et al. 2013; He et al. 2020; Wei et al. 2019; Wu et al. 2007). While some conclusions have been drawn by neglecting the non-Newtonian behavior of blood in certain hemodynamic studies, the accuracy of hemodynamic measurements may be compromised (Thondapu et al. 2018). Therefore, the viscosity of blood should be measured for individual animal models to perform more accurate simulations. Arteries are considered rigid and do not deform during blood flow, but they may undergo various structural changes in pulsatile flow, which can be overcome by using fluid-structure interaction models or establishing injury-induced growth models for further investigation. Fereidoonnezhad et al. (2017) and Cheng and Zhang (2019) proposed a microstructure-based injury-induced growth model to simulate in-stent restenosis, establishing an evolving equation for the abundant growth of neointima, considering smooth muscle cell migration and proliferation caused by stent-induced injury. Mechanical-biological growth models provide a link between in-stent restenosis and cellular-level changes after stent deployment (Fereidoonnezhad et al. 2017).

OCT has been widely used during interventional procedures because it provides unprecedented high-definition visualization of plaques and stent structures in real time during PCI (Prati et al. 2015). Typically, a combination of OCT and CTA has been used for 3D reconstruction of vascular models, accurately reproducing the geometric shape of the arterial wall. Migliori et al. (2017), Zhou et al. (2019a, b), Chiastra et al. used OCT and CTA to reconstruct coronary bifurcation vessels in patients and simulated changes in the microenvironment of blood flow after different stent implantations (Chiastra et al. 2013). These studies have shown that improper positioning of the stent can significantly alter the mechanical environment of the vessel, thus having a critical impact on ISR occurrence. However, the aforementioned studies did not validate the numerical models by comparing them with real animal pathological models.

In recent years, several research groups have conducted studies combining numerical simulations and animal experiments to explore factors influencing ISR development after stent implantation (Amatruda et al. 2014; Caputo et al. 2013; Morlacchi et al. 2011; Zun et al. 2019). Caputo et al. investigated the role of WSS and reduced oxygen concentration in geometric models reconstructed from real pig arteries treated with stents (Caputo et al. 2013). Their results demonstrated that disturbances in vascular curvature can induce hemodynamic conditions that stimulate ISR. Zu et al. reported a method for validating a computer 3D model of in-stent restenosis in pig coronary arteries, which included a multiscale model incorporating the stent structure obtained from microcomputed tomography (micro-CT), tissue growth within the blood flow and stent-vessel interface, including smooth muscle cell proliferation and extracellular matrix deposition. The validity of the modeling approach was confirmed by comparing the modeling results with in vivo data from pigs implanted with stents for 14 and 28 days (Zun et al. 2019). He et al. combined numerical simulations with animal experiments to study the relationship between hemodynamic parameters and stent position in a stenotic vessel and their effects on neointimal formation in rabbit carotid arteries. The study found that the TAWSS was associated with the stent implantation position, and the vessel segment with the lowest TAWSS had the highest neointimal growth (He et al. 2020).

By combining medical imaging with numerical simulations and animal models, further investigations can be conducted to explore the impact of future stent implantation on the reconstruction of narrowed blood vessels. This approach can also provide a means to study the effects of endothelial cell engulfment under abnormal hemodynamics on vascular fibrosis (Zhou et al. 2019a).

7.4.4 Mechanical Microenvironment of Arterial Atherosclerosis After Stent Implantation

Arterial atherosclerosis commonly occurs at geometrically complex sites within the blood vessels, such as the outer parts of arterial bifurcations and specific locations within the vessel lumen. On the one hand, the implantation of a vascular stent can lead to deformation of the arterial vessel, disrupting its natural curved state. On the other hand, the stent exerts long-term radial forces on the arterial vessel, resulting in significant adaptational mismatch between the vessel and the relatively rigid stent. The differences in adaptability at the stent ends can cause pressure wave reflections, leading to disturbed flow within the vessel containing the stent (Aoki & Tanabe 2021). Currently, there exist various types of stents with vastly different mechanical and biochemical properties, which influence the interaction between the stent and the vascular tissue. For instance, neointimal growth within the stent may occur in regions of low endothelial wall shear stress. Shishido et al. (2016) and Goh et al. focused on elucidating the biological responses following stent implantation and how these biochemical reactions contribute to stent failures in different generations of stents. Gori (2021) and Wang et al. provided a review of research on changes occurring in stents within the arterial mechanical environment and discussed their roles in stent restenosis and late thrombosis from three perspectives:

① The interaction between the stent and the host vessel, involving vascular wall responses, mechanical signal transduction mechanisms, re-endothelialization, and the processes of late thrombosis formation.
② Changes in blood flow dynamics within the lumen of the vessel segment and alterations in the mechanical microenvironment within the vessel segment wall due to stent implantation (Wang et al. 2018).

PCIs not only cause mechanical injury and vascular inflammation but also involve complex interactions between endothelial cells, smooth muscle cells, platelets, and inflammatory cells, including neutrophils, monocytes, and lymphocytes, accompanied by the presence of foreign bodies and biochemical processes triggered by drug release from stents (Chaabane et al. 2013). Suna et al. used proteomics to characterize extracellular matrix (ECM) remodeling in porcine coronary arteries with bare metal stents (BMS) and drug-eluting stents (DES). The study found significant differences in the ECM of the coronary arteries after implantation of bare metal stents and drug-eluting stents, specifically an upregulation of aggregating proteoglycans. The accumulation of aggregating proteoglycans correlated with changes in ADAMTS gene expression (ADAMTS is a metalloproteinase with thrombospondin motifs), indicating the involvement of aggregating proteoglycans and proteoglycans in the vascular injury response following stent implantation (Suna et al. 2018). Escuer et al. utilized a novel 2D continuous mathematical model to describe the complex process of restenosis after stent insertion in coronary arteries. They identified growth factors, matrix metalloproteinases, extracellular matrix, smooth muscle cells, and endothelial cells as crucial species involved in restenosis development. The results qualitatively captured key characteristics of arterial lesion growth and healing under nonphysiological mechanical forces (Escuer et al. 2019).

Due to the close relationship between restenosis, inflammatory reactions, cytokines, and the disease itself, reducing restenosis at the genetic level through methods from genetics and molecular biology may become mainstream in the future.

The development of biomedical engineering, materials science, and engineering technologies has provided broad prospects for the treatment of coronary heart disease, and the application of coronary stents, especially drug-eluting stents and bioresorbable stents, has been a shining chapter. Although the incidence of restenosis remains high and long-term survival rates have not significantly improved, it has provided a powerful means for the treatment of acute coronary syndrome. It is believed that with the advancement of medical science and continuous progress in clinical practice, new methods and approaches will be discovered in the near future to further reduce restenosis and achieve new milestones in the treatment of coronary heart disease.

References

AbuRahma AF. Predictors of perioperative stroke/death after carotid artery stenting: a review article[J]. Ann Vasc Dis. 2018, 11(1): 15–24.

Ali ZA, Maehara A, Généreux P, Shlofmitz RA, Fabbiocchi F, Nazif TM, Guagliumi G, Meraj PM, Alfonso F, Samady H, Akasaka T, Carlson EB, Leesar MA, Matsumura M, Ozan MO, Mintz GS, Ben-Yehuda O, Stone GW, ILUMIEN III: OPTIMIZE PCI Investigators. Optical coherence tomography compared with intravascular ultrasound and with angiography to guide coronary stent implantation (ILUMIEN III: OPTIMIZE PCI): a randomised controlled trial[J]. Lancet. 2016, 388(10060): 2618–2628.

Amatruda CM, Bona Casas C, Keller BK, Tahir H, Dubini G, Hoekstra A, Hose DR, Lawford P, Migliavacca F, Narracott AJ, Gunn J. From histology and imaging data to models for in-stent restenosis[J]. Int J Artif Organs. 2014, 37(10): 786–800.

Aoki J, Tanabe K. Mechanisms of drug-eluting stent restenosis[J]. Cardiovasc Interv Ther. 2021, 36(1): 23–29.

Borhani S, Hassanajili S, Ahmadi Tafti SH, Rabbani S. Cardiovascular stents: overview, evolution, and next generation[J]. Prog Biomater. 2018, 7(3): 175–205.

Buccheri S, Franchina G, Romano S, Puglisi S, Venuti G, D'Arrigo P, Francaviglia B, Scalia M, Condorelli A, Barbanti M, Capranzano P, Tamburino C, Capodanno D. Clinical outcomes following intravascular imaging-guided versus coronary angiography-guided percutaneous coronary intervention with stent implantation: a systematic review and bayesian network meta-analysis of 31 studies and 17,882 patients[J]. JACC Cardiovasc Interv. 2017, 10(24): 2488–2498.

Caputo M, Chiastra C, Cianciolo C, Cutrì E, Dubini G, Gunn J, Keller B, Migliavacca F, Zunino P. Simulation of oxygen transfer in stented arteries and correlation with in-stent restenosis[J]. Int J Numer Method Biomed Eng. 2013 Dec, 29(12): 1373–1387.

Chaabane C, Otsuka F, Virmani R, Bochaton-Piallat ML. Biological responses in stented arteries[J]. Cardiovasc Res. 2013, 99(2): 353–263.

Chen ZY, Xu Q, Liang ZH. New progress in mechanism and animal models of atherosclerosis[J]. Laboratory Animal Science, 2015, 32(04): 54–58.

Cheng J, Zhang LT. Simulation of vessel tissue remodeling with residual stress: an application to in-stent restenosis[J]. International Journal of Smart and Nano Materials, 2019, 10: 11–27.

Chiastra C, Morlacchi S, Gallo D, Morbiducci U, Cárdenes R, Larrabide I, Migliavacca F. Computational fluid dynamic simulations of image-based stented coronary bifurcation models[J]. J R Soc Interface. 2013, 10(84): 20130193.

Curcio A, Torella D, Indolfi C. Mechanisms of smooth muscle cell proliferation and endothelial regeneration after vascular injury and stenting: approach to therapy[J]. Circ J. 2011, 75(6): 1287–1296.

Dangas GD, Claessen BE, Caixeta A, Sanidas EA, Mintz GS, Mehran R. In-stent restenosis in the drug-eluting stent era[J]. J Am Coll Cardiol. 2010, 56(23): 1897–1907.

Darmoch F, Alraies MC, Al-Khadra Y, Moussa Pacha H, Pinto DS, Osborn EA. Intravascular ultrasound imaging-guided versus coronary angiography-guided percutaneous coronary intervention: a systematic review and meta-analysis[J]. J Am Heart Assoc. 2020, 9(5): e013678.

Escuer J, Martínez MA, McGinty S, Peña E. Mathematical modelling of the restenosis process after stent implantation[J]. J R Soc Interface. 2019, 16(157): 20190313.

Fan YB, Deng XY. Biomechanical modeling simulation and application[M]. Shanghai Jiao Tong University Press, 2017.

Fan J, Kitajima S, Watanabe T, Xu J, Zhang J, Liu E, Chen YE. Rabbit models for the study of human atherosclerosis: from pathophysiological mechanisms to translational medicine[J]. Pharmacol Ther. 2015, 146: 104–119.

Fan J, Chen Y, Yan H, Niimi M, Wang Y, Liang J. Principles and applications of rabbit models for atherosclerosis research[J]. J Atheroscler Thromb. 2018, 25(3): 213–220.

Fereidoonnezhad B, Naghdabadi R, Sohrabpour S, Holzapfel GA. A Mechanobiological model for damage-induced growth in arterial tissue with application to in-stent restenosis[J]. J. Mech. Phys. Solid, 2017, 101: 311–327.
Gao XF, Wang ZM, Wang F, Gu Y, Ge Z, Kong XQ, Zuo GF, Zhang JJ, Chen SL. Intravascular ultrasound guidance reduces cardiac death and coronary revascularization in patients undergoing drug-eluting stent implantation: results from a meta-analysis of 9 randomized trials and 4724 patients[J]. Int J Cardiovasc Imaging. 2019, 35(2): 239–247.
Garg S, Serruys PW. Coronary stents: current status[J]. J Am Coll Cardiol. 2010, 56(10 Suppl): S1–42.
Gastaldi D, Morlacchi S, Nichetti R, Capelli C, Dubini G, Petrini L, Migliavacca F. Modelling of the provisional side-branch stenting approach for the treatment of atherosclerotic coronary bifurcations: effects of stent positioning[J]. Biomech Model Mechan. 2010, 9(5): 551–561.
Ge S, Xi Y, Du R, Ren Y, Xu Z, Tan Y, Wang Y, Yin T, Wang G. Inhibition of in-stent restenosis after graphene oxide double-layer drug coating with good biocompatibility[J]. Regen Biomater. 2019, 6(5): 299–309.
Gori T. Vascular wall reactions to coronary stents-clinical implications for stent failure[J]. Life (Basel). 2021, 11(1): 63.
He S, Liu W, Qu K, Yin T, Qiu J, Li Y, Yuan K, Zhang H, Wang G. Effects of different positions of intravascular stent implantation in stenosed vessels on in-stent restenosis: An experimental and numerical simulation study[J]. J Biomech. 2020, 113: 110089.
He SC. Effects of different position of intravascular stent implantation in stenosed vessels on in-stent restenosis[D]. Chongqing University, 2019.
Hung OY, Molony D, Corban MT, Rasoul-Arzrumly E, Maynard C, Eshtehardi P, Dhawan S, Timmins LH, Piccinelli M, Ahn SG, Gogas BD, McDaniel MC, Quyyumi AA, Giddens DP, Samady H. Comprehensive assessment of coronary plaque progression with advanced intravascular imaging, physiological measures, and wall shear stress: a pilot double-blinded randomized controlled clinical trial of nebivolol versus atenolol in nonobstructive coronary artery disease[J]. J Am Heart Assoc. 2016, 5(1): e002764.
Iqbal J, Chamberlain J, Francis SE, Gunn J. Role of animal models in coronary stenting[J]. Ann Biomed Eng. 2016, 44(2): 453–465.
Jang TS, Lee JH, Kim S, Park C, Song J, Jae HJ, Kim HE, Chung JW, Jung HD. Ta ion implanted nanoridge-platform for enhanced vascular responses[J]. Biomaterials. 2019 Dec, 223: 119461.
Jones DA, Rathod KS, Koganti S, Hamshere S, Astroulakis Z, Lim P, Sirker A, O'Mahony C, Jain AK, Knight CJ, Dalby MC, Malik IS, Mathur A, Rakhit R, Lockie T, Redwood S, MacCarthy PA, Desilva R, Weerackody R, Wragg A, Smith EJ, Bourantas CV. Angiography alone versus angiography plus optical coherence tomography to guide percutaneous coronary Intervention: outcomes from the pan-london PCI cohort[J]. JACC Cardiovasc Interv. 2018, 11(14): 1313–1321.
Khosravi A, Akbari A, Bahreinizad H, Salimi Bani M, Karimi A. Optimizing through computational modeling to reduce dogboning of functionally graded coronary stent material[J]. J Mater Sci Mater Med. 2017, 28(9): 142.
Koskinas KC, Chatzizisis YS, Antoniadis AP, Giannoglou GD. Role of endothelial shear stress in stent restenosis and thrombosis: pathophysiologic mechanisms and implications for clinical translation[J]. J Am Coll Cardiol. 2012, 59(15): 1337–1349.
Kurogi K, Ishii M, Yamamoto N, Yamanaga K, Tsujita K. Optical coherence tomography-guided percutaneous coronary intervention: a review of current clinical applications[J]. Cardiovasc Interv Ther. 2021, 36(2): 169–177.
Li H, Liu T, Wang M, Zhao D, Qiao A, Wang X, Gu J, Li Z, Zhu B. Design optimization of stent and its dilatation balloon using kriging surrogate model[J]. Biomed Eng Online. 2017, 16(1): 13.
Li X, Zhang W, Lin W, Qiu H, Qi Y, Ma X, Qi H, He Y, Zhang H, Qian J, Zhang G, Gao R, Zhang D, Ding J. Long-term efficacy of biodegradable metal-polymer composite stents after the first and

the second implantations into porcine coronary arteries[J]. ACS Appl Mater Interfaces. 2020, 12(13): 15703–15715.

World health statistics 2020: monitoring health for the SDGs, sustainable development goals. Geneva: World Health Organization, 2020. Licence: CC BY-NC-SA 3.0 IGO.

Liu WZ, Tuo HY, Wu GT, Guo M, Ren Y. Research progress of hyperlipidemia and atherosclerosis models in rabbits[J]. Journal of Gansu University of Traditional Chinese Medicine, 2016, 33(01): 79–82.

Migliori S, Chiastra C, Bologna M, Montin E, Dubini G, Aurigemma C, Fedele R, Burzotta F, Mainardi L, Migliavacca F. A framework for computational fluid dynamic analyses of patient-specific stented coronary arteries from optical coherence tomography images[J]. Med Eng Phys. 2017, 47: 105–116.

Morlacchi S, Keller B, Arcangeli P, Balzan M, Migliavacca F, Dubini G, Gunn J, Arnold N, Narracott A, Evans D, Lawford P. Hemodynamics and in-stent restenosis: micro-CT images, histology, and computer simulations[J]. Ann Biomed Eng. 2011, 39(10): 2615–2626.

Nakajima A, Araki M, Kurihara O, Minami Y, Soeda T, Yonetsu T, Higuma T, Kakuta T, McNulty I, Lee H, Malhotra R, Nakamura S, Jang IK. Predictors for rapid progression of coronary calcification: an optical coherence tomography study[J]. J Am Heart Assoc. 2021, 10(3): e019235.

Nicolais C, Lakhter V, Virk HUH, Sardar P, Bavishi C, O'Murchu B, Chatterjee S. Therapeutic options for in-stent restenosis[J]. Curr Cardiol Rep. 2018, 20(2): 7.

Olinic DM, Spinu M, Homorodean C, Ober MC, Olinic M. Real-life benefit of OCT imaging for optimizing PCI indications, strategy, and results[J]. J Clin Med. 2019, 8(4): 437.

Opolski MP, Schumacher SP, Verouden NJW, van Diemen PA, Borucki BA, Sprengers R, Everaars H, de Winter RW, van Rossum AC, Nap A, Bom MJ, Knaapen P. On-site computed tomography versus angiography alone to guide coronary stent implantation: a prospective randomized study[J]. J Invasive Cardiol. 2020, 32(11): E268–E276.

Prati F, Romagnoli E, Burzotta F, Limbruno U, Gatto L, La Manna A, Versaci F, Marco V, Di Vito L, Imola F, Paoletti G, Trani C, Tamburino C, Tavazzi L, Mintz GS. Clinical impact of OCT findings during PCI: the CLI-OPCI II study[J]. JACC Cardiovasc Imaging. 2015, 8(11): 1297–1305.

Rykowska I, Nowak I, Nowak R. Drug-eluting stents and balloons-materials, structure designs, and coating techniques: a review[J]. Molecules. 2020, 25(20): 4624.

Samady H, Eshtehardi P, McDaniel MC, Suo J, Dhawan SS, Maynard C, Timmins LH, Quyyumi AA, Giddens DP. Coronary artery wall shear stress is associated with progression and transformation of atherosclerotic plaque and arterial remodeling in patients with coronary artery disease[J]. Circulation. 2011, 124(7): 779–788.

Shishido K, Antoniadis AP, Takahashi S, Tsuda M, Mizuno S, Andreou I, Papafaklis MI, Coskun AU, O'Brien C, Feldman CL, Saito S, Edelman ER, Stone PH. Effects of low endothelial shear stress after stent implantation on subsequent neointimal hyperplasia and clinical outcomes in humans[J]. J Am Heart Assoc. 2016, 5(9): e002949.

Stefanini GG, Taniwaki M, Windecker S. Coronary stents: novel developments[J]. Heart. 2014, 100(13): 1051–1061.

Sun CK, Shao PL, Wang CJ, Yip HK. Study of vascular injuries using endothelial denudation model and the therapeutic application of shock wave: a review[J]. Am J Transl Res. 2011, 3(3): 259–268.

Suna G, Wojakowski W, Lynch M, Barallobre-Barreiro J, Yin X, Mayr U, Baig F, Lu R, Fava M, Hayward R, Molenaar C, White SJ, Roleder T, Milewski KP, Gasior P, Buszman PP, Buszman P, Jahangiri M, Shanahan CM, Hill J, Mayr M. Extracellular matrix proteomics reveals interplay of aggrecan and aggrecanases in vascular remodeling of stented coronary arteries[J]. Circulation. 2018, 137(2): 166–183.

Tearney GJ, Regar E, Akasaka T, Adriaenssens T, Barlis P, Bezerra HG, Bouma B, Bruining N, Cho JM, Chowdhary S, Costa MA, de Silva R, Dijkstra J, Di Mario C, Dudek D, Falk E, Feldman MD, Fitzgerald P, Garcia-Garcia HM, Gonzalo N, Granada JF, Guagliumi G, Holm

NR, Honda Y, Ikeno F, Kawasaki M, Kochman J, Koltowski L, Kubo T, Kume T, Kyono H, Lam CC, Lamouche G, Lee DP, Leon MB, Maehara A, Manfrini O, Mintz GS, Mizuno K, Morel MA, Nadkarni S, Okura H, Otake H, Pietrasik A, Prati F, Räber L, Radu MD, Rieber J, Riga M, Rollins A, Rosenberg M, Sirbu V, Serruys PW, Shimada K, Shinke T, Shite J, Siegel E, Sonoda S, Suter M, Takarada S, Tanaka A, Terashima M, Thim T, Uemura S, Ughi GJ, van Beusekom HM, van der Steen AF, van Es GA, van Soest G, Virmani R, Waxman S, Weissman NJ, Weisz G, International working group for intravascular optical coherence tomography (IWG-IVOCT). Consensus standards for acquisition, measurement, and reporting of intravascular optical coherence tomography studies: a report from the international working group for intravascular optical coherence tomography standardization and validation[J]. J Am Coll Cardiol. 2012, 59(12): 1058–1072.

Thondapu V, Tenekecioglu E, Poon EKW, Collet C, Torii R, Bourantas CV, Chin C, Sotomi Y, Jonker H, Dijkstra J, Revalor E, Gijsen F, Onuma Y, Ooi A, Barlis P, Serruys PW. Endothelial shear stress 5 years after implantation of a coronary bioresorbable scaffold[J]. Eur Heart J. 2018, 39(18): 1602–1609.

Tonino PA, De Bruyne B, Pijls NH, Siebert U, Ikeno F, van' t Veer M, Klauss V, Manoharan G, Engstrøm T, Oldroyd KG, Ver Lee PN, MacCarthy PA, Fearon WF, FAME Study Investigators. Fractional flow reserve versus angiography for guiding percutaneous coronary intervention[J]. N Engl J Med. 2020, 360(3): 213–224.

Ullrich H, Münzel T, Gori T. Coronary stent thrombosis-predictors and prevention[J]. Dtsch Arztebl Int. 2020, 117(18): 320-326.

Wang YC, Shao M. Basic principle and numerical method of finite element. Second edition[M]. Tsinghua University Press, 1997.

Wang J, Jin X, Huang Y, Ran X, Luo D, Yang D, Jia D, Zhang K, Tong J, Deng X, Wang G. Endovascular stent-induced alterations in host artery mechanical environments and their roles in stent restenosis and late thrombosis[J]. Regen Biomater. 2018, 5(3): 177–187.

Wei L, Leo HL, Chen Q, Li Z. Structural and hemodynamic analyses of different stent structures in curved and stenotic coronary artery[J]. Front Bioeng Biotechnol. 2019, 7:366.

Wiesent L, Schultheiß U, Schmid C, Schratzenstaller T, Nonn A. Experimentally validated simulation of coronary stents considering different dogboning ratios and asymmetric stent positioning[J]. PLoS One. 2019, 14(10): e0224026.

Wu W, Wang WQ, Yang DZ, Qi M. Stent expansion in curved vessel and their interactions: a finite element analysis[J]. J Biomech. 2007, 40(11): 2580–2585.

Yang H, Wang C, Liu C, Chen H, Wu Y, Han J, Jia Z, Lin W, Zhang D, Li W, Yuan W, Guo H, Li H, Yang G, Kong D, Zhu D, Takashima K, Ruan L, Nie J, Li X, Zheng Y. Evolution of the degradation mechanism of pure zinc stent in the one-year study of rabbit abdominal aorta model[J]. Biomaterials. 2017, 145: 92–105.

Yang X, Yang Y, Guo J, Meng Y, Li M, Yang P, Liu X, Aung LHH, Yu T, Li Y. Targeting the epigenome in in-stent restenosis: from mechanisms to therapy[J]. Mol Ther Nucleic Acids. 2021, 23: 1136–1160.

Yelamanchili VS, Hajouli S. Coronary Artery Stents. 2020. In: StatPearls [Internet]. Treasure Island (FL): StatPearls Publishing, 2021.

Zhao Y, Du R, Zhou T, Yang D, Huang Y, Wang Y, Huang J, Ma X, He F, Qiu J, Wang G. Arsenic trioxide-coated stent is an endothelium-friendly drug eluting stent[J]. Adv Healthc Mater. 2018, 7(15): e1800207.

Zhou T, Zheng Y, Sun L, Badea SR, Jin Y, Liu Y, Rolfe AJ, Sun H, Wang X, Cheng Z, Huang Z, Zhao N, Sun X, Li J, Fan J, Lee C, Megraw TL, Wu W, Wang G, Ren Y. Microvascular endothelial cells engulf myelin debris and promote macrophage recruitment and fibrosis after neural injury[J]. Nat Neurosci. 2019a, 22(3): 421–435.

Zhou Y, Tong J, Li X, Li X, Wang G. Numerical simulation of haemodynamics of the descending aorta in the non-diabetic and diabetic rabbits[J]. J Biomech. 2019b, 91: 140–150.

Zun PS, Narracott AJ, Chiastra C, Gunn J, Hoekstra AG. Location-specific comparison between a 3D in-stent restenosis model and micro-CT and histology data from porcine in vivo experiments[J]. Cardiovasc Eng Technol. 2019, 10(4): 568–582.

Chapter 8
Mechanobiology Mechanism for the Inhibitory Effect of Vascular Restenosis by Drug Regulation of Cell Viability

Atherosclerosis (as one of the most common clinical diseases with the highest morbidity and mortality worldwide (Libby et al. 2019; Milutinović et al. 2020), is the pathological basis for the occurrence and development of a variety of cardiovascular diseases with high health risks, such as unstable angina pectoris, myocardial infarction and stroke (Benjamin et al. 2019; Libby et al. 2011; Niu et al. 2019). The phenotypic change of vascular smooth muscle cells (VSMCs) is key in the pathogenesis of AS, and inhibiting the transformation of VSMCs to a synthetic phenotype can effectively improve the development of AS (Bennett et al. 2016; Gomez & Owens 2012; Yang et al. 2018). Arsenic trioxide (ATO) is the main active ingredient of arsenic, showing broad-spectrum antitumor effects (Izdebska et al. 2014; Wu et al. 2020). The high remission rate of ATO in the treatment of acute promyelocytic leukemia has established a high pharmacological value. At present, it has been gradually used to prevent and treat the excessive proliferation of "benign tumors" of VSMCs (Yang et al. 2018). However, there are few reports on the mechanism of ATO in the phenotypic transformation of VSMCs. In this chapter, by constructing animal and cell models of AS lesions in vivo and VSMC phenotypes in vitro, using cytoskeletal regulators and inhibitors of yap, and focusing on the mechanical properties of cells, we will further study the involvement of F-actin and Yap in the regulation of VSMC phenotype by ATO and the reversal mechanism of AS. The results showed that ATO could regulate the phenotypic transformation of vscms through the f-actin/yap pathway to realize the organic unity of mechanical properties and biological regulatory mechanisms such as elastic modulus, migration force and contraction force. The results of this study further analyzed the pharmacological basis of ATO and improved the phenotypic regulatory mechanism of VSMCs to provide a new drug basis for the reversal of AS.

G. Wang, *The Mechanobiology in Atherosclerosis*,
https://doi.org/10.1007/978-981-97-6512-6_8

8.1 ATO is an Endothelium-Friendly Anti-Stent Restenosis Drug

Cardiovascular disease is the disease with the highest incidence rate and mortality in the world. At present, drug-eluting stents (DESs) are the main method of treating stenotic cardiovascular diseases and can greatly reduce the occurrence of in-stent restenosis. However, the drugs carried by DESs (rapamycin, paclitaxel, etc.) also inhibit the re-endothelialization of the injured site of stent implantation to a certain extent, resulting in catastrophic consequences such as intrastent thrombosis and late thrombosis. Several mechanisms are involved in the thrombosis induced by DES, among which the significant reason is delayed endothelialization of blood vessels. Therefore, the development of "endothelium-friendly" DESs to inhibit stent restenosis while promoting vascular endothelial repair may become one of the most effective strategies to solve this problem.

For stent drugs with adverse effects such as rapamycin and paclitaxel, the potential application of ATO is significant (Wu et al. 2020). ATO, the main component of traditional Chinese medicine arsenic, has been used in medicine for 2400 years in China. Currently, ATO has been recognized as the standard treatment for acute promyelocytic leukemia (APL). In addition, ATO also has definite therapeutic effects in treating malignant tumors such as gastric cancer and cervical cancer. The excessive proliferation of vascular smooth muscle cells (VSMCs) has a similar molecular biological mechanism to tumor cell proliferation. Based on these successful experiences in the treatment of tumor cell hyperplasia and the clear inhibitory effect of the drug on cell proliferation, ATO was used to treat the hyperproliferation of VSMCs.

Taking advantage of the high efficiency and low water solubility of ATO, the academician team of Ge Junbo from Zhongshan Hospital affiliated with Fudan University and Beijing Amsino Medical Co., Ltd. collaborated to independently develop arsenic trioxide-eluting stents (AES). The safety and effectiveness of AES have been confirmed by in vitro cell and animal experiments and clinical observations (Gong et al. 2010; Yang et al. 2006). Clinical evaluation shows that this type of stent can significantly inhibit stent restenosis and effectively reduce late thrombosis; however, its mechanism of action remains unclear. Late-stage thrombosis within the stent is largely related to the level of endothelial repair; therefore, AES is likely to promote vascular re-endothelialization at the site of stent implantation. Therefore, it can be speculated that "ATO may be an endothelial-friendly antirestenosis drug".

To verify the above argument, representative mainstream stents such as arsenic trioxide drug eluting stents (AES), bare metal stents (BMS), polymer-coated metal stents (PMS), and rapamycin eluting stents (RES) were implanted into New Zealand white rabbits and compared for analysis to study the safety and effectiveness of AES stents. The pathological section staining experiment results show that AES stents can significantly inhibit the occurrence of in-stent restenosis, and their performance is even better than that of commonly used RES stents. Using the EC surface-specific markers vWF and Ki67 labeled with nuclear proliferation, surface staining was performed on the RES and AES (associated vessels) implanted in the carotid artery after 1 week of longitudinal incision. The results indicate that compared to RES, AES can significantly promote the process of vascular re-endothelialization at the stent site. Further mechanistic research revealed that ATO may exert its endothelial-friendly effect by enhancing the structural integrity of mitochondria and promoting ATP production. In summary, it was confirmed that ATO is an endothelial-friendly anti-stent restenosis drug, and the differential influence concentration of ATO on ECs and VSMCs was screened (2–6 μM) to provide a solid foundation for further exploring the mechanism by which ATO inhibits stent restenosis in this chapter (Fig. 8.1).

8.2 ATO Inhibited the Proliferation of VSMCs and Showed an Effect on Their Phenotypic Transformation

Clinically, restenosis plays a role as a "silencer" in the intervention of atherosclerotic plaque and has a negative impact on prognosis. VSMCs exhibit extraordinary phenotypic plasticity during restenosis by regulating blood flow to target tissues through contraction and relaxation, which are considered important targets for restenosis treatment.

VSMCs are involuntary muscles that have no transverse stripes and are innervated by autonomic nerves. It has contractile properties and is mainly distributed in the walls of blood vessels, the trachea, the iris, the bladder, and the digestive tract. VSMCs are crucial for the function of blood vessels and maintain blood pressure primarily through contraction and relaxation. VSMCs also play a crucial role in maintaining and reconstructing the extracellular matrix (ECM) of blood vessels. In contrast to other muscle cells, VSMCs are not terminally differentiated cells and exhibit phenotypic plasticity. Under the induction of pathological factors, VSMCs can transform from contractile (differentiated) to synthetic (dedifferentiated), characterized by increased proliferation, migration, and extracellular matrix protein synthesis abilities along with decreased expression levels of contractile phenotype genes/proteins in smooth muscle cells (laconetti et al. 2015; Wang et al. 2018). According to morphological classification, VSMCs are further divided into two subgroups: fusiform VSMCs obtained from normal arterial media and epithelial-shaped VSMCs obtained from proliferative intima after arterial endothelial injury. The former grows in a typical alpine canyon pattern, while the latter grows in a single-layer manner. Epithelial VSMCs may play an important role in arterial intimal hyperplasia.

Recently, in studying the formation mechanism of endometrial hyperplasia after vascular injury, Li Zhang's team found that ectopic viral integration site 1 (EVI1) can directly cause phenotypic changes in VSMCs (Yang et al. 2018). Other studies have shown that EVI1 is a specific downstream target of ATO, suggesting that ATO can induce phenotypic differentiation of VSMCs. However, there is little direct evidence that ATO regulates the phenotypic transformation of VSMCs (Metallo & Heiden 2013). During research on the mechanism by which AES inhibits the formation of

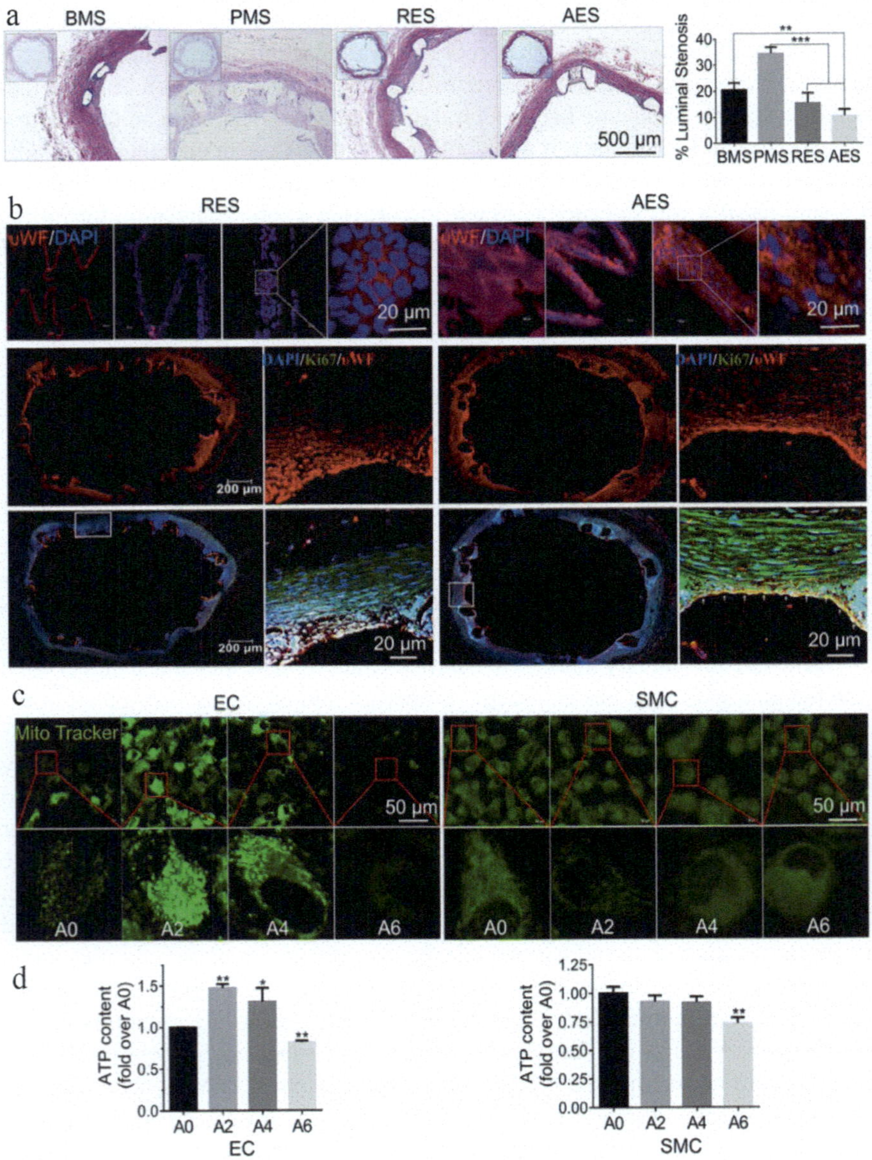

Fig. 8.1 ATO drug-eluting stent is an endothelium-friendly drug eluting stent. A0, A2, A4, and A6 represent ATO concentrations of 2, 4, 6, and 8 μM, respectively. Cited from Zhao et al. (2018)

restenosis in the stent, the research team found that VSMCs in the neointima underwent spindle-shaped phenotypic changes toward mature VSMCs earlier under the action of ATO, accelerating the speed of vascular injury repair. Further screening of appropriate concentrations of ATO in vitro cell experiments can promote the transformation of smooth muscle cells from dedifferentiated synthetic types to differentiated and mature contractile types. It was mainly manifested in the high expression of α-SMA (alpha-smooth muscle actin) and SM22α (smooth muscle 22 alpha), the cell aspect ratio was longer, and the cell was fusiform.

8.2.1 ATO in Vivo Promotes the Transformation of VSMCs from Synthetic to Contractile Types

After implanting different types of vascular stents (BMS: bare metal stents; PMS: polymer-coated metal stents; AES: ATO drug-eluting stents) into the carotid artery of New Zealand white rabbits for 1 and 3 months, the smooth muscle phenotype factors of the vessels at the stent site were evaluated (α-SMA) by HE staining and immunofluorescence staining (green). Its expression level was evaluated through fluorescence intensity, and the aspect ratio of VSMCs at the neointima was statistically analyzed to characterize their cell shape. It was found that ATO drug-eluting stents can significantly promote α-SMA expression, leading to the early completion of spindle-shaped phenotypic changes toward mature VSMCs (Fig. 8.2).

8.2.2 Cell Experiment of ATO Promoting the Transformation of VSMCs from Synthetic to Constrictive Type

The synthetic rat smooth muscle cell line A7r5 was selected as the study object. After ATO treatment, the results of immunofluorescence staining of smooth muscle contraction phenotype proteins and western blot analysis showed that ATO could increase the expression of the A7r5 contraction marker protein.

The above experiment was conducted using extracted porcine primary coronary artery smooth muscle cells (PCASMCs). Using primary PCASMCs (3–5 generations) as the research object, PCASMCs treated with PDGF (20 ng/mL) for 4 days were cultured under high serum conditions to obtain synthetic PCASMCs with high proliferation ability. PCASMCs that have not been treated with PDGF and cultured under low serum conditions are considered contractile with low proliferative capacity. Figure 8.3 shows that PCASMCs from the same source and concentration, cultured under the two different conditions mentioned above, exhibit different cell proliferation rates. In addition, the MTS method was used to determine the activity of synthetic and contractile PCASMCs. Differences in response concentration and mode to ATO were found between the contractile type (without PDGF treatment group) and the

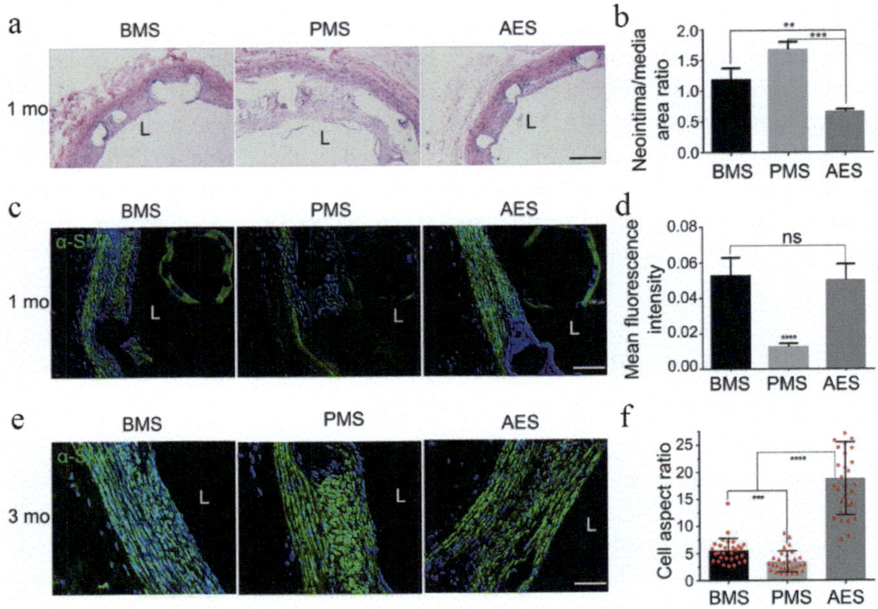

Fig. 8.2 In vivo, AES inhibits in-stent restenosis and induces differentiation of smooth muscle cells in neointima. **a** Hematoxylin and eosin (HE) staining of vascular after stent implantation for 1 mo, scar bar = 200 μm. **b** The statistical results of neointima/media area ratio for HE staining after stent implantation for 1 mo, n = 3. **c** Immunofluorescence staining with α-SMA (green) after stent implantation for 1 mo, scar bar = 100 μm. **d** The statistical results of mean fluorescence intensity of α-SMA after stent implantation for 1 mo in neointima, n = 6. **e** Immunofluorescence staining with α-SMA (green) after stent implantation for and 3 mo, scar bar = 50 μm. **f** The statistical results of cell aspect ratio (cell long/short axis ratio) of VSMCs after stent implantation for 3 mo in neointima, n = 30. "L" for the lumen. BMS (bare metal stent), PMS (polymer coating-metal stent), and AES (arsenic trioxide-drug eluting stent); 1 mo (1 month) and 3 mo (3 months), "ns" means no significance, and P values <0.01(**), <0.001(***) and <0.0001 (****). Cited from Zhao et al. (2021)

synthetic type (with PDGF treatment group). Contractile PCASMCs have a higher tolerance to ATO than synthetic PCASMCs, indicating that ATO tends to inhibit the proliferation of synthetic PCASMCs. We found that low concentrations of ATO (2 and 4 μmol/L) had a differential effect on the activity of synthetic and contractile PCASMCs.

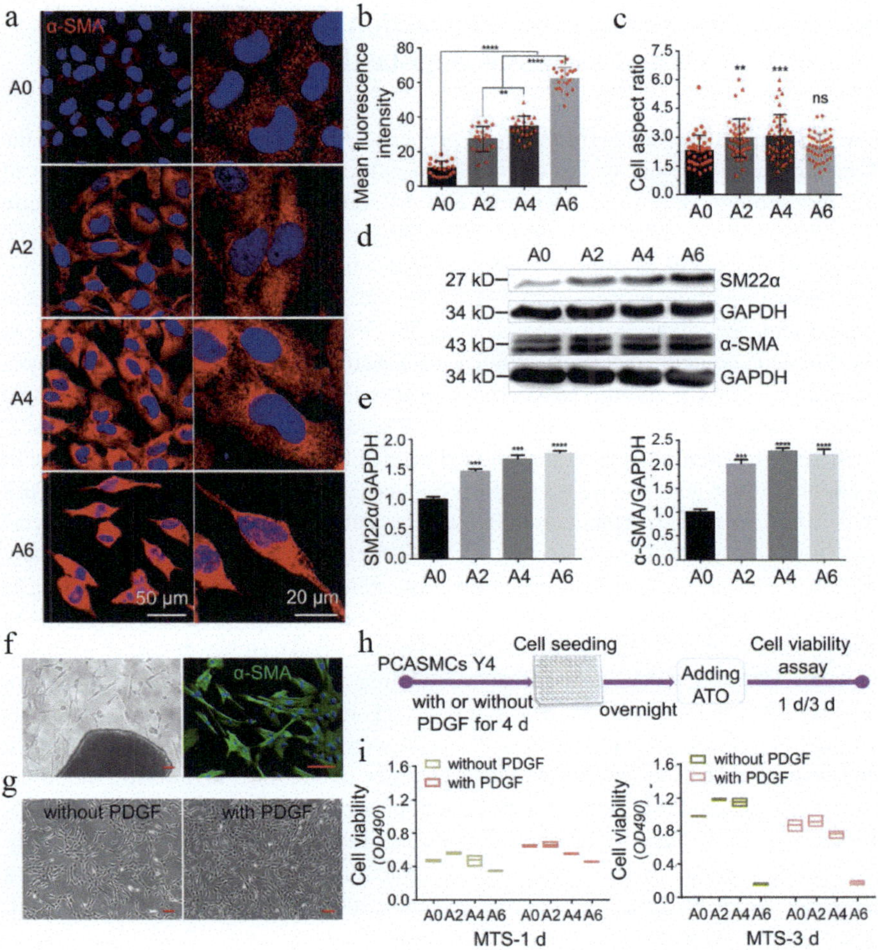

Fig. 8.3 In vitro, arsenic trioxide induces phenotype modulation, and inhibits cell viability of synthetic phenotype VSMCs. **a** α-SMA immunostaining of A7r5 with or without ATO treating for 1 d. **b** The statistical results of mean fluorescence intensity of α-SMA in (**a**). **c** The statistical results of cell aspect ratio (cell long/short axis ratio) of A7r5 with or without ATO treating for 1 d. (**d**) and (**e**), Protein levels of SM22α and α-SMA were determined by WB after ATO treating for 1 d. **f** The extraction and identification of primary porcine coronary artery smooth muscle cells (PCASMCs), scar bar = 100 μm. **g** The proliferation profile of PCASMCs, scar bar = 100 μm. **h** The schematic diagram for the phenotype induction of PCASMCs. **i** The viability of contractile- and synthetic-phenotype PCASMCs on different concentration of ATO for 1 and 3 d. A0, A2, A4 and A6 represent 0, 2, 4 and 6 μM of ATO, respectively. "ns" means no significance, P values <0.01(**), <0.001(***) and <0.0001 (****). Cited from Zhao et al. (2021)

8.3 F-Actin Enhances the Contraction Morphology of VSMCs and Increases the Elastic Modulus of Cells

The mechanical properties (biophysical properties) of cells are closely related to the dynamic rearrangement of the cytoskeleton, affecting the functional behaviors of cell morphology, adhesion, migration, and differentiation (Ni et al. 2019). F-actin can generate forces related to cellular morphological changes and the cellular elastic modulus (Fletcher & Mullins 2010; Rotsch & Radmacher 2000). During the phenotypic transformation process of VSMCs, their cell morphology and cytoskeleton system undergo changes. In general, synthetic VSMCs are flat in shape with low myofilament content and short bundles of F-actin. The morphology of contractile VSMCs cells is spindle-shaped or long spindle shaped, with abundant muscle filaments and F-actin arranged in longitudinal parallel, forming neat stress fibers.

Recent research has found that the cytoskeletal structure of actin regulates cell death and tumorigenesis and is considered a very attractive therapeutic target (Park et al. 2020). For ATO, proteins rich in cysteine and thiol groups are its main targets of action. The cytoskeleton is a target of ATO in cells due to its composition of many proteins with high levels of thiol groups. For example, cytoskeleton remodeling is involved in the apoptosis process of the Chinese hamster ovary cell line (CHO AA8) with p53 deficiency induced by ATO (Izdebska et al. 2014). However, the relationship between ATO and VSMC cytoskeletal remodeling and phenotype regulation has not been reported in the literature. The synthetic rat smooth muscle cell line A7r5 was selected as the research object, and the cells were treated with ATO to study the aggregation of the F-actin cytoskeleton, changes in cell size, and elastic modulus. The results showed that ATO (2 and 4 μM) induced the transformation of VSMCs from the synthetic type to the contraction type, and the cell morphology gradually changed to a spindle shape. The cytoskeleton texture was clearer, F-actin formed a dense and orderly stress fiber bundle, and the cell elastic modulus was enhanced. However, the 6 μM ATO treatment group showed a decrease in cell shrinkage, actin filament fluorescence intensity, and Young's modulus. From the above, it can be concluded that ATO promotes the aggregation of skeletal actin and the expression of contractile phenotype proteins while also increasing the elastic modulus of VSMCs (Fig. 8.4).

8.4 The HiPO-YAP Pathway Mediates ATO to Regulate the Phenotypic Transformation of VSMCs

Yes-associated protein (YAP) is widely present in the body and is the core downstream effector of typical Hippo signaling cascade reactions (Ma et al. 2017). It can regulate organ size, maintain stem cell self-renewal, regulate cell phenotype, induce epithelial mesenchymal transformation, promote cell proliferation, and promote tumor development. YAP and transcriptional coactivator with PDZ-binding motif

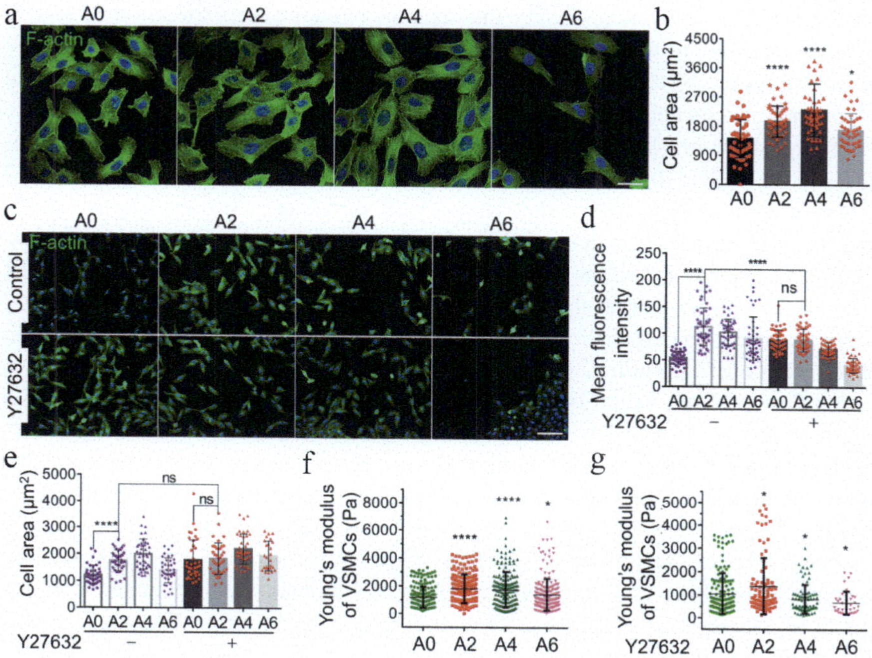

Fig. 8.4 ATO induces actin cytoskeleton organization and mechanical changes in VSMCs in vitro. **a** Immunostaining for cytoskeleton F-actin of A7r5 with or without ATO treating for 1 d, scar bar = 50 μm. **b** The statistical analysis of cell area. **c** Immunostaining for F-actin of A7r5 with or without Y27632 treating for 4 h, then treating ATO for 1 d, scar bar = 200 μm. The statistical analysis of fluorescence intensity (**d**) and cell area (**e**). **f** Young's modulus determined by atomic force microscope for A7r5 treating ATO for 1 d. **g** Young's modulus determined by atomic force microscope for A7r5 with Y27632 treating for 4 h, then treating ATO for 1 d. A0, A2, A4 and A6 represent 0, 2, 4 and 6 μM of ATO, respectively; "ns" means no significance, P values < 0.05 (*) and <0.0001 (****). Cited from Zhao et al. (2021)

(TAZ) are homologous chromosomes in mammals. Phosphorylated YAP/TAZ may bind to 14-3-3 proteins in the cytoplasm and remain in the cytoplasm, inhibiting the activity of YAP and leading to inhibition of proliferation and antiapoptotic effects, thereby promoting cell apoptosis. It may also continue to be phosphorylated by CK1 δ/ε phosphorylation, recruiting E3 ubiquitin ligase to degrade YAP due to ubiquitination. After dephosphorylation, YAP and TAZ can enter the nucleus and bind to transcription factors such as the SMAD (mother against-dependent regulatory protein) family members, ErB4, Runt-related transcription factor (RUNX2), P73, P53BP-2, and transcriptional enhancer domain (TEAD) family (Piccolo et al. 2014). Afterwards, the expression of target genes such as connective tissue growth factor (CTGF) and cysteine rich angiogenesis inducing factor (Cyr61) are induced to promote cell proliferation and survival. The Hippo pathway can be activated by cell density or mechanical signals, resulting in phosphorylation and deactivation of YAP/TAZ and degradation after ubiquitination. Conversely, the decrease in Hippo signaling causes

YAP and TAZ to activate and enter the nucleus, inducing the expression of related genes and thereby promoting cell survival and proliferation. YAP translocation into the nucleus is the core process of activation of this pathway. Research has found that cell polarity and mechanical factors such as extracellular matrix hardness, membrane stretching, shear stress, and low permeability shock can regulate the activity of this pathway (Dupont et al. 2011).

Recent studies have shown that Hippo/YAP signaling is involved in the development of restenosis in rodent models, but the regulatory mechanism is complex. MST1 can phosphorylate the YAP/TAZ protein and inhibit its activity, but MST1 is induced and activated at the site of carotid artery injury in rats. However, after 14 days of carotid artery balloon injury in rats, overexpression of MST1 can inhibit the formation of neointima, which is a contradiction. If we infer from the above assumptions, the activation of MST1 can inhibit the activity of YAP/TAZ, but in the rat carotid artery balloon injury model and proliferative SMCs cultured in vitro, the expression of YAP is also significantly increased, which is contradictory. The enhanced expression and activity of YAP mediate the transformation of the SMC phenotype to syndesmosis, promoting the formation of new intima. The simultaneous increase in YAP and MST1 levels in SMCs at the neointima suggests that the mechanism by which YAP and MST1 regulate neointima is still unclear and complex (Fig. 8.5). However, it is clear that activating Hippo or inhibiting YAP activity will be a potential treatment method for reducing neointimal formation.

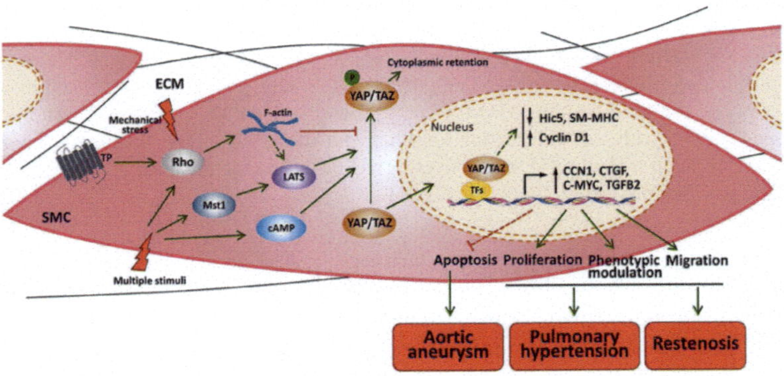

Fig. 8.5 Regulation of the Hippo pathway in smooth muscle cells. Several stimuli, such as mechanical stress, ECM stiffness and TP receptors regulate the nuclear translocation of YAP/TAZ by modulating the activity of Rho GTPases or remodeling the actin cytoskeleton. The target genes of proliferation, phenotypic modulation, migration and apoptosis in smooth muscle cells are regulated by YAP/TAZ. Cited from Zhao (2018) and He et al. (2018)

8.4.1 *Hippo/YAP is Predicted to Participate in ATO Regulation of VSMC Phenotype by RNA Transcriptome Sequencing*

The Hippo/YAP pathway plays an important role in many known signaling pathways that regulate VSMC phenotypic switch-related factors. Recent studies have reported that PI3Kγ regulates the phenotypic transformation and neointima formation of VSMCs through the cyclic adenosine monophosphate (cAMP) reaction element-binding protein CREB/YAP (Yu et al. 2019). In addition, cAMP can inhibit the expression of VSMC-related pro-division genes by inducing cytoskeletal remodeling to promote YAP nucleation and decrease its activity (Kimura et al. 2016).

Our research group extracted primary porcine coronary artery smooth muscle cells (PCASMCs) and used third-generation PCASMCs as the research object for RNA transcriptome sequencing. The untreated cells were contractile smooth muscle cells (SS), and those treated with PDGF (10 ng/ml) were synthetic smooth muscle cells (SH). After the synthetic and contracted VSMCs were treated with ATO (2 μM) for 8 h, total RNA samples were collected and sent to the company for transcriptome sequencing analysis. A Venn diagram was used to analyze the differential genes of the two types of smooth muscle cells after ATO stimulation, and the results showed that in addition to 2916 genes belonging to the same changing genes, there were 4837 genes that were different in the response of the two phenotypes of SMCs to ATO (Fig. 8.6b). Through Gene Ontology (GO) enrichment analysis, it was found that after ATO treatment, there were significant differences in the genes of contracted and formed PCASMCs in cell processes such as cell metabolism and the cell cycle (Fig. 8.6c).

Afterwards, taking synthetic PCASMCs as the research object, we further analyzed the differential expression of SH0 and SH2 before and after ATO treatment and found that ATO had an upregulatory effect on apoptosis, Hippo signaling, actin cytoskeleton and vasoconstriction in synthetic PCASMCs. It has downregulated effects on cellular processes such as the cell cycle, DNA replication and the TCA cycle (Fig. 8.6d). Compared with SH0, apoptosis-related genes, actin skeleton-related genes and vascular smooth muscle contraction-related genes in the SH2 group were upregulated, but cell migration-related genes were downregulated. In other words, ATO treatment can promote apoptosis of synthetic PCASMCs and inhibit cell migration. In addition, it promoted the expression of some actin skeleton genes and smooth muscle cell contraction phenotype genes. Both positive and negative implications suggest that ATO can induce the transformation of PCASMCs from the synthetic type to the contraction type.

RNA sequencing showed that the gene expression of VSMCs with the two phenotypes was significantly different after ATO treatment. ATO can upregulate Hippo signaling in synthetic VSMCs, promote actin cytoskeleton aggregation and induce vasoconstriction, which are closely related to YAP signaling. Then, qPCR was performed on A7r5 smooth muscle cells to detect the expression of genes associated with cell contraction, cytoskeleton and the Hippo signaling pathway (α-SMA,

a

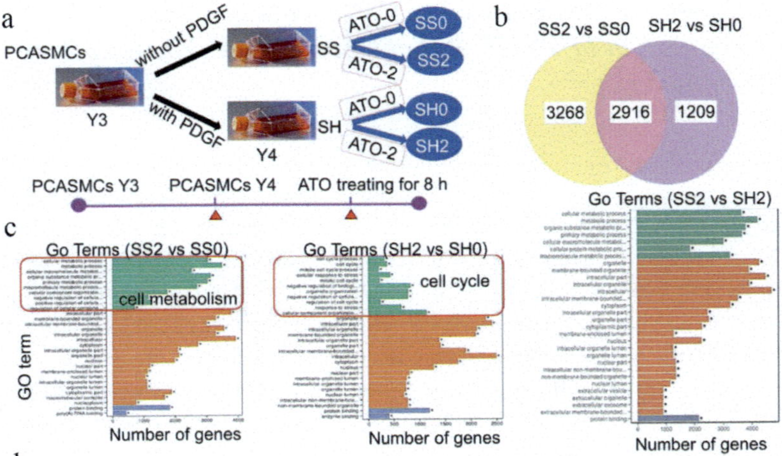

d

SH2 vs SH0	Up				Down		
Pathway	Apoptosis	Hippo signaling	Actin cytoskeleton	Vascular contraction	Cell cycle	DNA replication	TCA cycle
Gene#	18	22	30	18	55	27	13
P value	0.054	0.17	0.18	0.23	6.95E-9	2.76E-7	0.02

e

Pathway	Gene Symbol	SH2/SH0 $\log_2$ fold change	*P* value
Apoptosis	Bcl-XL	-0.43	00015
	PIDD1	-0.54	0.00014
	CASP3	1.16	7.41E-30
	CASP9	0.54	0.00011
Cell migration	MAPK3	-0.30	0.0029
	MRLC2	-0.26	0.0048
	RhoA	-0.26	0.0013
	SRC	0.41	0.00086
	VCL	-0.46	4.03E-9
Actin cytoskeleton	ACTN1	0.31	5.13E-3
	BAIAP2	0.32	0.0035
	BPK1	0.32	0.0031
	PFN2	0.36	0.0027
	TMSB4	0.35	8.78E-6
Vascular smooth muscle contraction	ACTA2	-0.58	2.12E-14
	ACTG2	0.24	0.0032
	CNN1	0.54	6.70E-8
	FAM129A	0.59	7.32E-10
	MYLK2	2.43	2.15E-6
	PPP1R15 A	1.43	8.66E-43
	PPP1R15 B	0.46	7.14E-7
	TAGLN	0.23	0.0025

f

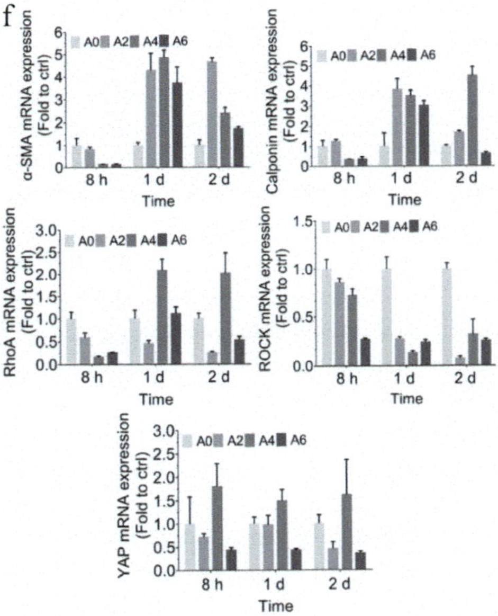

Fig. 8.6 RNA genome sequencing analysis of contractile and synthetic PCASMCs with or without ATO treating for 8 h, and qPCR validation of genes and signals. **a** The schematic diagram for the phenotype induction and ATO treating of PCSMCs. **b** Venn diagram. **c** Gene Ontology (GO) enrichment analysis. (**d**) and (**e**) are the statistical tables of changes in ATO treatment, related cell processes (**d**) and main gene expression (**e**) of synthetic PCASMCs (SH2 vs SH0). **f** qPCR of α-SMA, Calponin, RhoA, ROCK and YAP for A7r5. SS0: contractile PCASMCs without ATO treatment; SS2: contractile PCASMCs treated with 2 μM ATO for 8 h; SH0: synthetic PCASMCs without ATO treatment; and SH2: synthetic PCASMCs treated with 2 μM ATO for 8 h. A0, A2, A4 and A6 represent 0, 2, 4 and 6 μM of ATO, respectively. Cited from Zhao et al. (2021)

calponin, RhoA, ROCK and YAP). The data confirmed the predicted results of RNA sequencing.

8.4.2 *YAP Mediates ATO to Regulate the Phenotypic Transformation of VSMCs*

A7r5 rat smooth muscle cells were treated with different concentrations of ATO (0, 2, 4 and 6 μM) for 24 h in vitro, and YAP immunofluorescence staining was performed. The results in Fig. 8.7a show that the total amount of YAP and the amount of YAP entering the nucleus gradually decrease as the concentration of the drug increases. Similarly, in vivo experiments (stents implanted for 1 week) further confirmed that ATO (AES group: ATO drug-eluting stents) reduced YAP induction while inhibiting VSMC proliferation (Fig. 8.7b). WB analysis results showed (Fig. 8.7c) that ATO inhibited YAP expression by promoting the expression of the systolic phenotype factor SM22α, and the expression of the synthetic phenotype factor osteopontin (OPN) decreased accordingly. The results showed that the amount of YAP entering the nucleus and the expression of YAP protein decreased in the process of ATO regulating the transformation of the cell phenotype to the contraction type. It is suggested that the Hippo/YAP signaling pathway may respond to mechanical changes in cells and mediate ATO-induced VSMC transformation from synthetic to contractile.

Interestingly, it can be seen from Fig. 8.8a that phosphorylated YAPp-YAP and cytoskeletal F-actin colocalize, the distribution of the two in the cell is highly correlated, and the degree of cytoskeletal polymerization is consistent with the distribution and arrangement of p-YAP. In conclusion, the expression of p-YAP in A7r5 smooth muscle cells treated with ATO was enhanced with increasing drug concentration, and the distribution of p-YAP was closely related to cytoskeletal arrangement. WB experiments were further used to verify the correlation between p-YAP and ROCK and phenotype-related factors. The results showed that YAP-mediated phenotypic transformation of VSMCs is regulated by or partly by the ROCK pathway.

8.5 Research Progress and Prospects

The transformation of contractile VSMCs into syntaxis is the premise of VSMC proliferation. Effectively controlling this phenotypic transformation is one of the keys to preventing cardiovascular disease. However, there are still some difficulties in treating atherosclerotic vascular diseases by inhibiting related factors. The study of the mechanism of VSMC superficial conversion has potential guiding value for reversing vascular remodeling and clinical prevention.

M22α is an important cytoskeleton-related protein and a classic VSMC marker protein that is highly expressed in differentiated VSMCs and is one of the markers

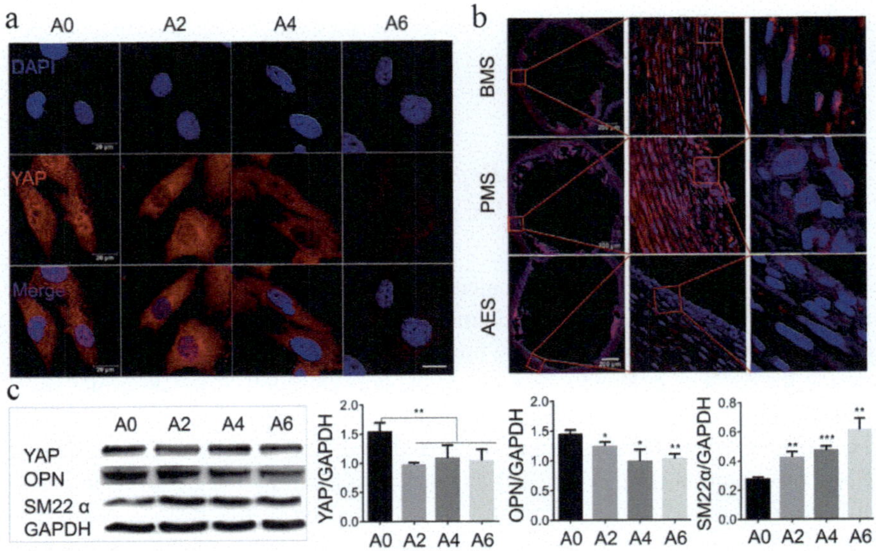

Fig. 8.7 ATO induces VSMCs phenotype modulation. **a** Immunostaining of YAP after ATO treating A7r5 for 1 d, scar bar = 20 μm. **b** Immunostaining for YAP in cross-sections of carotid arteries postimplanting stent for 1 week. **c** Protein expression levels of YAP, SM22α and Osteopontin (OPN) were determined by WB after ATO treating for 1 d. A0, A2, A4 and A6 represent 0, 2, 4 and 6 μM of ATO, respectively; BMS (bare metal stent), PMS (polymer coating-metal stent), and AES (arsenic trioxide-drug eluting stent); P values < 0.05 (*), <0.01(**) and <0.001 (***). Cited from Zhao et al. (2021)

of the early differentiation of smooth muscle cells. SM22α can regulate the smooth muscle cell phenotype by interacting with actin and participating in microfilament polymerization and skeletal remodeling of vascular smooth muscle cells. In this study, the expression of α-SMA and SM22α in smooth muscle cells was used to study the effect of ATO on the phenotypic changes of VSMCs. The results showed that the expression of SM22α in immunofluorescence staining of VSMCs could better represent the phenotypic change of SMCs.

PDGF-BB (platelet-derived growth factor) is a mitogen that plays an important role in animal development and the maintenance of homeostasis. PDGF can bind to the PDGF receptor (tyrosine kinase receptor, PTK) on the cell membrane and improve the proliferation and migration ability of VSMCs by inhibiting the expression of specific genes of smooth muscle cells to promote the transformation of the smooth muscle cell phenotype from contractile to synthetic. Therefore, our study further compared the different performances of contractile and synthetic primary porcine coronary artery smooth muscle cells (PCASMCs) in response to ATO by using PDGF to stimulate the transformation of PCASMCs into syntaxis. ATO selectively inhibits the viability, proliferation and migration of synthetic PCASMCs. These results directly prove that ATO inhibits neointimal hyperplasia.

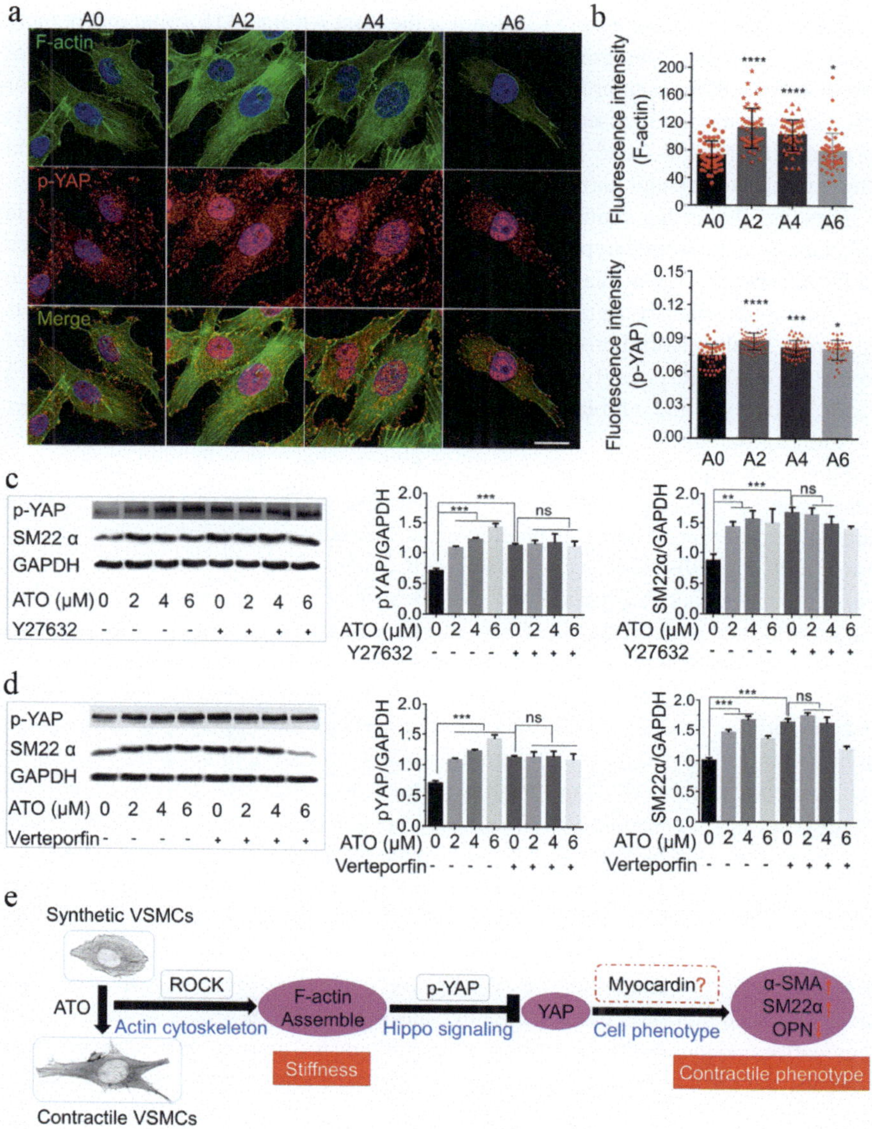

Fig. 8.8 The activity of YAP is associated with ATO regulating the phenotype transformation of VSMCs in vitro. **a** Immunofluorescence costaining of F-actin and p-YAP after ATO treating A7r5 for 1 d, scar bar = 20 μm. **b** The fluorescence intensity of F-actin and p-YAP. **c** VSMCs (A7r5 cell lines) were treated with control or ATO for 1 d with or without ROCK inhibitor Y27632 and WB experiments for indicated the protein levels of p-YAP and SM22α were performed. **d** VSMCs (A7r5 cell lines) were treated with control or ATO for 1 d with or without YAP inactivator verteporfin and WB experiments for indicated the protein levels of p-YAP and SM22α were performed. **e** Working hypothesis which elaborates role of YAP and ROCK in ATO modulation the differentiation of VSMC. A0, A2, A4 and A6 represent 0, 2, 4 and 6 μM of ATO, respectively; "ns" means no significance, P values < 0.05 (*), <0.01(**), <0.001 (***) and <0.001 (****). Cited from Zhao et al. (2021)

In addition, Professor Zhang Li's team at the First Affiliated Hospital of Zhejiang University School of Medicine recently studied the formation mechanism of endometrial dysplasia after vascular injury and found that miR-22, as a therapeutic target for vascular dysremodeling diseases, can change the phenotype of smooth muscle cells by regulating MECP-2 and EVI-1 to reduce pathological endometrial hyperplasia after vascular injury and narrowing of blood vessels.

These results further confirm the positive and negative roles of the smooth muscle cell phenotype in intimal hyperplasia and vascular stenosis. On the other hand, since EVI-1 is a specific downstream target of ATO, we speculated that ATO also has a role in regulating the phenotype of VSMCs. Thus, the regulation of VSMC phenotypes is a complex, multifactorial process. In addition to ATO/EVI-1 being a possible axis of regulation of VSMC phenotype, other possible mechanisms by which ATO regulates VSMC phenotype and inhibits the occurrence of restenosis in the scaffold are another question for our study.

Yap is a downstream core effector molecule of the Hippo signaling cascade that can regulate organ size, promote cell proliferation, inhibit apoptosis, and maintain stem cell self-renewal and cell phenotype. It has been reported that when arterial injury and other diseases occur, SMCs will change from the contractile type to the zygotic type, and Yap expression increases in this process. Studies have shown that endometrial hyperplasia after arterial injury or in the course of atherosclerosis shares some common biological mechanisms with tumorigenesis, namely, excessive proliferation and migration of vascular smooth muscle cells and loss of expression of differentiation markers. Studies have confirmed that the hippo/yap pathway plays an important role in many known signaling pathways that regulate SMC phenotype switch-related factors. The Yap signaling pathway has received extensive attention in recent years and may become an effective therapeutic or drug target. Our study found that the inactivation of Yap was related to phenotypic changes in SMCs. However, the change in the total amount of Yap is not the main factor that determines the phenotype of SMCs. The change process of Yap through phosphorylation inactivation determines the contractile phenotype of SMCs. It is further presumed that p-yap is also positively correlated with the cytoskeleton and cell stiffness of SMCs. Studies have confirmed that ATO can increase the elastic modulus of SMCs through F-actin aggregation and induce the transition from a synthetic to a contractile phenotype. ATO regulates the change in SMC phenotype partly depending on the phosphorylation degree of yap.

As is an intimal proliferative cardiovascular disease, the phenotypic transformation of VSMCs is the main pathological basis of its occurrence. Arsenic trioxide (ATO), an antiproliferative traditional Chinese medicine, has been used as an alternative drug for the coating of vascular stents and is of great significance for the prevention and treatment of as. This chapter focuses on the key scientific question "does hippo/yap pathway mediate the effect of ATO on the phenotypic transformation of VSMCs and then inhibit the occurrence of intimal hyperplasia?". Taking the phenotype regulation of VSMCs as the main line or monitoring index, different phenotype models of VSMCs were constructed by intravascular stent implantation in white rabbits and in vitro, and the mediating role of the yap and hippo pathways,

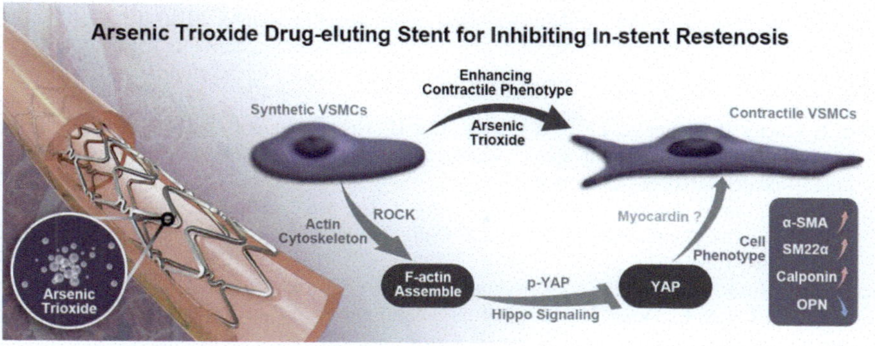

Fig. 8.9 Effects of ATO and Rapamycin (Rapa) on the cell viability of PCAECs and PCASMCs. Cited from Zhao et al. (2021)

which have key indicators in the process of vascular intimal hyperplasia, in the phenotypic transformation of VSMCs induced by ATO was systematically explored. The results showed that ATO could promote the phenotypic differentiation of VSMCs, during which cytoskeleton remodeling and elastic modulus increased, and yes-related protein (Yap) and rock pathway partially mediated the above process. It can be seen from the above that ATO is subject to the interaction of F-actin and Yap in regulating the phenotypic transformation of VSMCs, and the mechanical properties of the cells themselves are changed accordingly (Fig. 8.9). The results of this part are expected to enrich the medical scope of ATO, extend the regulatory mechanism of VSMC phenotype transformation, deepen the mechanism of ATO inhibiting vascular remodeling diseases such as in stent restenosis, and provide new ideas and strategies for the clinical prevention and treatment of intimal hyperplasia (vascular stenosis and restenosis).

References

Benjamin EJ, Muntner P, Alonso A, Bittencourt MS, Callaway CW, Carson AP, Chamberlain AM, Chang AR, Cheng S, Das SR , Delling FN , Djousse L, Elkind MSV, Ferguson JF , Fornage M, Jordan LC, Khan SS, Kissela BM, Knutson KL, Kwan TW, Lackland DT, Lewis TT, Lichtman JH, Longenecker CT, Loop MS, Lutsey PL, Martin SS, Matsushita K, Moran AE, Mussolino ME, O'Flaherty M, Pandey A, Perak AM, Rosamond WD, Roth GA, Sampson UKA, Satou GM, Schroeder EB, Shah SH, Spartano NL, Stokes A, Tirschwell DL, Tsao CW, Turakhia MP, VanWagner LB, Wilkins JT, Wong SS, Virani SS. Heart disease and stroke statistics-2019 update: a report from the American Heart Association[J]. Circulation, 2019, 139(10): e56–e528.

Bennett MR, Sinha S, Owens GK. Vascular smooth muscle cells in atherosclerosis[J]. Circ Res, 2016, 118(4): 692–702.

Dupont S, Morsut L, Aragona M, Enzo E, Giulitti S, Cordenonsi M, Zanconato F, Digabel JL, Forcato M, Bicciato S, Elvassore N, Piccolo S. Role of YAP/TAZ in mechanotransduction[J]. Nature, 2011, 474(7350): 179–183.

Fletcher DA, Mullins RD. Cell mechanics and the cytoskeleton[J]. Nature, 2010, 463(7280): 485–492.

Gomez D, Owens GK. Smooth muscle cell phenotypic switching in atherosclerosis[J]. Cardiovasc Res, 2012, 95(2): 156–164.

Gong F, Cheng X, Wang S, Zhao Y, Gao Y, Cai H. Heparin-immobilized polymers as non-inflammatory and non-thrombogenic coating materials for arsenic trioxide eluting stents[J]. Acta Biomater, 2010, 6(2): 534–546.

He J, Bao Q, Yan M, Liang J, Zhu Y, Wang C, Ai D. The role of Hippo/yes-associated protein signalling in vascular remodelling associated with cardiovascular disease. Br J Pharmacol, 2018, 175(8):1354–1361.

Iaconetti C, Rosa SD, Polimeni A, Sorrentino S, Gareri C, Carino A, Sabatino J, Colangelo M, Curcio A, Indolfi C. Down-regulation of miR-23b induces phenotypic switching of vascular smooth muscle cells in vitro and in vivo[J]. Cardiovasc Res, 2015, 107(4): 522–533.

Izdebska M, Klimaszewska-Wiśniewska A, Lewandowski D, Nowak JM, Gagat M, Grzanka A. Arsenic trioxide preferentially induces nonapoptotic cell deaths as well as actin cytoskeleton rearrangement in the CHO AA8 cell line. Postep Hig Med Dosw, 2014, 68: 1492–1500.

Kimura TE, Duggirala A, Smith MC, White S, Sala-Newby GB, Newby AC, Bond M. The hippo pathway mediates inhibition of vascular smooth muscle cell proliferation by cAMP[J]. J Mol Cell Cardiol, 2016, 90: 1–10.

Libby P, Ridker PM, Hansson GK. Progress and challenges in translating the biology of atherosclerosis[J]. Nature, 2011, 473(7347): 317–325.

Libby P, Buring JE, Badimon L, Hansson GK, Deanfield J, Bittencourt MS, Tokgözoğlu L, Lewis EF. Atherosclerosis[J]. Nat Rev Dis Primers, 2019, 5(1): 56.

Ma X, Zhang H, Xue X, Shah YM. Hypoxia-inducible factor 2alpha (HIF-2alpha) promotes colon cancer growth by potentiating Yes-associated protein 1 (YAP1) activity[J]. J Biol Chem, 2017, 292(41): 17046–17056.

Metallo CM, Heiden MGV. Understanding metabolic regulation and its influence on cell physiology[J]. Mol Cell, 2013, 49(3): 388–398.

Milutinović A, Šuput D, Zorc-Pleskovič R. Pathogenesis of atherosclerosis in the tunica intima, media, and adventitia of coronary arteries: an updated review[J]. Bosn J Basic Med Sci, 2020, 20(1): 21–30.

Ni F, Yu WM, Wang X, Fay ME, Young KM, Qiu Y, Lam WA, Sulchek TA, Cheng T, Scadden DT, Qu CK. Ptpn21 controls hematopoietic stem cell homeostasis and biomechanics[J]. Cell Stem Cell, 2019, 24(4): 608–620.

Niu N, Xu S, Xu Y, Little PJ, Jin ZG. Targeting mechanosensitive transcription factors in atherosclerosis[J]. Trends Pharmacol Sci, 2019, 40(4): 253–266.

Park JS, Burckhardt CJ, Lazcano R, Solis LM, Isogai T, Li L, Chen CS, Gao B, Minna JD, Bachoo R, DeBerardinis RJ, Danuser G. Mechanical regulation of glycolysis via cytoskeleton architecture[J]. Nature, 2020, 578(7796): 621–626.

Piccolo S, Dupont S, Cordenonsi M. The biology of YAP/TAZ: hippo signaling and beyond[J]. Physiol Rev, 2014, 94(4): 1287–1312.

Rotsch C, Radmacher M. Drug-induced changes of cytoskeletal structure and mechanics in fibroblasts: an atomic force microscopy study[J]. Biophys J, 2000, 78(1): 520–535.

Wang Y, Cao W, Cui J, Yu Y, Zhao Y, Shi J,Wu J, Xia Z, Yu B, Liu J. Arterial wall stress induces phenotypic switching of arterial smooth muscle cells in vascular remodeling by activating the YAP/TAZ signaling pathway[J]. Cell Physiol Biochem, 2018, 51(2): 842–853.

Wu Q, Chen X, Wang P, Wu Q, Qi X, Han X, Chen L, Meng X, Xu K. Delivery of arsenic trioxide by multifunction nanoparticles to improve the treatment of hepatocellular carcinoma[J]. ACS Appl Mater Interfaces, 2020, 12(7): 8016–8029.

Yang W, Ge J, Liu H, Zhao K, Liu X, Qu X, Li W, Huang Y, Sun A, Zou Y. Arsenic trioxide eluting stent reduces neointima formation in a rabbit iliac artery injury model[J]. Cardiovasc Res, 2006, 72(3): 483–493.

Yang F, Chen Q, He S, Yang M, Maguire EM, An W, Afzal TA, Luong LA, Zhang L, Xiao Q. MiR-22 is a novel mediator of vascular smooth muscle cell phenotypic modulation and neointima formation[J]. Circulation, 2018, 137(17): 1824–1841.

Yu Q, Li W, Jin R, Yu S, Xie D, Zheng X, Zhong W, Cheng X, Hu S, Li M, Zheng Q, Li G, Song Z. PI3Kγ (Phosphoinositide 3-Kinase γ) regulates vascular smooth muscle cell phenotypic modulation and neointimal formation through CREB (Cyclic AMP-Response element binding protein)/YAP (Yes-associated protein) signaling[J]. Arterioscler Thromb Vasc Biol, 2019, 39(3): e91–e105.

Zhao YP. Biological mechanism of arsenic trioxide for promoting vascular endothelialization and inhibiting in-stent restenosis[D]. Chongqing University, 2018.

Zhao Y, Du R, Zhou T, Yang D, Huang Y, Wang Y, Huang J, Ma X, He F, Qiu J, Wang G. Arsenic trioxide-coated stent is an endothelium-friendly drug eluting stent[J]. Adv Healthc Mater, 2018, 7(15): e1800207.

Zhao Y, Zang G, Yin T, Ma X, Zhou L, Wu L, Daniel R, Wang Y, Qiu J, Wang G. A novel mechanism of inhibiting in-stent restenosis with arsenic trioxide drug-eluting stent: enhancing contractile phenotype of vascular smooth muscle cells via YAP pathway[J]. Bioact Mater, 2021, 6(2): 375–385.

Glossary of Terms

AF Actin filament
Angiogenesis
ANG2 Angiopoietin-2
ATO Arsenic trioxide
AES Arsenic trioxide-eluting stents
AS Atherosclerosis
BMS Bare metal stents
BBB Blood-brain barrier
CVDs Cardiovascular diseases
CVP Caudal vein plexus
CMTA Cell membrane tension accumulation
CSVD Cerebral small vessel disease
CTA Computed tomography angiography
CXCR4 CXC motif chemokine receptor type 4
DSA Digital subtraction angiography
DES Drug eluting stent
eNOS Endothelial nitric oxide synthase
EPCs Endothelial progenitor cells
ET-1 Endothelin-1
ECM Extracellular matrix
ECM Extracellular matrix
ERK Extracellular signal-regulated kinase
FAK Focal adhesion kinase
FA Focal adhesion
Tg(flk1:GFP) Green fluorescent protein transgenic zebrafish flk 1
GRO Growth regulatory oncogene
HO Heme oxygenase
h.p.f. Hours post-fertilization
HBMECs Human brain microvascular endothelial cells
Id Inhibitors of DNA binding/differentiation
ISR In-stent restenosis

G. Wang, *The Mechanobiology in Atherosclerosis*,
https://doi.org/10.1007/978-981-97-6512-6

ICAM-1 Intercellular adhesion molecule-1
IL-8 Interleukin-8
IEL Internal elastic lamina
ITRAQ Isobaric tags for relative and absolute quantification
LSCM Laser scanning confocal microscoy
LCA Left carotid artery
LPS Lipopolysaccharides
LDLR Low density lipoprotein receptor
MMP Matrix metalloproteinase
MCP-1 Monocyte chemoattractant protein-1
MO Morpholino oligonucleotides
NF-κB Nuclear factor kappa-B
OSS Oscillatory shear stress
OPN Osteopontin
PPFC Parallel plated flow chamber
PIV Particle image velocimetry
PCI Percutaneous coronary intervention
PTCA Percutaneous transluminal coronary angioplasty
PBS Phosphate buffer saline
PI3K phosphatidylinositol 3 kinase-Akt Protein kinase B
PECAM-1 Platelet-endothelial cell adhesion molecule-1
PGI2 Prostacyclin I2
RCA Right carotid artery
SSRE Shear stress response element
SREBP1 Sterol-regulatory element binding protein-1
TNF-α Tumor necrosis factor-α
VECs Vascular endothelial cells
VEGF Vascular endothelial growth factor
VSMCs Vascular smooth muscle cells
Vasculogenesis
Zinc finger Kruppel-like transcription factor

The manufacturer's authorised representative in the EU is Springer Nature Customer Service Centre GmbH, Europaplatz 3, 69115 Heidelberg, Germany. If you have any concerns regarding our products, please contact ProductSafety@springernature.com

Printed and bound by CPI Group (UK) Ltd, Croydon, CR0 4YY
30/09/2026
02225304-0004